POLYCRYSTALLINE SILICON FOR INTEGRATED CIRCUITS AND DISPLAYS

Second Edition

POLYCRYSTALLINE SILICON FOR INTEGRATED CIRCUITS AND DISPLAYS

Second Edition

by

Ted Kamins
Hewlett-Packard Laboratories

KLUWER ACADEMIC PUBLISHERS
Boston / Dordrecht / London

Distributors for North, Central and South America:
Kluwer Academic Publishers
101 Philip Drive
Assinippi Park
Norwell, Massachusetts 02061 USA

Distributors for all other countries:
Kluwer Academic Publishers
Distribution Centre
Post Office Box 322
3300 AH Dordrecht, THE NETHERLANDS

Library of Congress Cataloging-in-Publication Data

A C.I.P. Catalogue record for this book is available
from the Library of Congress.

Copyright © 1998 by Kluwer Academic Publishers

All rights reserved. No part of this publication may be reproduced, stored in a retrieval system or transmitted in any form or by any means, mechanical, photo-copying, recording, or otherwise, without the prior written permission of the publisher, Kluwer Academic Publishers, 101 Philip Drive, Assinippi Park, Norwell, Massachusetts 02061

Printed on acid-free paper.

Printed in the United States of America

Contents

1	**Preface**	xi
1	**Deposition**	1
	1.1 Introduction	1
	1.2 Thermodynamics and kinetics	2
	1.3 The deposition process	3
	1.4 Gas-phase and surface processes	5
	1.4.1 Convection	5
	1.4.2 The boundary layer	6
	1.4.3 Diffusion through the boundary layer	8
	1.4.4 Reaction	10
	1.4.5 Steady state	11
	1.5 Reactor geometries	15
	1.5.1 Low-pressure, hot-wall reactor	15
	1.5.2 Single-wafer, cold-wall reactor	24
	1.5.3 Cold-wall batch reactor	33
	1.6 Reaction	34
	1.6.1 Decomposition of silane	35
	1.6.2 Surface adsorption	37
	1.6.3 Deposition rate	39
	1.6.4 Rate-limiting surface process	42
	1.7 Deposition from disilane	44
	1.8 Deposition of doped films	45
	1.8.1 n-type films	46
	1.8.2 p-type films	49
	1.8.3 Electrostatic models	51
	1.9 Conformal deposition	51
	1.10 Enhanced deposition techniques	53

| | 1.11 Summary | 56 |

2 Structure — 57

- 2.1 Nucleation — 57
  - 2.1.1 Amorphous surfaces — 58
  - 2.1.2 Single-crystal surfaces — 64
- 2.2 Surface diffusion and structure — 65
- 2.3 Evaluation techniques — 70
- 2.4 Grain structure — 72
- 2.5 Grain orientation — 76
  - 2.5.1 Films formed by thermal CVD — 79
  - 2.5.2 Effect of plasma on structure — 84
  - 2.5.3 Evaporated and sputtered films — 85
  - 2.5.4 Other mechanisms controlling structure — 86
- 2.6 Optical properties — 86
  - 2.6.1 Index of refraction — 87
  - 2.6.2 Absorption coefficient — 88
  - 2.6.3 Ultraviolet surface reflectance — 90
  - 2.6.4 Use of optical properties for film evaluation — 91
- 2.7 Thermal conductivity — 96
- 2.8 Mechanical properties — 97
- 2.9 Oxygen contamination — 99
- 2.10 Etching — 100
- 2.11 Structural stability — 102
  - 2.11.1 Recrystallization mechanisms — 103
  - 2.11.2 Undoped or lightly doped polycrystalline films — 103
  - 2.11.3 Heavily doped polycrystalline films — 105
  - 2.11.4 Amorphous films — 110
- 2.12 Hemispherical-grain (HSG) polysilicon — 116
- 2.13 Epitaxial realignment — 117
- 2.14 Summary — 121

3 Dopant Diffusion and Segregation — 123

- 3.1 Introduction — 123
- 3.2 Diffusion mechanism — 124
  - 3.2.1 Diffusion along a grain boundary — 124
  - 3.2.2 Diffusion in polycrystalline material — 128
- 3.3 Diffusion *in* polysilicon — 129
  - 3.3.1 Arsenic diffusion — 134

CONTENTS

	3.3.2	Phosphorus diffusion	137
	3.3.3	Antimony diffusion	138
	3.3.4	Boron diffusion	138
	3.3.5	Limits of applicability	139
	3.3.6	Heavy doping	140
	3.3.7	Nitrogen	140
	3.3.8	Implant channeling	140
3.4	Diffusion *from* polysilicon		141
3.5	Interaction with metals		144
	3.5.1	Aluminum	145
	3.5.2	Other metals	146
	3.5.3	Silicides	147
3.6	Dopant segregation at grain boundaries		149
	3.6.1	Theory of segregation	149
	3.6.2	Experimental data	152
3.7	Computer modeling of diffusion		159
3.8	Summary		162

4 Oxidation   163

4.1	Introduction		163
4.2	Oxide growth on polysilicon		164
	4.2.1	Oxidation of undoped films	164
	4.2.2	Oxidation of doped films	166
	4.2.3	Effect of grain boundaries	174
	4.2.4	Effect of device geometry	177
	4.2.5	Oxide-thickness evaluation	180
4.3	Conduction through oxide on polysilicon		181
	4.3.1	Interface features	184
	4.3.2	Deposition conditions	185
	4.3.3	Oxidation conditions	188
	4.3.4	Dopant concentration and annealing	189
	4.3.5	Carrier trapping	191
	4.3.6	CVD dielectrics	192
4.4	Summary		193

5 Electrical Properties   195

5.1	Introduction	195
5.2	Undoped polysilicon	196
5.3	Amorphous silicon	198

5.4		Moderately doped polysilicon	199
	5.4.1	Carrier trapping at grain boundaries	200
	5.4.2	Carrier transport	205
	5.4.3	Trap concentration and energy distribution	213
	5.4.4	Thermionic-field emission	218
	5.4.5	Grain-boundary barriers	219
	5.4.6	Limitations of models	221
	5.4.7	Segregation and trapping	224
	5.4.8	Summary: Moderately doped polysilicon	225
5.5		Grain-boundary modification	225
	5.5.1	Grain-boundary passivation	226
	5.5.2	Recrystallization	228
5.6		Heavily doped polysilicon	230
	5.6.1	Solid solubility	231
	5.6.2	Method of doping	232
	5.6.3	Stability	235
	5.6.4	Mobility	237
	5.6.5	Trends	237
5.7		Minority-carrier properties	238
	5.7.1	Lifetime	238
	5.7.2	Switching characteristics	240
5.8		Summary	243

6 Applications — 245

6.1		Introduction	245
6.2		Silicon-gate MOS transistor	246
	6.2.1	Complementary MOS	248
	6.2.2	Threshold voltage	249
	6.2.3	Silicon-gate process	251
	6.2.4	Polysilicon interconnections	254
	6.2.5	Gate-oxide reliability	255
	6.2.6	Limitations	257
	6.2.7	Process compatibility	259
	6.2.8	New structures	260
6.3		Nonvolatile memories	261
6.4		Polysilicon resistors	262
6.5		Fusible links	267
6.6		Gettering	268
6.7		Polysilicon contacts	268

	6.7.1	Reduction of junction spiking	268
	6.7.2	Diffusion from polysilicon	269
6.8	Vertical *npn* bipolar transistors		270
	6.8.1	Fabrication: Polysilicon contacts	270
	6.8.2	Physics of the polysilicon-emitter transistor . . .	274
6.9	Lateral *pnp* bipolar transistors		279
6.10	Device isolation		281
	6.10.1	Dielectric isolation	281
	6.10.2	Poly-buffered LOCOS	283
	6.10.3	Trench isolation	284
6.11	Dynamic random-access memories		285
	6.11.1	Trench capacitor	287
	6.11.2	Stacked capacitor	289
6.12	Polysilicon diodes		292
6.13	Polysilicon thin-film transistors		296
	6.13.1	Device physics	298
	6.13.2	Methods of improving polysilicon for TFTs . . .	304
	6.13.3	TFTs for active-matrix, liquid-crystal displays .	308
	6.13.4	TFTs for static random-access memories	310
6.14	Microelectromechanical Systems		311
	6.14.1	Integrated sensors	312
	6.14.2	Polysilicon for MEMS	313
6.15	Summary		315

Preface

Polycrystalline silicon has played an important role in integrated-circuit technology for two decades. It was first used in self-aligned, silicon-gate, MOS ICs to reduce capacitance and improve circuit speed. In addition to this dominant use, polysilicon is now also included in virtually all modern bipolar ICs, where it improves the basic physics of device operation. The compatibility of polycrystalline-silicon with subsequent high-temperature processing allows its efficient integration into advanced IC processes. This compatibility also permits polysilicon to be used early in the fabrication process for trench isolation and dynamic random-access-memory (DRAM) storage capacitors.

In addition to its integrated-circuit applications, polysilicon is becoming vital as the active layer in the channel of thin-film transistors in place of amorphous silicon. When polysilicon thin-film transistors are used in advanced active-matrix displays, the peripheral circuitry can be integrated onto the same substrate as the pixel transistors. Recently, polysilicon has been used in the emerging field of microelectromechanical systems (MEMS), especially for microsensors and microactuators. In these devices, the mechanical properties, especially the stress in the polysilicon film, are critical to successful device fabrication.

The properties of polysilicon differ in important ways from those of single-crystal silicon, with significant effects on device performance. During the past two decades, a great deal of information has been published about polysilicon. A wide range of deposition conditions has been used to form films exhibiting markedly different properties. Seemingly contradictory results can often be explained by considering the details of the crystal structure formed with the particular deposition conditions used.

This monograph is an attempt to synthesize much of the available knowledge about polysilicon. It represents an effort to interrelate the deposition, properties, and applications of polysilicon. By properly understanding the properties of polycrystalline silicon and their relation to the deposition conditions, polysilicon can be designed to ensure the most optimum device and integrated-circuit performance. As feature sizes become smaller and intrinsic device delays decrease, however, some of the fundamental properties of polysilicon can restrict the overall performance of an integrated circuit. Understanding the basic limitations of polycrystalline silicon is essential to optimize process and circuit design and minimize these limitations.

Because of the limited size of this monograph (constrained by publishing economics), the total scope of polysilicon deposition, properties, and applications could not be covered as thoroughly as I had hoped. Since the first edition was published in 1988, polysilicon has been incorporated in more demanding IC applications and extended to the field of active-matrix, liquid-crystal displays. The resulting increase in the breadth and depth of understanding has produced an increasing number of publication in the field. Although this second edition is appreciably longer than the first edition, the amount of material recently published is too great to be covered exhaustively in a volume of reasonable size.

To provide a generally useful treatment, I have tried to emphasize trends and models and place specific experimental data in the context of these models. The intent is to provide a framework that can be used to plan experiments pertinent to specific deposition and processing conditions so that the detailed data needed for practical device fabrication can be rapidly obtained.

In addition to updating the material presented in the first edition, this second edition includes a significant amount of discussion of devices — and the related deposition technology and materials — that have become important since the first edition was published. Substantial material has been added to describe polysilicon deposition and application for the thin-film transistors used in active-matrix displays. The use of polysilicon in dynamic random-access memories is also emphasized. Depositing polysilicon in cold-wall, single-wafer reactors is also included in this second edition.

PREFACE

The material covered in this book is divided into six chapters, which are briefly described below:

Chapter 1: DEPOSITION
In a continuous-flow reactor, kinetics, as well as thermodynamics, affect the deposition process. Either gas-phase or surface mechanisms can limit the overall deposition process, and the choice of the limiting step should be compatible with the reactor geometry being used. Both cold-wall and hot-wall reactors can be used to deposit polysilicon, with the most controllable operating regime depending especially on the reactor geometry. Deposition in a single-wafer reactor allows control of interfaces between polysilicon and related layers deposited without intermediate air exposure in other chambers of a cluster tool. Introducing arsenic- or phosphorus-containing dopant gases during the deposition can decrease the deposition rate, while adding a boron-containing dopant gas can increase it. Using more reactive species or plasma-enhancement can increase the deposition rate.

Chapter 2: STRUCTURE
Initial nucleation of a polysilicon deposit on an amorphous surface depends on the deposition conditions and the nature of the surface. Surface migration of adsorbed atoms during the deposition of the film also influences its structure. Surface migration can be affected by deposition temperature, rate, and pressure. Complementary structural information is provided by various evaluation techniques, such as transmission-electron microscopy (for grain size and detailed microstructure) and x-ray diffraction (for quantitative comparison of films deposited under different conditions). The structure of undoped polycrystalline films is stable at high temperatures, while marked grain growth can occur in highly doped films, even at moderate temperatures. Large grains can be obtained by depositing amorphous silicon and subsequently crystallizing it. Epitaxial regrowth of polysilicon on single-crystal silicon can also occur after deposition.

Chapter 3: DOPANT DIFFUSION AND SEGREGATION
Rapid diffusion of dopant atoms along grain boundaries dominates the diffusion of dopant in polysilicon. Diffusion of dopant *within* polysilicon is differentiated from diffusion *from* polysilicon into underlying single-crystal silicon; in the latter case, residual oxide layers at the interface or (in the opposite extreme) epitaxial regrowth can retard the diffusion of

dopant from polysilicon. The n-type dopants phosphorus, arsenic, and probably antimony segregate to grain boundaries, especially at lower temperatures, limiting the amount of dopant available to contribute carriers to the conduction process; boron does not appear to segregate. Computer-aided modeling allows prediction of the complex behavior of dopant atoms in polysilicon and the underlying single-crystal silicon.

Chapter 4: OXIDATION

The oxidation of undoped polysilicon is controlled by the different oxidation rates of differently oriented crystallites. Oxidation of doped films depends on the dopant concentration near the surface and the ability of the dopant to diffuse away from the surface during oxidation. Conduction through oxide grown on polysilicon depends on the surface morphology of the film and the device structure.

Chapter 5: ELECTRICAL PROPERTIES

The resistivity of undoped polysilicon films is similar to that of intrinsic silicon and depends only weakly on the deposition conditions. The electrical properties of moderately doped films are determined both by dopant segregation to the grain boundaries and by carrier trapping at the grain boundaries. Trapping creates potential barriers, impeding carrier transport between grains and reducing the "effective mobility." Current flow is normally modeled by thermionic emission, but tunneling can be important in some cases. The nonuniformity of the grain size restricts detailed modeling of conduction in polysilicon. The conductivity of heavily doped films is limited by the solid solubility of the dopant in crystalline silicon. The rate of cooling, as well as the time at the processing temperature, can affect the active dopant concentration. Hydrogen can passivate some of the grain-boundary defects, while laser melting can increase the grain size.

Chapter 6: APPLICATIONS

The main application of polysilicon is as the gate electrode in silicon-gate MOS integrated circuits, in which the high-temperature compatibility of polysilicon allows fabrication of self-aligned transistors with low parasitic capacitances. The polysilicon can also serve as an additional partial level of circuit interconnection. Lightly doped polysilicon can provide the high-value resistors needed for static memories and other applications. Polysilicon is used in high-performance, bipolar integrated circuits for contact to the base and emitter of the transistor. Using

PREFACE

polysilicon for the emitter contact, markedly improves the transistor gain. In dynamic random-access memories polysilicon is widely used for trench capacitors in the substrate and stacked capacitors above the substrate. When polysilicon replaces amorphous silicon in the channel of a thin-film transistor, the higher mobility allows integrating peripheral circuitry onto the same substrate as the pixel transistors of an active-matrix display.

The first edition was prepared during a year I spent at Stanford University as Hewlett-Packard's resident industrial representative. I would like to express my appreciation to Stanford University's Center for Integrated Systems, which provided an environment conducive to writing the first edition. At Stanford Professor John Linvill provided welcome encouragement to pursue this work, and Ms. Louise Peterson and Ms. Joyce Pelzl provided logistical assistance. During preparation of the monograph, presenting a course at Stanford covering the same material provided me with the opportunity to refine the content and its organization. I would especially like to thank Dr. John Moll, formerly of Hewlett-Packard, for encouragement in selecting this monograph as one of the projects to pursue while at Stanford.

I would also like to acknowledge those at Hewlett-Packard, Stanford and elsewhere, who shared information about polysilicon with me. At Stanford the research groups of Professors J. Plummer and R. Dutton and Professor J. Bravman, E. Crabbé, E. Demirlioglu, J. Hoyt, M. Ghannam and others provided useful discussion. Dr. John Andrews, formerly of Bell Laboratories, kindly allowed me to use a number of figures that he had drawn for a course we taught together through the University of California. I also appreciate the loan of original micrographs by Professor J. Bravman of Stanford University, Dr. Y. Wada of Hitachi, Dr. R. B. Marcus of Bell Communications Research, and Dr. H. Oppolzer of Siemens and the permission of Silvaco International to allow adaptation of their treatment of dopant diffusion in polysilicon.

Ted Kamins
Hewlett-Packard Laboratories

Chapter 1

Deposition

1.1 Introduction

Silicon integrated circuits continue to play an increasingly important role in the electronics industry. The content of integrated circuits in electronic products has increased continuously over the past three decades until today it can dominate the value, as well as the cost, of an electronic system such as a computer. One of the critical factors leading to this rapid increase in the use of integrated circuits has been the development of high-density, complementary metal-oxide-semiconductor (CMOS) integrated circuits, which allow complex logic or high-density memories to be built on a single silicon chip. Key to the fabrication of these dense CMOS chips is the use of polycrystalline silicon as the gate-electrode material. The use of polysilicon allows formation of a self-aligned structure, greatly improving the device characteristics by reducing parasitic capacitance. It also permits more complex structures to be fabricated because of its compatibility with high-temperature silicon integrated-circuit processing.

Although the first widespread application of polysilicon was in MOS integrated circuits, its availability led to its use in bipolar circuits as well, where it has similarly improved device performance and allowed increased density. Today, virtually all advanced bipolar integrated circuits include one or two layers of polysilicon. Polycrystalline silicon is increasingly being employed in novel ways within both CMOS and bipolar integrated circuits. For example, the storage regions of dynamic, random-access memories (DRAMs) are placed above the access transistors in complex structures composed of several layers of polysilicon. Alternatively, the storage regions and even the access transistors can be placed in wells or trenches etched in the silicon surface and refilled with polysilicon. These trenches are also used for isolation between elements of the integrated circuit.

In addition to integrated-circuit applications, polysilicon is being increasingly used as the channel of thin-film transistors for switching pixels of liquid crystal displays (LCDs). Because of the reasonable carrier mobility in polysilicon, polysilicon thin-film transistors can be used in the peripheral addressing circuitry, as well as in the individual pixel switch. Polysilicon thin-film transistors are also used as the load elements of silicon static random-access memories (SRAMs).

In addition, polysilicon can form mechanical elements of microelectromechanical systems (MEMS), where its mechanical properties are critically important. Residual stress in the polysilicon can deform structural elements and prevent operation of the device.

A considerable amount of understanding of the properties of polysilicon has been gained by research motivated by solar-cell applications. However, the material used for solar cells often differs substantially in grain size and method of formation from that important for integrated-circuit applications. Although most of the basic considerations in this volume are common to the two types of material, our discussion will be directed toward the type of polysilicon used in integrated circuits and active matrix displays.

1.2 Thermodynamics and kinetics

Polycrystalline silicon is generally deposited by chemical vapor deposition (CVD), utilizing the thermal decomposition of silane (SiH_4) or disilane (Si_2H_6) to form elemental silicon and molecular hydrogen by the overall reactions

$$SiH_4 \text{ (g)} \rightarrow Si \text{ (s)} + 2H_2 \text{ (g)} \tag{1.1}$$

and

$$Si_2H_6 \text{ (g)} \rightarrow 2Si \text{ (s)} + 3H_2 \text{ (g)} \tag{1.2}$$

In the open-flow reactors used for chemical vapor deposition of thin films for electronic applications, reactant and carrier gases are continuously introduced into the reactor and unused gases and gaseous reaction byproducts are continuously removed. Because of the continuous gas flow, the deposition process is influenced by kinetic factors, as well as by the thermodynamics of the decomposition reaction itself.

By considering the possible species present and thermochemical data, the *equilibrium* concentrations of the various gaseous species can be determined. Available computer programs calculate these quantities by minimizing the free energy of the overall system. The equilibrium constant K of a reaction can be calculated from the relation

$$\ln K(T) = -\frac{\Delta G(T)}{RT} \tag{1.3}$$

where G is the Gibb's free energy, and ΔG is found from the sum of the free energies of formation of the products minus that of the reactants. The amount of solid formed from a given amount of reactant can be determined if the reaction proceeds fully to equilibrium. The deposition rate (solid product per unit time) can similarly be determined from the input gas flow rate (reactant per unit time). In actual reactors, however, the gases are not in the reaction zone long enough for equilibrium to be established, and *kinetic* factors limit the amount of material deposited to less than that expected solely from equilibrium considerations. For good uniformity of the deposit across a group of wafers and within a wafer, the deposition rate is usually much less than suggested by thermodynamics; thermodynamic considerations only set an upper bound on the amount deposited. Kinetic factors are often of more practical concern because they can depend strongly on the geometry of the reactor and the operating conditions, over which we have control.

1.3 The deposition process

The overall deposition process can be viewed as the sum of a series of individual steps, as shown in Fig. 1.1. First, the silicon-containing source gas enters the reaction chamber by forced convection and flows to the vicinity of the wafer. It approaches the wafer by diffusing through a *boundary layer* near the wafer and may partially decompose in the gas phase. When the source gas and reaction intermediates reach the wafer surface, they adsorb and may be able to diffuse on the surface before completely decomposing to silicon and hydrogen. The resulting silicon atoms diffuse to stable sites, generally at steps formed by

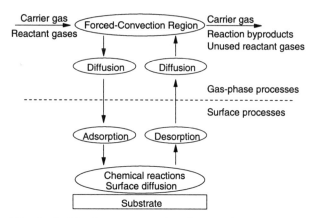

Figure 1.1: The individual steps in the chemical vapor deposition process can be divided into those occurring primarily in the gas phase and those occurring primarily on the wafer surface.

partially completed layers of silicon previously deposited. Subsequently arriving silicon atoms surround the first atom, complete bonds, and bind it firmly into the deposited layer. (The initial stage of deposition, in which nucleation occurs on an amorphous layer, will be discussed in Chapter 2.) Simultaneously, hydrogen atoms or molecules liberated by the decomposition reaction diffuse on the surface. Two hydrogen atoms can encounter each other, combine, and desorb as a hydrogen molecule.

The overall deposition process thus includes the following individual steps (and possibly others):

- Forced convection
- Boundary-layer diffusion
- Surface adsorption
- Surface diffusion
- Decomposition
- Surface diffusion
- Nucleation
- Incorporation
- Byproduct desorption

Any of these individual steps can limit the overall deposition process, and the temperature and pressure dependences of the deposition process reflect those of the rate-limiting step. Proper process design often requires selecting the rate-limiting step to optimize the deposition

1.4. GAS-PHASE AND SURFACE PROCESSES

uniformity; the limiting step is often chosen to be the one which can be best controlled in a specific type of reactor. For ease of discussion, the entire series of individual steps can be divided into a group of gas-phase processes and a group of surface processes, as shown in Fig. 1.1, although this division is somewhat idealized.

1.4 Gas-phase and surface processes

1.4.1 Convection

Most CVD reactions important in the electronics industry are carried out in *open-flow reactors*. The reactor may operate at atmospheric pressure, with the gas forced through it by a slightly higher pressure of the incoming gases, or it may operate at reduced pressure, with the gas flow resulting from pumping at the outlet (exhaust) end of the reaction chamber.[1]

At low velocities temperature gradients can determine the nature of the gas flow, and *free convection* established by the temperature gradients must be considered. At higher gas velocities, *forced convection* caused by the gas flowing through the reactor is usually more important [1.1, 1.2].

When free convection dominates at low input gas flow rates, the gas motion is established by the largest thermal gradients in the system, with the gas flowing from the hottest region to the coolest. When a heated plate forms one boundary of the reaction zone (as in the cold-wall reactor shown in Fig. 1.2 and to be discussed in Sec. 1.5.2) the gas flows in an approximately circular path from the center of the heated plate to the center of the opposite cool wall and then along the relatively cool walls and back across the heated plate. Obtaining uniform deposition in a reactor dominated by the irregular gas flow of free convection is difficult, and reactors are usually operated so that free convection does not control the gas flow. As the gas velocity through the deposition chamber increases, forced convection along the gas-flow direction is superposed on free convection to produce a spiral gas flow. At higher velocities, forced convection dominates, and the gas flows primarily parallel to the

---

[1]Pressure is usually expressed in units of *Torr* or *pascals* (pa): Atmospheric pressure = 760 Torr = 1.013×10^5 pa; 1 Torr = 133 pa. A less frequently used unit is the *bar*: 1 bar = 1 dyne/cm^2 = 0.987 atmosphere = 750 Torr.

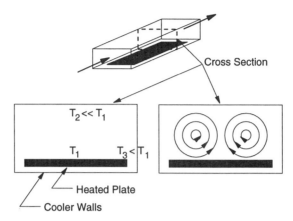

Figure 1.2: Cross section of reactor showing free-convection flow, which dominates at low input gas flow rates. Thermal gradients cause free convection, with the gas flowing from the hottest to the coldest parts of the reactor. (The input gas flow is perpendicular to the plane of the paper.)

reactor walls. If the walls are uniformly heated, as in the hot-wall, low-pressure reactor to be discussed in Sec. 1.5.1, little thermal gradient exists within the deposition chamber, and forced convection dominates even at low gas velocities.

Gas flow in the forced-convection regime is described by the *Reynold's number* Re

$$\mathrm{Re} = \frac{D_T v \rho}{\eta} \quad (1.4)$$

where D_T is a characteristic dimension of the reactor cross section, v is the linear gas velocity, ρ is the gas density, and η is its viscosity. Low Reynold's numbers (≤ 2000) correspond to laminar flow, while turbulent flow is obtained at high Reynold's numbers.

1.4.2 The boundary layer

Although forced convection influences the overall deposition process, it is generally not the limiting step. Of the gas-phase processes, diffusion through a *boundary layer* often limits. The concept of a boundary layer can be discussed when the distance traveled by gas molecules between collisions with other gas molecules (the *mean free path*) is much less than the shortest dimension of the deposition chamber. Most CVD reactors operate in this *viscous flow* regime.

1.4. GAS-PHASE AND SURFACE PROCESSES

The mean free path λ can be written as

$$\lambda = \frac{1}{\sqrt{2}\,\pi\,d_0^2\,n} \tag{1.5}$$

where d_0 is the effective diameter of the gas molecules and n is the density of atoms. For air λ (cm) = 0.66/P(pa) = 0.005/P(Torr). For other gases, the mean free path can differ by about a factor of 3, being shorter for lighter gases such as H_2 or He and longer for heavier gases such as HCl or Cl_2. At a deposition pressure of 1-10 Torr, the mean free path is of the order of 10 μm, much smaller than the reactor dimensions, but larger than the device feature lengths and heights. Because the mean free path is much shorter than the reactor dimensions, the gas molecules mainly collide with other gas molecules, rather than with the walls of the chamber. The microscopic momentum transfer between molecules when they collide can be described on the macroscopic scale as friction or *viscous* forces.

When we also consider collisions between the molecules and the walls of the reactor, we see that the velocity of the gas is not uniform over a cross section of the reactor. Momentum transfer from the molecules near the walls of the reactor to the walls slows these molecules. The reduced velocity of molecules near the walls, in turn, slows other nearby molecules (by momentum transfer on the microscopic scale or viscous forces on the macroscopic scale). This friction force of the walls on the flowing gas creates a velocity gradient between gas near the walls and the free-flowing gas in the forced-convection region far from the walls. The concept of a low-velocity *boundary layer* separating the forced-convection region from the stationary surfaces (*eg*, the walls of the reactor chamber or the wafers) is useful in understanding both the hot-wall, low-pressure reactor and the cold-wall reactor to be discussed in Sec. 1.5.

We illustrate the idea of gas-phase diffusion through the boundary layer by considering a flat plate parallel to the gas flow, as shown in Fig. 1.3, before applying the concept to CVD. The velocity within the boundary layer gradually increases from a low value (often ∼0) at the plate to its value in the forced-convection region, as shown in Fig. 1.3. A parabolic increase in velocity with distance y from the plate is sometimes assumed [1.3]. In some analyses the calculations are simplified by taking the velocity to be zero throughout the boundary layer (*ie*, by assuming that the layer is stationary or *stagnant* [1.4]); consequently,

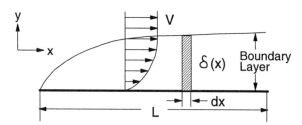

Figure 1.3: A *boundary layer* separates the rapidly moving gas in the *forced-convection region* from the stationary surfaces in the deposition chamber.

the boundary layer is sometimes (although incorrectly) called a *stagnant layer*. With the widespread availability of high-speed computers, detailed modeling of the gas flow in the boundary layer, as well as in the forced-convection region, is feasible, reducing the need for simplifying assumptions. The treatment in this section is meant to provide a physical picture of the process, rather than exact solutions.

The boundary layer also develops along the length of the reactor so that the gas flow is a function of the longitudinal position x, as well as the distance y from the plate. The thickness δ of the boundary layer can be approximated by the expression

$$\delta(x) = K_{BL}\sqrt{\frac{\eta x}{\rho v}} \qquad (1.6)$$

where v is the gas velocity in the forced-convection region, η is the gas viscosity, ρ is its density, and K_{BL} is a numerical coefficient of the order of one. Equation 1.6 shows that, as the gas velocity in the forced-convection region increases, the viscous forces exerted on the boundary layer by the flowing gas increase, and the thickness of the boundary layer decreases.

1.4.3 Diffusion through the boundary layer

The reactant gases reach the wafers by diffusing through the boundary layer. In steady state the flux of reactant F_1 across the boundary layer is given by the expression

$$F_1 = D\frac{dC}{dy} \qquad (1.7)$$

1.4. GAS-PHASE AND SURFACE PROCESSES

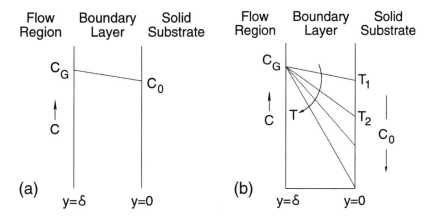

Figure 1.4: Diffusion through the boundary layer to the wafer surface supplies the reactant needed for the reaction to proceed. (a) The concentration gradient across the boundary layer provides the driving force for this diffusion. (b) As the temperature increases, the concentration gradient needed to supply the reactant for the surface reaction increases, and the surface concentration decreases.

where D is the gas-phase diffusion coefficient of the reactant gas in the boundary layer, C is the concentration of the reactant, and y is the distance along the direction of gas diffusion (assumed perpendicular to the direction of forced-convection flow). If no reaction occurs in the gas phase, F is constant, and the concentration varies linearly with position in the boundary layer, as shown in Fig 1.4a. Equation 1.7 can then be written as

$$F_1 = D \frac{C_G - C_0}{\delta} \tag{1.8}$$

where C_G is the reactant concentration at the edge of the boundary layer near the forced-convection region, C_0 is the reactant concentration in the gas phase adjacent to the wafer surface, and δ is the thickness of the boundary layer. The concentration gradient $(C_G - C_0)/\delta$ provides the *driving force* for the transport of the reactant through the boundary layer.

The gas-phase diffusion coefficient describes the ease with which the reactant molecules can move in response to the driving force. From the elementary kinetic theory of gases, the gas-phase diffusion coefficient in a mixture containing gas A and gas B (for example, a reactant gas A

in a carrier gas B) is given by [1.5]

$$D_{AB} = \frac{1}{3} \frac{n_A \lambda_B V_B + n_B \lambda_A V_A}{n_A + n_B} \qquad (1.9)$$

where n is the molecular density of each species, λ is the mean free path, and V is the average molecular velocity.

In macroscopic terms, the gas-phase diffusion coefficient can be written as

$$D = K_D \frac{\eta}{\rho} \qquad (1.10)$$

where K_D is an empirical coefficient that varies slightly for different gases (typically $1.3 \leq K_D \leq 1.6$), η is the viscosity, and ρ is the density of the gas, which is proportional to the pressure. From the temperature and pressure dependences of the quantities in Eq. 1.10, D can be written as

$$D = D_0 \left(\frac{T}{T_0}\right)^\alpha \frac{P_0}{P} \qquad (1.11)$$

where D_0 is commonly between 0.1 and 1.2, $T_0 = 273$ K (0°C), and $P_0 = 760$ Torr. The diffusion coefficient is only a weak function of temperature, with α typically between 1.75 and 2. This relatively slow variation of the *gas-phase* diffusion coefficient with temperature contrasts with the rapidly varying *solid-phase* diffusion coefficient

$$D_{\text{solid}} = D_0 \exp\left(-\frac{E_{aD}}{kT}\right) \qquad (1.12)$$

with its apparent activation energy E_{aD}, which is often several eV.[2]

1.4.4 Reaction

The flux F_2 of reactant entering into the chemical reaction at or near the wafer surface can be expressed as

$$F_2 = k(C_0 - C_{\text{eq}}) \qquad (1.13)$$

where k is the reaction-rate coefficient describing the relevant reaction and C_{eq} is the concentration of the gas in equilibrium with the solid.

---

[2] 1 eV/atom = 23 kcal/mole

1.4. GAS-PHASE AND SURFACE PROCESSES

The reaction-rate coefficient is a strong function of temperature, with an apparent activation energy E_a:

$$k = k_0 \exp\left(-\frac{E_a}{kT}\right) \qquad (1.14)$$

For deposition of silicon, the apparent activation energy is typically in the range 1.6-2.0 eV.

1.4.5 Steady state

In steady state the flux of reactant reaching the vicinity of the wafer equals the amount reacting (assuming no gas-phase reaction), and the fluxes F_1 and F_2 are equal. The gas-phase reactant concentration near the wafer surface can then be written as

$$C_0 = \frac{C_G}{1 + (k\delta/D)} \qquad (1.15)$$

For $D/\delta \gg k$, diffusion through the boundary layer occurs rapidly compared to the reaction, and the overall deposition process is said to be *reaction-rate limited*. The deposition rate is proportional to k, which (from Eq. 1.14) is a strong function of temperature. Thus, the deposition rate increases rapidly with increasing temperature when the overall deposition process is reaction-rate limited. For the 1.6-2.0 eV activation energy frequently observed for the deposition of silicon, a 10°C uncertainty in temperature near 600°C causes a 25-35% uncertainty in the deposition rate and, therefore, the thickness. Consequently, operating in the reaction-rate-limited regime requires a reactor in which the temperature is very well controlled.

When the reaction rate limits the deposition process, little driving force (concentration gradient) is needed to bring to the wafer surface the small amount of reactant consumed by the slow surface reaction, and $C_0 \approx C_G$ ($T = T_1$ in Fig. 1.4b). As the temperature increases, the reaction rate increases rapidly while the diffusion coefficient only increases slowly. Consequently, a significant concentration gradient is needed across the boundary layer to create the driving force for diffusion to supply the needed reactant, and C_0 decreases significantly below C_G (eg, $T = T_2$ in Fig. 1.4b). In this region, both reaction and diffusion affect the deposition rate.

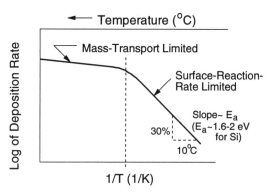

Figure 1.5: The deposition rate is a rapidly varying function of temperature in the surface-reaction-rate-limited regime of operation (low temperatures), while it changes only slowly with temperature in the mass-transport-limited regime (higher temperatures).

As the temperature continues increasing, the reaction rate increases further, and the concentration gradient needed to supply the reactant must also increase. C_0 continues decreasing until it approaches C_{eq}, the gas concentration which would be in equilibrium with the solid. When $C_0 \approx C_{eq}$, additional increases in temperature cannot reduce C_0 further, and the concentration gradient remains approximately constant. The overall deposition process is then limited by the rate at which the reactant can be transported to the surface of the wafer, and the deposition process is said to be diffusion limited or *mass-transport limited;* in this operating regime $D/\delta \ll k$. [Note, that over a considerable range of temperatures ($\geq 100°C$), $C_G > C_0 > C_{eq}$. In this range both mass transport and surface reaction influence the overall deposition process, and both must be considered.]

In the mass-transport limited regime C_0 in Eq. 1.8 is often taken to be zero, and the concentration gradient in the boundary layer is approximately C_G/δ. The parameters in the expression DC_G/δ describing the flux of reactant reaching the surface (and, consequently, the deposition rate) all vary relatively slowly with temperature. Therefore, the overall deposition rate varies much less rapidly with temperature in the mass-transport-controlled regime of operation than in the surface-reaction-controlled regime, as shown in the Arrhenius plot in Fig. 1.5.

The group of parameters $k\delta/D$ describes the limiting mechanism of

1.4. GAS-PHASE AND SURFACE PROCESSES

the overall deposition process and is defined as the *Sherwood number* Sh:

$$\text{Sh} = k\delta/D \tag{1.16}$$

When $\text{Sh} \gg 1$, the deposition process is mass-transport limited, while it is reaction controlled for $\text{Sh} \ll 1$.

To obtain the best thickness uniformity by minimizing the number of variables that must be controlled, the operating regime of a reactor is usually chosen to be completely dominated either by the chemical reaction or by mass transport. The particular operating regime is selected to make the overall deposition process least sensitive to variables which are poorly controlled in a given reactor. However, other factors (such as the need for high deposition rate, high wafer capacity, or small grain size) sometimes make operating near the transition region necessary. In this case, small changes in the operating conditions can change the relative importance of surface reaction and mass transport.

Although we are primarily concerned with the deposition of silicon films from silane and disilane, similar transitions between mass-transport and reaction control occur for deposition from other silicon-containing gases. The chlorinated silanes, SiH_2Cl_2, $SiHCl_3$, and $SiCl_4$, have been studied in most detail. The activation energy observed at low temperatures is similar to that seen with silane. The transition between the two limiting regions occurs at higher temperatures with the more chlorinated compounds, for which the reaction proceeds less readily. (The chlorinated silanes are only used for polysilicon deposition in specialized applications because of poor nucleation on oxide when chlorine is present.)

Temperature is not the only variable determining whether the deposition process is limited by reaction or by mass transport; the nature of the diluent gas or *carrier gas* can also influence the operating regime. A carrier gas, which does not enter directly into the deposition process, is often added to the reactant gases, especially at higher pressures, to increase the gas velocity in the forced-convection region. Hydrogen is often used as the carrier gas. However, hydrogen is also a product of the reaction, so it suppresses the silane decomposition reaction so that the transition between the two regimes occurs at a higher temperature than when nitrogen is used. The total gas pressure in the reactor also influences the limiting mechanism. Because the gaseous diffusion coefficient is inversely proportional to the gas density (Eq. 1.10), and

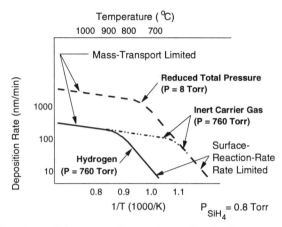

Figure 1.6: The deposition rate depends on both the type of carrier gas and the pressure. Hydrogen suppresses the reaction (important at lower temperatures) while lowering the total pressure increases the gas-phase diffusion coefficient (important at higher temperatures). Adapted from [1.7]. Reprinted with permission.

consequently the pressure (Eq. 1.11), the ability of the reactant gas to diffuse through the boundary layer increases as the pressure decreases. Therefore, by decreasing the pressure, the limitation on the overall deposition process can change from mass transport to surface reaction. As the pressure is reduced from atmospheric pressure to the fraction of a torr (tens of pascals) typically used in the low-pressure CVD reactors to be discussed below, the diffusion coefficient increases by approximately a factor of 1000. From Eq. 1.6 we see that the boundary-layer thickness also varies with the pressure; however, the thickness of the boundary layer only increases by a factor of about 3-10 as the pressure decreases by a factor of 1000 [1.6]. Therefore, the ease of diffusion, represented by the quantity D/δ, increases substantially as the pressure decreases; and reaction, rather than mass transport, is likely to limit the overall deposition process at low pressures.

Figure 1.6 [1.7] illustrates the influence of both the type of carrier gas and the pressure. At high temperatures, where the overall deposition process is limited by mass transport, decreasing the total pressure increases the gas-phase diffusion coefficient, and therefore the deposition rate. In this mass-transport-limited regime, the nature of the carrier gas has only a minor effect because the gas-phase diffusion coefficient

1.5. REACTOR GEOMETRIES

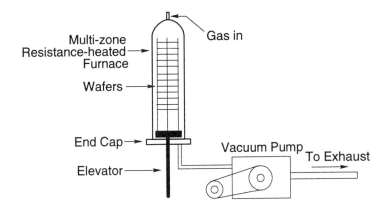

Figure 1.7: The hot-wall, low-pressure reactor is used for routine deposition of polysilicon because of its high wafer capacity and simplicity. The vertical reactor configuration is shown here.

depends only moderately on the diffusing species. At low temperatures, where the reaction limits the overall deposition process, the rate is higher in nitrogen, which does not suppress the silane decomposition reaction and the subsequent hydrogen desorption. In hydrogen the rate increases as the total pressure decreases (while keeping the silane partial pressure constant). In the intermediate temperature range, where both mass transport and reaction influence the overall deposition process, changes in either the total pressure or the nature of the carrier gas affect the deposition rate.

1.5 Reactor geometries

1.5.1 Low-pressure, hot-wall reactor

For production of high-volume integrated circuits, the number of wafers processed simultaneously should be as large as possible. This requirement for high wafer capacity led to the development of the commercially important, hot-wall, low-pressure CVD (LPCVD) reactor shown in Fig. 1.7. For maximum wafer capacity, the wafers are stacked adjacent to each other with only a narrow space of a few millimeters between adjacent wafers, and the wafers are supported at several locations around their edges. (Because the wafers are only supported at their edges in the low-pressure CVD reactor, polysilicon is deposited on the back of each wafer, as well as on the front.) As in a diffusion or oxidation furnace, the wafers are placed perpendicular to the main gas flow in a fused-silica[3] tube with a circular cross section. Fused silica is used for high-temperature compatibility and because its high purity reduces contamination of the depositing film.

The gases flow by forced convection through the annular space between the wafers and the walls of the deposition chamber and then must diffuse along the narrow space between the wafers (Fig. 1.8). The close wafer spacing makes diffusion of the gases to the center of the wafer difficult — especially for larger wafers. If the overall deposition process is limited by mass transport in this reactor, the difficult diffusion of the reactant gas leads to a much thinner deposit at the center of the wafer than near the edges. Therefore, the LPCVD reactor must be operated in the reaction-limited regime to obtain deposited films with thickness uniform from center to edge [1.8, 1.9].

We have already seen that decreasing the deposition temperature reduces the reaction rate greatly but changes the diffusion rate only slowly, increasing the relative ease of diffusion compared to reaction. However, making the reaction more difficult by decreasing the temperature is not adequate to achieve operation in the reaction-limited regime when the space between wafers is narrow; making diffusion easier is also necessary. We saw from Eq. 1.11 that the gas-phase diffusion coefficient increases as the pressure decreases, reducing the mass-transport limitation on the deposition process. The combination of low-temperature and low-pressure operation moves the deposition process into the reaction-rate-limited regime, allowing deposition of uniform layers of polysilicon with closely spaced wafers in a hot-wall, tube-type reactor. As the wafer diameter increases, the distance that the reactant must diffuse increases, and mass transport becomes more difficult. Therefore, a deposition process that is reaction limited with smaller wafers can also be influenced by mass transport with larger wafers, and different operating conditions may be needed to obtain uniform layers.

Although the requirements for gas-flow control imposed by mass transport are reduced by operating in the reaction-limited regime, excellent temperature control is required to operate in this temperature-

---

[3]The amorphous fused silica used here is commonly, but incorrectly, called "quartz." Quartz is a crystalline form of silicon dioxide.

1.5. REACTOR GEOMETRIES

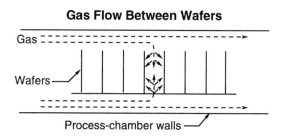

Figure 1.8: The narrow space between wafers in the low-pressure reactor makes gas diffusion to the centers of the wafers difficult. (Shown for a horizontal reactor.)

sensitive regime. Near the commonly used 625°C deposition temperature, the deposition rate changes approximately 2-2.5% for every degree change in temperature [1.8]. Fortunately, the necessary temperature control is available in the multi-zone, resistance-heated furnace originally developed for oxidation and diffusion.

To obtain reasonable deposition rates in the low-pressure, hot-wall reactor, the silane partial pressure must comprise all, or at least a significant fraction, of the deposition pressure, and little or no carrier gas is used. The maximum ratio of carrier-gas flow to silane flow is about four to one, but many processes operate with no added carrier gas. Because of the low pressure, the velocity of even the small amount of gas used is adequate to transport the gas through the reactor before it reacts completely. The deposition chamber is effectively at a constant pressure because of its large cross section. Pressure changes occur primarily in the gas supply system and perhaps in the exhaust line leading to the pump.

Because of the method of heating, the walls of the reaction chamber are as hot as the wafers during the deposition. Consequently, silicon is deposited on the walls, as well as on the wafers, making this reactor suitable only for deposition of thin films. The thermal coefficient of expansion of silicon is much higher than that of fused silica (3×10^{-6} cm^{-3} for silicon compared to 0.5×10^{-6} for fused silica). Therefore, after a thick silicon deposit accumulates on the fused-silica walls of the chamber, particles can form as the film flakes during thermal cycles. After deposition of a specified thickness of silicon, the fused-silica chamber is generally removed from the furnace, and the silicon is removed by wet-chemical etching, or the section of the chamber coated with a thick silicon deposit is replaced.

The deposition chamber of the low-pressure reactor may be oriented either vertically or horizontally. Most newer reactors have the vertical configuration shown in Fig. 1.7, with the closely spaced wafers oriented horizontally. The chamber of older reactors was often horizontal, with the wafers oriented vertically. The vertical chamber arrangement has several advantages. The horizontal wafer orientation makes automated loading of wafers into the fused-silica wafer carrier ("boat") easier; the wafers can be readily loaded from a cassette with a robot rotating on only one axis.

The gas flow in the vertical furnaces also creates a purer ambient than in a horizontal furnace and probably leads to less oxygen contamination in the deposited layer. During loading and unloading of the horizontal reactor, nitrogen can be forced through the chamber toward the open reactor door in an attempt to prevent oxygen and moisture contamination from entering the chamber. However, the nitrogen rises toward the top of the horizontal chamber as it heats while flowing through the hot chamber. Air then flows into the lower part of the chamber, allowing oxygen and moisture to be adsorbed on the walls and fixtures. By contrast, in the vertical configuration, nitrogen can be forced more uniformly through the chamber while it is open for loading, reducing air and moisture intrusion and adsorption and leading to a purer ambient during the subsequent deposition process. If necessary, the loading area can be totally sealed in a nitrogen-containing enclosure, eliminating oxygen and moisture contamination in the chamber. However, improved device yield would have to be demonstrated to justify the added cost and complexity of the equipment [1.10].

The floor space occupied by vertical and horizontal reactors is configured differently. For both arrangements, the actual deposition chamber is usually located "through the wall" of the *clean room*, into the less-clean (and less-expensive) *gray room*. The ability to stack several horizontal furnaces on top of each other, reduces the total floor space per horizontal furnace, while the floor space increases linearly with the number of vertical furnaces. While horizontal furnaces require a large loading area within the clean room itself, the loading area of vertical furnaces is below the furnace so that it does not occupy clean-room floor space. Vertical furnaces do, however, require a high ceiling in the gray room.

1.5. REACTOR GEOMETRIES

In the horizontal configuration of the LPCVD reactor, the gases are usually inserted at the end of the chamber where the wafers are loaded. Although a purer ambient can be obtained when the gases are introduced into the end of the chamber opposite the loading door, the loosely adherent material deposited on the walls of the chamber near the gas outlet can fall onto the wafers as they are being loaded and unloaded. Inserting the gases and loading the wafers at the same end of the chamber reduces particles, but requires maintaining a leak-free door closure to deposit high-quality films.

Gas depletion

As the gases flow along the length of the reaction chamber, a portion of the reactant gas is consumed. The partial pressure of reactant gas is lower near the outlet end of the reactor, and the deposition rate there is lower if all other deposition parameters remain constant. In addition, as the silane reacts, two moles of hydrogen are produced for every mole of silane consumed. The additional hydrogen dilutes the silane, further decreasing its partial pressure along the length of the reactor. To illustrate the importance of gas depletion, Fig. 1.9 shows the deposition-rate variation along the length of a hot-wall, low-pressure reactor for temperatures ranging from 525° to 725°C (with constant temperature along the reaction chamber). At 525°C, the deposition rate does not decrease appreciably along the length of the reactor because only a very small fraction of the incoming silane is consumed. Near the entrance the rate is slightly lower because the incoming gases have not yet heated to the deposition temperature. However, the deposition rate of about 1 nm/min is lower than needed for economic deposition.

As the deposition temperature increases to 625°C, the deposition rate increases by approximately one order of magnitude to about 10 nm/min. However, the deposition rate decreases moderately along the length of the reactor as a significant fraction of the silane reactant gas is consumed. For these deposition conditions and reactor dimensions, about 30% of the silane is consumed by deposition on the 100-wafer load in the reactor, and about another 15-20% is deposited on the walls of the reactor. In the hot-wall, low-pressure reactor, both sides of each wafer are coated with silicon, increasing the reactant gas depletion. (Deposition on the backs of the wafers can also require extra processing to remove this polysilicon before further device processing.)

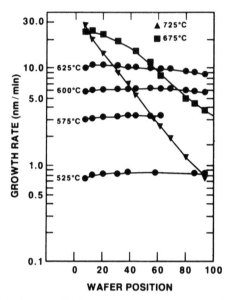

Figure 1.9: In the hot-wall, low-pressure reactor, the deposition rate decreases along the length of the reactor when a significant fraction of the reactant gas is consumed. This gas depletion is more important at higher deposition temperatures.

Several approaches can be used to achieve a uniform deposition rate along the length of the deposition zone in the hot-wall, low-pressure reactor. In one approach, a deposition parameter is varied along the reactor length to compensate for gas depletion. In the reaction-limited operating regime the deposition rate $R_D(x)$ at a position x along the reaction chamber can be written

$$R_D(x) = A \exp\left(-\frac{E_a}{kT}\right) C(x) \qquad (1.17)$$

where $C(x)$ is the silane concentration. Because of the strong temperature dependence of the deposition rate, a small temperature increase along the length of the reaction chamber can be used to compensate for *moderate* gas depletion, so that the thickness of the deposited films remains constant along the length of the reaction chamber. A temperature difference of 20-40°C is typically used to compensate for gas depletion.

1.5. REACTOR GEOMETRIES

At an even higher temperature of 725°C, the maximum deposition rate increases. However, a large fraction of the silane is consumed by deposition on the walls of the reactor before the gases reach the wafer-containing region, and most of the remaining silane is consumed on the first few wafers in the reactor. This almost complete depletion of silane cannot be compensated by a temperature gradient. Because gas depletion increases rapidly with increasing temperature, the low-pressure reactor can only be operated over a narrow temperature range. For deposition of polysilicon films, the operating temperature is usually as high as possible to increase the deposition rate, but low enough to avoid severe silane depletion.

Obtaining the desired structure in the silicon film may further constrain the deposition temperature. As we will discuss in Sec. 2.5, the crystal structure in the deposited film depends strongly on the deposition temperature. In some applications the electrical properties depend sensitively on the film structure, and variations caused by a temperature gradient cannot be tolerated.

To achieve a uniform thickness without a temperature gradient, the reactant gases can be injected into the reaction chamber at several points along its length, rather than only at one end. However, the extra fixtures needed for multi-point injection complicate the deposition system. Because the additional injectors are at approximately the deposition temperature, deposits can form within the injectors, increasing the required maintenance.

In the tube-type reactor the gas flow and pressure, as well as the local temperature, can affect the thickness uniformity along the length of the reaction chamber, although the effect of gas flow is not as significant as is that of temperature. Some improvement in uniformity can be obtained by increasing the gas velocity so that less of the silane has time to react before being transported to downstream wafers. The effect of varying the gas flow while keeping the reactant partial pressure constant is shown schematically in Fig. 1.10 [1.8]. If the flow rate increases while the pressure remains constant, the deposition rate is unchanged near the inlet end of the reactor. The same amount of reactant is consumed at the upstream end of the reactor, but the fraction consumed is less. More reactant gas is then available to travel to downstream wafers, and the thickness becomes more uniform along the entire wafer load. (This is equivalent to saying that the partial pressure of the reactant gas does not decrease as rapidly with position.) A similar trend is

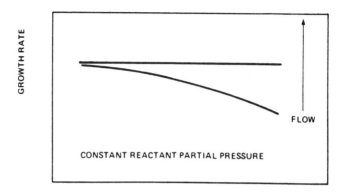

Figure 1.10: As the gas flow increases, the reactant gas can reach the downstream wafers more readily, and the deposition rate there increases. From [1.8]. Reprinted with permission of Solid State Technology.

found if the area on which the deposit is formed is reduced; *eg*, by spacing the wafers farther apart. If the pumping speed of the reactor is increased while keeping the reactant flow constant, the deposition rate also becomes more uniform along the entire wafer load, although at the expense of a lower deposition rate. The increased gas velocity lowers the partial pressure of the gas near the inlet end of the reactor so that a smaller fraction of the gas is consumed, again allowing more to reach downstream wafers. In both cases, however, the deposition efficiency decreases; *ie*, the fraction of the reactant gas deposited on the wafers decreases, increasing gas usage and operating costs.

An alternate, hot-wall, reactor geometry was suggested to eliminate the long length of the forced-convection region [1.11, 1.12]. In this *vertical-flow reactor,* as in the tube-type, low-pressure reactor, the wafers were closely spaced in a stacked configuration. However, the gases were injected along the entire length of the wafer load. They flowed between a pair of adjacent wafers and were then removed from the deposition chamber without passing other wafers. Because of the short path of the reactant gases, less gas depletion occurred. In the tube-type reactor the thickness changes by a factor of 35 over a 100-wafer load at 725°C (Fig. 1.9) [1.11], while it varied only by a factor of 2 or 3 in the vertical-flow reactor over a load containing two 50-wafer boats. However, this reactor did not become popular, partially because of the complicated fused-silica process chamber.

1.5. REACTOR GEOMETRIES

Operating sequence

Depositing layers requires a carefully controlled sequence of overhead procedures, in addition to the actual deposition. In the hot-wall, low-pressure reactor, the wafers are transferred from cassettes to a fused-silica wafer carrier, usually by robots in the vertical reactor and manually in the horizontal reactor. The fused-silica boat is inserted into the hot zone of the reactor, and the door is sealed. When the cold boat and wafers are inserted into the reactor, the reactor walls cool as heat is transferred to the wafers. While the wafers are heating and the reactor temperature is stabilizing, the reactor is pumped to base pressure for a specified length of time or until it reaches a specified pressure. Nitrogen may then be inserted to increase the pressure, and the reactor is pumped to its base pressure again. After reaching its base pressure, the reactor is isolated from the pump, and the increase of pressure in the chamber is monitored for several minutes to detect large leaks in the chamber (often caused by the loading door not sealing properly). If the rate of pressure increase is below a specified limit, the deposition process can start. The reactant gases are introduced, often slowly at first to avoid turbulence, which can move particles onto the wafer surface. The deposition continues for a given time calculated from the predetermined deposition rate and the desired thickness. At the end of the deposition time, the reactor is again pumped to base pressure to remove all the reactant gases, and then nitrogen is inserted to increase the pressure to atmospheric pressure before the reactor door is opened and the wafers are unloaded.

Details of the operating procedure vary with the particular equipment being used and with the device being processed. For example, to avoid unwanted oxide formation on exposed silicon regions during loading, the reactor may be cooled somewhat ($\sim 100°C$) before loading the wafers and then heated to the deposition temperature before starting the deposition. Because of the large thermal mass that must be heated, this change in temperature takes a considerable amount of time and decreases throughput. It is, therefore, only used when critical [for example, for deposition of the polysilicon emitter of bipolar transistors (see Sec. 6.8)]. For most applications, the entire surface of the wafer is covered with oxide, and the furnace is kept at the deposition temperature during loading.

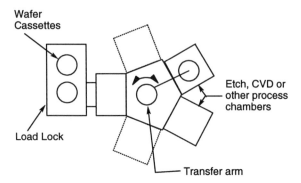

Figure 1.11: Top view of the arrangement of the chambers of a cluster tool, showing load locks and process chambers *clustered* around a central wafer handler.

Because of the large number of overhead steps in addition to the actual deposition step, the total cycle time can be significantly greater than the actual deposition time. However, depositing on a large number of wafers reduces the overhead time per wafer to a reasonable value.

1.5.2 Single-wafer, cold-wall reactor

Most polysilicon layers are deposited in the hot-wall, low-pressure, large-batch reactors. However, with increasing wafer size, processing one wafer at a time becomes practical. Single-wafer processing is especially useful when closely related process steps can be performed in different chambers of a *cluster tool,* in which wafers can be moved from one chamber to another without exposing the freshly formed surface to air (Fig. 1.11). Using a cluster tool allows excellent control of the interface properties, and is critical in applications such as the sequential deposition of polysilicon and tungsten silicide. Any oxygen on the polysilicon surface interferes with the uniform deposition and properties of the silicide. Transporting the wafer from the polysilicon deposition chamber to the tungsten silicide deposition chamber in the oxygen-free, controlled ambient of a cluster tool can greatly improve the interface quality and device yield.

Another potential application of *cluster processing* is the gate stack of an MOS integrated circuit. As the gate dielectric thickness decreases for advanced devices, it becomes more sensitive to contamination and degradation from handling. Cleaning the wafer surface in one chamber

1.5. REACTOR GEOMETRIES

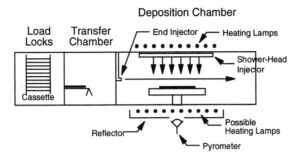

Figure 1.12: Cross section showing the basic elements of a typical cold-wall, single-wafer reactor, including the wafer support plate and the lamp heating of the deposition chamber, as well as the load lock and wafer transfer chamber.

of a cluster tool, forming the gate dielectric in another chamber, and then depositing the polysilicon in a third chamber can improve interface control and potentially increase yield by avoiding air exposure and extra handling. Because of likely contamination from an oxygen ambient, the polysilicon is best deposited in a different chamber than used to form the gate dielectric.

Single-wafer processing requires very rapid deposition [1.13] and also low overhead time to obtain a wafer throughput that can justify the cost of the equipment. Because the overhead time is not divided by the large number of wafers in a batch reactor, reducing the overhead time is especially important, as will be discussed below.

Deposition chamber

Most single-wafer deposition chambers have cold-walls and are heated with lamps. The deposition conditions appropriate for such chambers are markedly different from those in the hot-wall, low-pressure, batch reactors discussed above. The basic elements of the deposition chamber and associated portions of the reactor are shown in Fig. 1.12.

In a single-wafer reactor, the wafer is usually placed on a support plate with moderate thermal mass to improve uniformity of temperature and gas flow. Improved temperature control is especially important when multiple layers are present on the wafer surface, changing the reflectivity of the wafer. However, the moderate mass of the support plate precludes extremely rapid heating and cooling of the wafer.

The support plate also provides better control of the gas flow. Cold-wall, single-wafer reactors usually operate at higher pressures ($\sim$10 Torr to atmospheric pressure) than tube-type, hot-wall, low-pressure reactors ($\sim$0.1-1 Torr). The partial pressure of silane in the single-wafer reactor is usually only moderately higher than that used in the hot-wall, low-pressure reactor. To obtain an adequate gas velocity at the higher pressures a large flow of carrier gas is usually added to the small flows of reactant gases. Hydrogen is often used as the carrier gas because of its purity, even though it is a product of the deposition reaction and, therefore, decreases the deposition rate. The reactant and carrier gases are usually introduced at one end of the process chamber so that the forced-convection region is approximately parallel to the wafer surface. The support plate is larger than the wafer and is surrounded by coplanar fixtures so that the gas flows uniformly in the forced-convection region. The reactant gas then diffuses through the boundary layer between the forced-convection region and the wafer surface. Alternatively, the reactant gases can be introduced through a *shower-head* type gas injector near the top of the chamber to distribute the gas more uniformly over the wafer surface. However, this type of gas injector can interfere with the lamp heating above the wafer.

The support plate is usually graphite covered with an impervious coating of CVD silicon carbide. Graphite was first used in induction-heated reactors where good electrical conduction was needed for adequate electrical coupling of energy from the heating coils located outside the fused-silica process chamber to the support plate, which served as an electrical *susceptor*. The impure graphite was coated with silicon carbide to prevent impurities from moving into the silicon wafer being processed. The conducting graphite is not needed in lamp-heated reactors, but the technology to make silicon-carbide-coated graphite support plates was well developed for induction-heated reactors and is convenient to use in other reactors.

Most single-wafer reactors have cold walls to be compatible with the requirements of cluster tools. The wafer is usually heated by arrays of tungsten-halogen lamps operating with very high temperature filaments above and below the the fused-silica deposition chamber. The light passes through the clear chamber walls without significant absorption and heats the wafer and support plate. The walls remain fairly

1.5. REACTOR GEOMETRIES

cool. However, they are heated by radiation and convection from the hot parts of the system, so they must be cooled adequately – usually by forcing air across the outside of the fused-silica chamber and then cooling the air by transferring the heat to cooling water in a heat exchanger. By controlling individual lamps or small sets of lamps, good temperature uniformity can be obtained across the wafer. Arc lamps and even furnace heating have also been used in single-wafer reactors.

For uniform gas flow across the wafer in the direction perpendicular to the gas flow, the distance from the wafer surface to the top wall of the chamber is kept as constant as practical. However, the top wall must be curved or reinforced to provide mechanical strength if the reactor operates at reduced pressure. A thin, flat fused-silica wall cannot support the large force that occurs when the pressure inside the reaction chamber is reduced.

Process

While the basic deposition kinetics in a single-wafer are the same as in other reactors, the region of optimum operation may be different. Because of the limited thermal mass and cool walls of the deposition chamber, the forced-convection region remains fairly cool, allowing flexibility in the deposition process. Because the reactant gases flow along the length of the reaction chamber primarily in this relatively cool region forced-convection region, the cold-wall reactor can be used over a wide temperature range. The major portion of the temperature gradient, as well as the reactant-gas concentration gradient, occurs across the boundary layer, and the gases do not usually react until they diffuse through the boundary layer to approach the hot wafer surface.

In any single-wafer reactor the deposition rate must be high to justify the cost of the system. Because the deposition rate of polysilicon increases rapidly with increasing temperature, the deposition temperature in the single-wafer reactor is usually higher than in the hot-wall, low-pressure reactor. Although the reactor can operate in the higher-temperature, mass-transport-limited regime of operation because the gas flow is well controlled, the deposition temperature is usually limited by the need to control the crystal structure of the polysilicon. As we will discuss in Chapter 2, at high temperatures the increased surface diffusion of the silicon atoms decreases the nucleation density and increases the grain size. To obtain smooth, continuous films, thin layers of polysilicon are usually deposited in the lower-temperature, reaction-rate-limited region of operation. Operating in this temperature-sensitive region requires excellent control of temperature across the wafer and reproducibility from cycle to cycle. As discussed above, in the reaction-rate-limited regime, a 10°C temperature variation causes about a 25-30% thickness nonuniformity near the 700°C temperature often used to deposit polysilicon in this reactor, while a similar temperature variation in the mass-transport-limited regime, causes only about a 2% thickness nonuniformity. For thicker layers the temperature may be chosen to be high enough for operation in the mass-transport-limited region, where control of the gas transport is more critical than temperature control.

The cool forced-convection region also allows a higher reactant partial pressure to be used without severe gas-phase reaction and creation of silicon particles. As we will see in Sec. 1.6.3, the deposition rate initially increases linearly with the partial pressure of SiH_4, and then saturates when all available adsorption sites are covered with unreacted SiH_4 or with reaction products that have not desorbed. This restriction is more severe at lower temperatures, again favoring deposition at a higher temperature in the single-wafer reactor.

Because the forced-convection region of the cold-wall reactor remains relatively cool, reactant gases can flow along the deposition chamber in the forced-convection region without reacting significantly, thereby reducing gas depletion. The limited surface area on which the silicon is being deposited, also reduces gas depletion. Therefore, gas depletion is not as severe as in the hot-wall, batch reactor, allowing a higher deposition temperature to be used. Adequate uniformity is obtained (assuming uniform temperature) by rotating the wafer slowly (~10s of rpm) during deposition to compensate for the decreasing partial pressure of the reactant gas as it moves across the wafer toward the exhaust end of the reactor.

Using a cold-wall reactor provides the flexibility to obtain the needed high deposition rate. However, the high deposition rate and generally higher deposition temperature can affect the structure of the deposited polysilicon, requiring careful consideration of the suitability of the new crystal structure for the desired device. For example, as we will see in Chapter 3, dopant diffusion is greatly affected by the crystal structure of the polysilicon. When polysilicon is used as the gate electrode of an MOS transistor, the electrical performance of the transistor is severely

1.5. REACTOR GEOMETRIES

degraded if the dopant does not penetrate the entire thickness of the polysilicon, making control of the crystal structure important.

While both sides of the wafer are accessible to the reactant gas in the low-pressure, hot-wall reactor, the film is only deposited on one side of the wafer in the single-wafer, cold-wall reactor because the wafer is placed on the support plate.

In a single-wafer reactor the temperature is often measured with an optical pyrometer, which senses black-body radiation, and is corrected for the lower emissivity of the surface on which it is focused. Because the reflectivity of the wafer surface changes during polysilicon deposition, the pyrometer is positioned to sense radiation from the back of the support plate. However, any deposition on the back of the support plate can change its emissivity and produce erroneous temperature measurements. If polysilicon deposits on the back of the support plate, the emissivity and indicated temperature can vary during the deposition. Coating the back of the support plate with a thick layer of polysilicon before loading the wafer can reduce further variations in emissivity during deposition on the wafer.

Rapid thermal processors

Although the wafer usually sits on a support plate with moderate thermal mass, the support plate can be eliminated so that the wafer is supported on fused-silica pins with low thermal conductivity, as in a *rapid thermal processor*. However, the requirements on such a system for chemical vapor deposition are severe. The temperature across the entire wafer must be unform, and the local temperature must not be affected by the varying local reflectivity arising from different oxide thicknesses. In addition, to obtain a uniform deposition rate across the wafer, the gas flow must not be perturbed when the gases flows onto the edges of the wafer. These severe additional requirements have impeded the acceptance of rapid thermal processors for chemical vapor deposition.

The change in reflectivity, and consequently power absorption, during deposition is especially important in a rapid thermal processor, which does not have the stabilizing thermal mass of a support plate. The reflectivity of a silicon wafer covered with various layers depends on constructive and destructive interference of light from all the interfaces (Fig. 1.13), which in turn depends on the thickness and index of refraction of each layer. As the thickness of the polysilicon layer changes

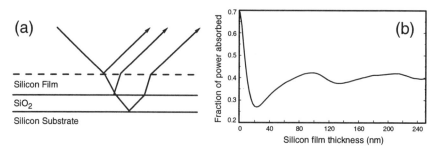

Figure 1.13: The reflectivity (a) and power absorbed (b) is determined by interaction of transmitted and reflected light from each interface between regions with different refractive indices.

during the deposition, the reflectivity and consequently the fraction of the incident lamp heating power absorbed also changes. If the heating power is kept constant, the temperature of the wafer can change markedly, causing major changes in the deposition rate and possibly even the structure of the polysilicon.

An extreme case is illustrated in Fig. 1.14 [1.14]. Polysilicon was deposited on an oxidized wafer using SiH_4 in a low-thermal mass, lamp-heated reactor while keeping the lamp power constant. At the start of the deposition, a polycrystalline structure formed. However, as the polysilicon thickness increased, interference between the light reflected from the top surface and from the underlying interfaces increased the overall reflectivity. Consequently, the amount of power absorbed by the wafer decreased, as did the temperature. The temperature of the wafer was no longer high enough to allow deposition of a polycrystalline structure, and the next portion of the silicon layer deposited was amorphous. As the thickness of the deposited silicon layer increased, the interference continued changing. Eventually the reflectivity decreased, and the amount of power absorbed increased. The temperature then increased and the structure of the depositing layer again became polycrystalline.

A temperature sensor, which measures a part of the system with constant reflectivity and feedback control circuitry can minimize the effect of reflectivity changes on the wafer surface. However, temperature variations across the wafer and the associated nonuniform reflectivity can degrade thickness uniformity [1.15].

1.5. REACTOR GEOMETRIES

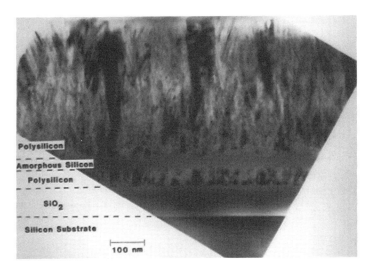

Figure 1.14: (a) Transmission electron micrograph, showing the change from the polycrystalline structure of the layer initially deposited on the oxide-covered substrate to an amorphous structure, and the subsequent renucleation of a polycrystalline structure as the reflectivity and absorbed power change. (b) Calculated variation with deposited thickness of the fraction of incident power absorbed.

Overhead time

The economics of single-wafer reactors require low overhead time, as well as a very short deposition time. The low overhead time requires integration of different steps peripheral to the deposition and imposes requirements on the construction of the wafer-transfer chamber and the wafer loading mechanism. However, measures needed to reduce the overhead time are basically compatible with the idea of controlling the ambient that the wafer sees between related process steps in a cluster tool. Key features of single-wafer reactors that allow reduced overhead time include a load lock and transfer chamber compatible with the hydrogen ambient often used as the carrier gas, a separate cooling chamber, automated wafer handling at reduced pressure, and reduced thermal mass.

The load lock and transfer chamber used with most single-wafer reactors and all cluster tools allow the wafer to be moved from the storage cassette into the chamber and between chambers in a nitrogen ambient.

Thus, during wafer loading and unloading, the deposition chamber can be kept in a hydrogen ambient at an elevated temperature and usually at the deposition pressure. In addition to reducing the overhead time associated with loading, purging, and pumping the chamber and heating the wafer, keeping the deposition chamber hot and in an oxygen-free ambient also improves purity of the ambient and the deposited film.

The wafer is loaded into the deposition chamber and removed at the deposition temperature (or as close to it as practical) to avoid the time needed to heat or cool the thermal mass of the support plate. Therefore, the wafer transport system must be capable of operating up to about 700°C. Second, the ambient in the deposition chamber should be the same as the carrier gas used for the deposition to eliminate the time needed to change gases in the deposition chamber. Because some of the carrier gas from the deposition chamber enters the transfer chamber during loading and unloading, the transfer chamber must be constructed to be compatible with hydrogen (*ie*, compatible with an explosive gas) even though the bulk of the gas in the transfer chamber is usually nitrogen. Some nitrogen enters the deposition chamber while the wafer is being inserted, but it is quickly pushed out of the chamber by the high hydrogen flow. To further reduce the overhead time, the transfer chamber should be at the same pressure as the deposition chamber. Because the wafer is hot when it is removed from the process chamber, it may need to be cooled by placing it on a heat sink in a cooling chamber before it is moved back into the plastic cassette in the load lock.

Although the automated wafer handling and controlled ambient in the transfer chamber reduces the oxygen impurity content in the deposition chamber, possible transfer of unwanted materials (either in the gas-phase or adsorbed on the wafer surface) from chamber to chamber still needs to be considered.

Because polysilicon is deposited on the parts of the support plate not covered by the wafer and some residue forms on the walls of the process chamber, *in situ* cleaning of the chamber in a mixture of gaseous H_2 and HCl is needed frequently — after every wafer or small group of wafers. Because the support plate must be heated to a high temperature ($\sim$1150°C) during the *in situ* cleaning, the thermal mass of the support plate should be fairly low to reduce the time needed for heating and cooling. Removing any deposit on the walls of the chamber is critical so that the heating radiation from lamps outside the process chamber reaches the wafer. The light from the lamps is not appreciably absorbed

1.5. REACTOR GEOMETRIES

by clean fused silica, so the walls remain fairly cool. However, if the walls become coated with silicon, a significant fraction of the heating power from the lamps is absorbed in the walls, heating them further and causing more deposition. Deposits on the walls also make temperature sensing with an optical pyrometer difficult.

1.5.3 Cold-wall batch reactor

The cold-wall, single-wafer reactor discused in Sec. 1.5.2 was derived in the late 1980s and early 1990s from an older cold-wall, horizontal, batch reactor that usually operated at atmospheric pressure. This older reactor was developed in the late 1960s and dominant in the 1970s. It was used in much of the initial exploration and development of polysilicon deposition, and the films deposited in it were extensively characterized. Because the forced-convection region remained fairly cool, this reactor could operate over a wider range of deposition temperatures than the low-pressure, hot-wall reactor, allowing polysilicon deposited over a wide range of temperatures to be investigated. However, in the older reactor, the wafers were placed in one plane on the support plate, so its wafer capacity was limited and also decreased as the wafer diameter increased.

For high-volume applications, a larger number of wafers needs to be processed at one time, and the reactor capacity should not depend strongly on the wafer diameter. Therefore, after the range of suitable operating parameters was determined in the versatile cold-wall reactor, deposition variables consistent with the limited temperature range of the hot-wall, low-pressure reactor were chosen so that this high-volume reactor could be used for routine manufacturing starting in the late 1970s.

Because we want to explore the properties of films deposited over a wide range of deposition conditions in subsequent chapters, we will necessarily be discussing films deposited in the older, cold-wall reactor, as well as those deposited in the low-pressure reactor and the single-wafer reactor.

Although conceptually similar to the cold-wall, single-wafer reactor, the older cold-wall reactor did not have a wafer transfer chamber, and air entered the deposition chamber while wafers were being loaded and unloaded. (The chamber was cooled and purged in nitrogen before being opened to air.) The air entering the chamber between depositions and the limited system integrity possible in the early 1970s restricted the purity of the ambient in the process chamber during deposition. The limited purity, in turn, required a higher temperature to obtain the same crystal structure obtained at a lower temperature in a purer ambient. The deposition temperature used to obtain a given structure in the older reactors is considerably higher than that needed to obtain the same structure in newer reactors. Therefore, the temperatures used in some of the earlier studies cited in subsequent chapters do not correspond to temperatures used in newer reactors to obtain the same crystal structure. However, the same trends are observed in the different reactors as the temperature varies, and these trends are the main theme of our discussion.

1.6 Reaction

In Sec. 1.3 we saw that the overall deposition process consists of a number of individual steps. Thus far, we have considered primarily gas flow in the forced-convection region and in the boundary layer and have discussed how diffusion through the boundary layer interacts with reactions occurring at or near the wafer surface. Some discussion of the microscopic chemical processes occurring is needed for a better understanding of the deposition process and its limitations so that the deposition conditions can be optimized. Because we are dealing with the chemistry of the deposition process, this basic discussion applies to any of the reactors used to deposit polycrystalline silicon. Gaining an understanding of the surface chemistry involved is important with any reactor operating in the reaction-rate-limited regime, but it is most important for the tube-type, hot-wall, low-pressure reactor, in which slight changes in the chemistry can move the deposition process out of the reaction-rate-limited regime, significantly degrading the uniformity.

As we mentioned in Sec. 1.3, the processes occurring at or near the surface include gas-phase or surface decomposition of the silicon-containing gas, adsorption of the silicon-containing gas or reaction intermediates on the surface, reaction to form elemental silicon, diffusion of the silicon to stable sites, incorporation into the depositing layer, and desorption of reaction byproducts. We now discuss these processes in more detail. To simplify our discussion, deposition from silane is considered first, and then disilane is discussed, although many characteristics of the processes are common to both gases.

1.6.1 Decomposition of silane

Silane can decompose either on the surface (*heterogeneous decomposition*) or in the gas phase (*homogeneous decomposition*). To form the dense, low-defect layers needed for silicon integrated circuits, heterogeneous decomposition should dominate. At least the final stage of the decomposition reaction should occur from species adsorbed on the wafer surface, even though intermediate species may be formed in the gas phase.

In certain ranges of temperature and pressure, partial decomposition of silane in the gas phase to form the gaseous intermediate silylene (SiH_2)

$$SiH_4 \rightarrow SiH_2 + H_2 \tag{1.18}$$

occurs during the deposition of silicon films [1.16, 1.17]. Thermodynamic data show that reaction of SiH_4 to form SiH_2 at atmospheric pressure is favored at higher temperatures, as shown in Fig. 1.15 [1.17]. However, at lower-temperatures and at pressures typical in low-pressure reactors, the rate constant associated with forming SiH_2 decreases several orders of magnitude [1.18], and forming SiH_2 is less likely. Furthermore, kinetic factors, such as the short residence time of the gas in the reactor (about 0.1 s for the low-pressure reactor), impede approaching the thermodynamic equilibrium concentrations of SiH_2 and other reaction intermediates. Therefore, for deposition of undoped layers in the low-pressure reactor the dominant reaction occurs through the adsorption of silane itself on the wafer surface, where it decomposes.

Homogeneous decomposition

Under adverse conditions, however, extensive gas-phase reaction and particle formation can occur in the gas phase. Decomposition of silane to form elemental silicon in the gas phase allows silicon atoms to agglomerate into clusters or larger particles in the gas phase; when the particles reach the surface, they are likely to form isolated defects, as well as a rough, porous layer unsuitable for integrated-circuit applications.

Homogeneous reaction occurs especially (1) at high gas temperatures, (2) at high concentrations of the silicon-containing gas, and (3) with the more reactive silicon-containing gases, such as disilane, which react at lower temperatures. Homogeneous reaction also occurs more readily in inert carrier gases, such as nitrogen and argon, than in hy-

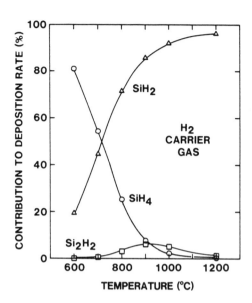

Figure 1.15: The relative contributions to the deposition process by the different chemical species present in an atmospheric-pressure reactor as a function of temperature. From [1.17]. Reprinted by permission of the publisher, The Electrochemical Society, Inc.

drogen. Hydrogen suppresses the decomposition reaction because it is one of the reaction products. Because the forced-convection region of the hot-wall, low-pressure reactor is at approximately the same temperature as the wafers, gas-phase reaction occurs more readily in hot-wall reactors than in cold-wall reactors, where the forced-convection region is much cooler than the wafers. The lack of a hydrogen carrier gas also promotes gas-phase reaction in the hot-wall, low-pressure reactor.

The gas-phase nucleation of silicon particles has been proposed to occur through polymerization reactions in which silylene reacts with a silicon polymer $Si_j H_2$ to eliminate a hydrogen molecule and form a higher polymer [1.19]:

$$SiH_2 + Si_j H_2 \rightarrow Si_{j+1} H_2 + H_2 \qquad (1.19)$$

Because larger polymers diffuse less rapidly than the monomers SiH_4 and SiH_2 in the gas phase, they do not readily move into the narrow spaces between wafers in the low-pressure reactor. Consequently, their incorporation into the depositing silicon layer in parallel with deposition

1.6. REACTION

from the monomers tends to degrade the thickness uniformity, as well as the film quality.

Measurable gas-phase reaction is seen above about 0.2 torr at a deposition temperature of 625°C in the hot-wall, low-pressure reactor, although large silicon particles are not visible until the temperature is increased to about 750°C at this pressure. However, even at 625°C and higher silane partial pressures, the surface of the depositing film can become rough because of the inclusion of silicon clusters formed in the gas phase. Therefore, the partial pressure and temperature must be limited to avoid decomposition of the reactant gas to elemental silicon in the gas phase, although partial decomposition to the gas-phase intermediate SiH_2 is sometimes useful.

1.6.2 Surface adsorption

After possible gas-phase reactions which partially decompose the silane, the remaining SiH_4 and any reaction intermediates (assumed to be SiH_2 in our discussion) travel to the wafer surface, where they have a high probability of adsorbing according to the reactions

$$SiH_4 + * \rightarrow SiH_4{}^* \qquad (1.20)$$

and

$$SiH_2 + * \rightarrow SiH_2{}^* \qquad (1.21)$$

where * represents a free surface site, and $SiH_4{}^*$ and $SiH_2{}^*$ are the adsorbed species.

In determining the dominant species adsorbed on the surface, the relative adsorption energies of the species should be considered, along with the amount of each in the gas phase. Silane is a fully coordinated molecule with four hydrogen atoms bonded to the silicon in a tetrahedral configuration. If the silane adsorbs without dissociating, one of the peripheral hydrogen atoms must be in contact with the substrate, forming a H–Si bond. On the other hand, SiH_2 is more strongly adsorbed; either of its two hydrogen atoms can form Si–H bonds or its silicon atom can form a stronger Si–Si bond with the silicon surface. Silicon atoms with their uncompleted bonds should also be strongly adsorbed [1.20]. Thus, the higher partial pressure of SiH_4 in the gas phase is partially compensated by the stronger adsorption of reaction intermediates on the surface. The behavior seen when dopant is added during the deposition (Sec. 1.8) suggests that SiH$_4$ plays the dominant role during deposition of undoped films in the low-pressure reactor.

After adsorbing, the reacting species may be firmly bound where they first adsorb and remain fixed until they re-evaporate or decompose, or they may diffuse on the surface. The surface diffusion coefficient D_S of adsorbed atoms is given by

$$D_S = D_{S0}\, \theta \exp\left(-\frac{E_D}{kT}\right) \qquad (1.22)$$

where E_D is the energy required for surface diffusion, D_{S0} is a constant, and θ is the fraction of surface sites which are unoccupied [1.21–1.23]. The associated *diffusion length* $L = \sqrt{Dt}$, where t is the time during which the adsorbed species can diffuse.

If the adsorbed silicon species decompose into Si and adsorbed H atoms, the decomposition reactions can be written

$$\mathrm{SiH_4^* + 4\,^* \rightarrow Si + 4H^*} \qquad (1.23)$$

and

$$\mathrm{SiH_2^* + 2\,^* \rightarrow Si + 2H^*} \qquad (1.24)$$

As the adsorbed silicon species diffuse on the surface, they encounter steps formed at the edges of partially completed layers of silicon previously deposited, as shown in Fig. 1.16, and also irregularities at grain boundaries and at defects within grains. Because of the greater number of silicon-silicon bonds possible, a *ledge* (step) has a lower energy than does a uniform silicon surface, and a *kink-site* (the intersection of two ledges) is still more energetically favorable. If the adsorbed molecules have not completely decomposed earlier, the final stages of the decomposition can occur at the ledge or kink sites, with reactions analogous to Eqs. 1.23 and 1.24. An adsorbed silicon atom remains, on the average, at a ledge or kink site longer than at a surface site. In addition, when an atom fills a kink site, it does not annihilate this low-energy site; the kink site is just displaced by one atomic position, providing an adjacent favorable position for another silicon atom. Both the longer residence time and the adjacent, energetically favorable site, favor subsequently arriving silicon atoms to surround an atom at a kink site. This step completes the bonding of the first atom, incorporating it firmly into the

1.6. REACTION

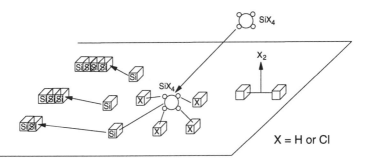

Figure 1.16: The incoming gas-phase species adsorb on the surface and decompose. The resulting silicon species diffuse on the surface to a stable ledge or kink site or other irregularity, where they are covered by subsequently arriving silicon atoms and firmly bound into the film.

depositing film and contributing to grain growth. (The initial stage of deposition — nucleation — will be discussed in Sec. 2.1.)

A hydrogen atom produced by the reaction can diffuse on the surface until it encounters another hydrogen atom with which it combines to form an adsorbed hydrogen molecule, which can then desorb, diffuse through the boundary layer to the forced-convection region, and be removed from the reactor. Because two hydrogen atoms are released at nearby sites by the decomposition of SiH_4 or SiH_2, adsorbed hydrogen molecules readily form.

1.6.3 Deposition rate

The deposition rate can be determined by quantitatively considering the mechanisms described above. One model [1.24] shows that the rate increases linearly with the SiH_4 partial pressure at low partial pressures and saturates at high SiH_4 partial pressures.

The silicon deposition rate is found by considering the sequential adsorption and reaction of the local gas-phase concentration of silane and the desorption of hydrogen molecules. To simplicity our discussion, the overall reaction $SiH_4 \rightarrow Si + 2H_2$ is assumed to proceed on the surface, and possible partial decomposition in the gas phase is neglected.

The rate at which SiH_4 molecules adsorb on the surface is proportional to the incoming flux F of SiH_4 molecules, the *reactive sticking coefficient* S_c (the probability that an impinging molecule adsorbs on the surface), and the number of sites available for adsorption $n_S \theta$:

$$R_{\text{adsorption}} = F_{\text{SiH}_4} S_c n_S \theta \qquad (1.25)$$

where n_S is the total number of surface sites per unit area, and θ is the fraction of surface sites available for adsorption. The fraction of surface sites available for adsorption is

$$\theta = 1 - \theta_S - \theta_H \qquad (1.26)$$

where θ_S is the fraction of surface sites occupied by silane molecules, and θ_H is the fraction of surface sites occupied by hydrogen. The rate at which adsorbed silane molecules are removed from the surface is the sum of the desorption rate and the rate at which adsorbed SiH$_4$ reacts to form solid silicon firmly bound into the depositing film; both terms are proportional to $n_S \theta_S$. Analogous expressions can be written for the adsorption, creation, and desorption of hydrogen. In steady state, the rate of addition of each species to the surface equals its rate of removal.

The deposition rate $R_D = dn_{\text{Si}}/dt$ is given by the reaction rate coefficient k for the decomposition reaction times the number of adsorbed SiH$_4$ molecules. An appropriate normalization constant converts from the number of deposited atoms per unit area n_{Si} to the thickness of deposited silicon.

$$R_D = \frac{1}{N_{\text{Si}}} \frac{dn_{\text{Si}}}{dt} = \frac{k}{N_{\text{Si}}} n_S \theta_S \qquad (1.27)$$

where N_{Si} is the atomic density of solid silicon.

To determine the deposition rate quantitatively, the local concentration of silane molecules in the gas phase must be known. Deposition of silicon on upstream wafers decreases the local concentration of silane available downstream. In a constant-pressure reactor, with two moles of H$_2$ reaction product created for each mole of SiH$_4$ consumed, dilution of the silane by hydrogen decreases the local concentration, as previously discussed. This dilution is significant in the low-pressure reactor where little or no carrier gas is used, and usually insignificant in reactors that use a large flow of carrier gas.

From the above expressions and the relation between the decomposition of silane and the creation of hydrogen, expressions for the surface coverage of silane and, consequently, the deposition rate can be obtained. Comparison with experimental data shows that the adsorption

1.6. REACTION

of silane and the desorption of hydrogen are generally the most important steps in the deposition process. The general expression for the deposition rate R_D at a given position x can be written in the form

$$R_D = \frac{A_1(1 - X_S)n_0}{(1 + X_S) + B_1 n_0} \tag{1.28}$$

where X_S is the fraction of silane consumed upstream of position x, n_0 is the input concentration of silane, and A and B are coefficients related to the reaction rate and equilibrium coefficients of the adsorption, decomposition, and desorption processes.

When the silane consumption is low ($X_S \approx 0$), Eq. 1.28 reduces to

$$R_D \approx \frac{A_1 n_0}{1 + B_1 n_0} \tag{1.29}$$

and the deposition rate is independent of position. Replacing the input silane concentration n_0 by the input partial pressure of SiH_4, we obtain the expression

$$R_D \approx \frac{A_2 P_{SiH_4}}{1 + B_2 P_{SiH_4}} \tag{1.30}$$

At low input silane partial pressures, the deposition rate increases linearly with increasing silane partial pressure. At high input silane partial pressures, a significant fraction of the adsorption sites are filled by unreacted SiH_4 and adsorbed H, and the deposition rate increases less than linearly with increasing SiH_4, until it finally saturates. This type of behavior has been observed in both the hot-wall, tube-type, low-pressure reactor (Fig.1.17) [1.24], in a vertical-flow, low-pressure reactor [1.11], and in cold-wall reactors operating at higher pressures, lending support to the model.

Adding a moderate amount of either hydrogen or nitrogen as a diluent gas can also influence the deposition rate [1.16]. With low partial pressures of SiH_4, where the deposition rate increases with increasing silane concentration, the deposition rate is lower when hydrogen is added than when nitrogen is used. However, the deposition rate becomes nearly the same for both diluent gases when the rate saturates at higher silane concentrations.

Figure 1.18 shows the effect of total pressure P_T (mainly hydrogen) on the deposition rate with constant SiH_4 partial pressure (not constant SiH_4 flow) for deposition at 675°C in a cold-wall, single-wafer reactor.

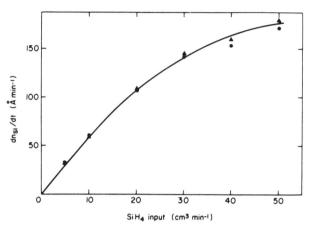

Figure 1.17: Deposition rate as a function of input silane flow in a type-type, hot-wall reactor, showing the linear dependence at low flows and saturation at high flows. From [1.24]. Reprinted with permission.

Over much of the pressure range, the deposition rate depends on total pressure approximately as $R \propto P_T^{-n}$, where n is ~ 0.5. The $P_T^{-0.5}$ dependence is consistent with hydrogen competing for surface sites at lower pressures where the deposition is reaction-rate limited. At higher pressures mass transport can start to limit the deposition process, and the $1/P_T$ dependence of the diffusion coefficient becomes important [1.25]. Other studies find n varying from small values to 1 [1.7, 1.25, 1.26], depending on the limiting mechanism of the deposition process.

1.6.4 Rate-limiting surface process

As we saw in Sec. 1.5.1, the deposition process in the hot-wall, low-pressure reactor is adjusted so that one of the surface processes is rate limiting, consistent with the strong temperature dependence observed (Eq. 1.17). Silane adsorption, surface diffusion, silane decomposition, silicon incorporation, or hydrogen desorption could limit the deposition process, and determining which individual process dominates is not straightforward.

For deposition of undoped films at low silane partial pressures, only a small fraction of the surface sites on which adsorption can take place are covered, and the rate is proportional to the partial pressure, as discussed above. As the partial pressure increases, the available sites

1.6. REACTION

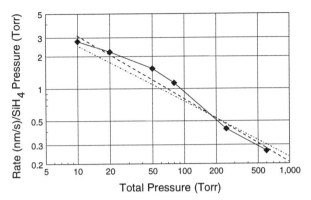

Figure 1.18: Decrease in deposition rate with increasing total pressure at 675°C in a cold-wall, single-wafer reactor. The deposition rate is normalized by the SiH_4 partial pressure to isolate the effect of the total pressure, which is > 94% H_2 over the entire pressure range shown. The dashed line has a slope of -0.5.

can become filled, and the rate becomes independent of the silane partial pressure. These observations suggest that the decomposition reaction on the wafer surface is the rate-limiting step. However, alternative rate-limiting mechanisms must also be considered.

Within a limited temperature range the deposition rate has the same temperature dependence for deposition of either polycrystalline silicon or amorphous silicon. In the latter case, surface diffusion should be minimal, while it can be significant for polysilicon deposition. Therefore, surface diffusion is unlikely to be the rate-limiting step.

In some cases, adding a strongly adsorbed species can limit the number of adsorption sites available for silane adsorption and decrease the deposition rate. We will consider this adsorption-limited deposition when we discuss the deposition of doped films in Sec. 1.8.

It has been proposed that at low temperatures, desorption of hydrogen from the surface is the rate-limiting step [1.27]. In this model, adding hydrogen to the gas phase impedes desorption of hydrogen from the surface, and therefore decreases the overall reaction rate by limiting desorption, rather than by suppressing the decomposition reaction directly. The decomposition is thought to be irreversible and insensitive to the hydrogen partial pressure under ordinary conditions [1.28].

1.7 Deposition from disilane

As the deposition temperature decreases, the deposition rate from silane decreases rapidly because of the high ($\sim$2 eV) activation energy for the surface reaction (Sec. 1.4.4). In large-area applications, such as active-matrix addressing of liquid-crystal displays, the substrate is often a low-cost, moderate-temperature glass (Sec. 6.13.3). The limited stability of the glass restricts the deposition temperature of the silicon layer used as the polysilicon channel of the thin-film transistors to $\leq$600°C. Obtaining the desired crystal structure further limits the deposition temperature to the mid-500°C range. Temperatures in the mid-500°C range are also needed for deposition of amorphous silicon for integrated circuits. In this temperature range the deposition rate from silane is impractically low, especially for cost-sensitive, large-area applications. The more reactive silicon source disilane Si_2H_6 is an attractive alternative silicon source at these lower temperatures.

Disilane fragments into SiH_4, SiH_3, and SiH_2 (silylene) in the gas phase [1.29, 1.30] by the reactions

$$Si_2H_6 \rightarrow SiH_4 + SiH_2 \qquad (1.31)$$

and

$$Si_2H_6 \rightarrow 2\ SiH_3 \qquad (1.32)$$

These gas-phase reaction intermediates adsorb on the surface and decompose there to form silicon and hydrogen. SiH_2 and SiH_3 decompose readily in the 500°C range, increasing the silicon deposition rate well above that of layers deposited from SiH_4; alternatively, the same rate can be obtained at a temperature about $\sim$100°C lower with Si_2H_6 than with SiH_4 [1.31].

However, relying on the more reactive Si intermediate species can cause other difficulties with the deposition process. In the hot-wall, low-pressure reactor with closely spaced wafers, the deposition conditions were carefully chosen so that the surface reaction was slow and was the limiting step in the overall deposition process. When silane is replaced by the more reactive disilane, the deposition process may no longer be limited solely by the surface reaction rate. If the ability of the gas to react on the surface becomes comparable to the ability to diffuse to the surface, the deposition rate near the center of the wafer can be markedly lower than the rate near the wafer edges, seriously degrading the thickness uniformity across the wafer.

1.8. DEPOSITION OF DOPED FILMS

In either the cold-wall or hot-wall reactor, the use of disilane can also be limited by premature decomposition of the disilane to elemental silicon in the forced-convection region. The silicon particles formed in the gas phase can reach the wafer surface, creating high particle densities and a porous deposit. This gas-phase nucleation is more serious at higher reactant partial pressures and in the hot-wall reactor, and can limit the practical deposition rate achievable with disilane.

In addition to the importance of disilane as an alternate deposition source, its behavior provides further insight into the deposition mechanism of undoped films from silane. If SiH_2 and SiH_3 played an important role in the deposition of undoped films from silane, as well as from disilane, the markedly different behavior of doped films deposited from the two sources (to be discussed in Sec. 1.8) would be unlikely. Therefore, formation of SiH_2 and SiH_3 in the gas phase probably does not play a dominant role in the deposition of undoped silicon films from silane in the low-pressure reactor under typical deposition conditions.

1.8 Deposition of doped films

In the previous section we saw that the deposition rate of polysilicon films is influenced by the number of adsorption sites on which the impinging molecules can adsorb. In the simplest case of undoped films, we only need to consider the adsorption of silicon- and hydrogen-containing species. When dopant gases are added to the system, we must consider the competition of additional species for the available adsorption sites to determine the deposition rate.

Adding the n-type dopant gases phosphine (PH_3) and arsine (AsH_3) during the deposition decreases the deposition rate significantly, while the rate increases when the p-type dopant gas diborane (B_2H_6) is added, as shown in Fig. 1.19 [1.32]. The changes are especially pronounced at low deposition temperatures where surface reactions dominate the deposition process. The deposition rate is scarcely affected at high temperatures where diffusion across the boundary layer limits the deposition rate [1.33]. The dependence of deposition rate on dopant-gas flow is seen in both atmospheric-pressure [1.33] and low-pressure [1.34] deposition systems. The behavior has been studied most thoroughly for deposition from silane, but a similar effect of adding arsine has been seen for films deposited from silicon tetrachloride [1.35]. In addition to its practical

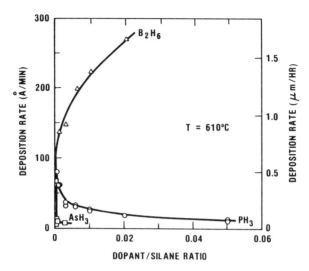

Figure 1.19: The deposition rate decreases when phosphine or arsine is added to the gas flow, while adding diborane increases the deposition rate. From [1.32]. ©1983, Bell Telephone Laboratories, Inc. Reprinted by permission.

importance, the effect of adding dopant gas on the deposition rate also provides information about the deposition mechanism of undoped films.

1.8.1 n-type films

At 680°C in an atmospheric-pressure reactor, the deposition rate decreases rapidly as the gas-phase concentration of either arsine or phosphine increases, until the effect saturates at high concentrations. The ratio of arsine to silane at saturation is about 3×10^{-4}, corresponding to a decrease in the deposition rate by about a factor of 7. For phosphine the deposition rate also decreases, but only by a factor of about 2.5 at saturation, and the ratio of phosphine to silane is about 2.5×10^{-3}. As we saw in Sec. 1.5.1, the deposition rate R_D can be described by the expression $R_D = R_{D0} \exp\left(-E_a/kT\right)$. For the n-type dopants arsenic and phosphorus, the apparent activation energy E_a is the same as for the deposition of undoped films, as shown in Fig. 1.20 [1.33]. The deposition rate is reduced by a change in the pre-exponential factor R_{D0}, which depends on the density of available adsorption sites.

1.8. DEPOSITION OF DOPED FILMS

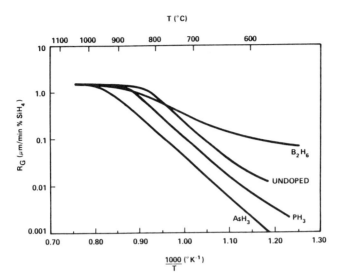

Figure 1.20: The activation energy of the deposition rate remains the same when phosphine and arsine are added but decreases when diborane is added. From [1.33]. Reprinted by permission of the publisher, The Electrochemical Society, Inc.

Of the dopant species considered, phosphorus has been investigated in most detail. Phosphorus or its precursors are adsorbed much more strongly than is silane (by at least 40 times). The adsorbed phosphorus-containing species block adsorption of silane by occupying adsorption sites, which are then unavailable for silane adsorption. The excess valence electrons on the phosphine molecule enable it to compete successfully for available surface sites with silane, which is a fully coordinated molecule. At low temperatures phosphine can adsorb without dissociating (ie, losing hydrogen) to form a strongly bound, adsorbed layer, while the silane molecule must dissociate before it can bond strongly to the silicon surface [1.36]. At higher temperatures the hydrogen desorbs from phosphine, vacating additional surface sites, on which more phosphorus can be adsorbed; at 600°C the phosphorus coverage can increase to about 4 times its room temperature value. At still higher temperatures the phosphorus desorbs.

When undoped polysilicon is deposited under typical conditions, most of the deposition occurs by the direct adsorption of SiH_4 on the surface and its decomposition into solid Si and adsorbed H. The adsorbed H combines with other adsorbed H and desorbs as H_2. When the adsorption of SiH_4 is blocked by phosphorus-containing species, the deposition rate decreases markedly.

However, the strong bonding of phosphorus to the silicon surface has been predicted to decrease the deposition rate even more than observed [1.37], leading to speculation that other silicon species are present to compete successfully with phosphine for adsorption sites. Normally both gas-phase and surface reactions of SiH_4 occur in parallel, but the deposition conditions are chosen so that adsorption and decomposition of SiH_4 on the surface dominate. However, when the adsorption of silane is blocked by the adsorbed phosphine, silicon deposition may occur from alternate, secondary species, such as silylene SiH_2, disilane Si_2H_6, and monosilylphosphine SiH_3PH_2, formed by reaction of silane in the gas phase. These alternate species are strongly adsorbed and can compete effectively with phosphine for adsorption sites, allowing deposition to occur. Because the concentrations of these alternate species are low, the deposition rate is markedly lower than that of undoped films.

In addition to the overall decrease in the deposition rate when phosphine is added, the film thickness across a wafer can become very nonuniform, especially in the low-pressure reactor [1.38, 1.39]. Surface reaction of the highly reactive, alternate silicon species occurs rapidly so that the deposition process is no longer limited by the surface reaction. When the ease of surface reaction becomes comparable to the ease of mass transport, the gases cannot readily reach the centers of the closely spaced wafers, and the deposited layer is thicker near the outer edges than near the center of the wafer. In addition to the degradation of the thickness uniformity, the phosphorus concentration also becomes nonuniform.

Deposition from disilane

When silane is replaced by an alternate silicon source, such as disilane Si_2H_6, that readily decomposes into more strongly adsorbed intermediate species, the deposition rate becomes less sensitive to the competitive adsorption of dopant species on the surface. The deposition rate of polysilicon from disilane decreases much less when phosphine is added than does the deposition rate from silane, leading to the use of disilane as a source gas for deposition of doped films. When mid-10^{20} cm^{-3} phosphorus atoms are incorporated, the decrease in deposition rate is about twenty times less for disilane than for silane [1.29]. The weak depen-

1.8. DEPOSITION OF DOPED FILMS

dence for disilane is attributed to the highly reactive intermediates SiH_2 (silylene), and SiH_3 formed as major constituents in the gas phase and subsequently strongly adsorbed on the surface to provide parallel paths for deposition. Deposition from silane is almost completely blocked by the strong adsorption of phosphine and phosphorus-related species. However, SiH_2 and SiH_3 are unsaturated, highly reactive species, which are very strongly adsorbed on the surface, so that deposition is impeded less by the competing adsorption of phosphine. Because these reactive species are major constituents when disilane is used, the overall deposition rate decreases only moderately when phosphorus is added.

The rapid surface reaction of SiH_2 and SiH_3 is consistent with the temperature dependence observed for deposition from Si_2H_6. Over a limited temperature range the deposition of undoped polysilicon from disilane is an activated process [1.29], with an apparent activation energy similar to that for deposition from silane [1.40]. When phosphine is added, the deposition path via silane is blocked, and deposition from SiH_2 dominates. In this case, a transition to a less-temperature-dependent, mass-transport-limited regime is seen in the same temperature range. The relatively difficult diffusion of the reactant gas between the closely spaced wafers degrades the uniformity of the deposited thickness across the wafer. In horizontal, low-pressure reactors, a *wafer cage*, made of fused silica rods or a fused-silica cylinder with holes in it, can be placed around the wafers to remove some of the highly reactive species before they reach the wafer, improving the thickness uniformity somewhat [1.41]. However, the wafer cage is not compatible with the automated wafer handling used in vertical, low-pressure reactors.

1.8.2 *p*-type films

The deposition rate is influenced quite differently when the *p*-type dopant gas diborane is added. The deposition rate *increases* by about a factor of two at a temperature of 680°C in an atmospheric-pressure reactor, saturating for diborane-to-silane ratios greater than about 2.5×10^{-3} [1.33, 1.42]. This increase in deposition rate with diborane can be advantageously used to deposit silicon films at very low temperatures, at which deposition of undoped films is too slow to be practical. In addition to the deposition rate increasing, it also becomes less sensitive to temperature when diborane is added. At intermediate temperatures, the activation energy of the deposition rate for heavily boron-doped films is about 0.9 eV, while it decreases to about 0.3 eV at lower temperatures. At even lower temperatures, an apparent activation energy of 0.54 eV is observed [1.43].

The cause of the increasing deposition rate for boron-doped films is more complex than the site-blocking mechanism proposed for the n-type films. It can be explained by the addition of a parallel deposition mechanism with a lower activation energy [1.44]. Normal deposition is attributed to the decomposition of silane on surface silicon atoms, while the additional parallel mechanism is thought to be the decomposition of silane on deposited boron surface atoms, which act as a catalyst. According to this model, the deposition rate of the parallel reaction path should be proportional to the number of surface boron atoms, $N_B^{2/3}$, in close agreement with the $N_B^{0.7}$ dependence observed. The diborane catalysis of silane pyrolysis is thought to be analogous to the use of diborane and hydrogen to produce other higher boranes; in these reactions, borine BH_3 has been suggested as an intermediate [1.43]. In addition to diborane, boron tribromide also increases the deposition rate [1.45].

Boron enhancement of the deposition rate is analogous to the increased deposition rate of silicon in a silicon-germanium alloy [1.46]. Hydrogen desorption can limit the overall deposition process, and hydrogen is less firmly bound to germanium than to silicon. Therefore, the deposition rate increases markedly when germanium is present to enhance the desorption of hydrogen, freeing more surface sites for adsorption of the silicon species. Similarly, surface boron may aid the desorption of hydrogen.

Deposition from disilane

When disilane is used as the silicon source, the deposition rate increases with increasing boron concentration, just as when silane is used [1.47]. This similar behavior for boron-doped films contrasts strongly with the limited effect on deposition rate when phosphorus is added during deposition of polysilicon from disilane. This difference suggests that the mechanism of deposition-rate modification by diborane is markedly different from that by phosphine and arsine. For disilane, as for silane, the addition of boron alters the surface reaction. The decomposition reaction appears to be enhanced at adsorption sites above boron atoms. As for boron-doped films deposited from silane, the activation energy of the overall deposition rate decreases when boron is added during deposition from disilane.

1.8.3 Electrostatic models

In addition to possible blocking of adsorption sites by dopant species or catalysis of parallel reaction paths, possible modification of the electrical surface potential by high dopant concentrations has been suggested as the cause of the changes in deposition rate with dopant addition [1.45]. The opposite influence of n-type and p-type dopants suggests that electrical factors may play a role in the deposition process. Another model, which relates the number of active adsorption sites for silane to the Fermi level in the polysilicon, has also been suggested to explain the decrease in deposition rate in heavily arsenic-doped films [1.48]. However, the site-blocking model appears to explain the behavior more readily than do models relying on surface potential changes.

1.9 Conformal deposition

In advanced integrated-circuit structures, irregular surface features must be covered with polysilicon, and in extreme cases, such as the trench structures to be discussed in Secs. 6.10.3 and 6.11.1, the polysilicon must be able to fill narrow, deep wells. Covering these irregular features depends on the ability of silicon-containing species to diffuse to all regions of the surface as the film grows. This diffusion can occur either in the gas phase or on the surface.

We saw in Sec. 1.5.1 that uniform deposition on closely spaced wafers is favored by operating in the surface-reaction-limited regime so that gas-phase diffusion occurs readily compared to surface reactions. Similar behavior applies to microscopic dimensions. Operating in the surface-reaction-limited regime promotes conformal coverage of device features, and undoped polysilicon films deposited under typical low-pressure CVD operating conditions are conformal.

The slow surface reaction is consistent with a low effective *sticking coefficient* (sometimes called the *reactive* sticking coefficient). With a low sticking coefficient the impinging gas molecules are likely to be reflected from the surfaces they strike. They can collide with the surfaces many times before being adsorbed and reacting, allowing them to penetrate into fine device features. The effective sticking coefficient for silane for typical low-pressure deposition of undoped polysilicon layers is $\sim 5 \times 10^{-5} - 1 \times 10^{-4}$ [1.49], allowing excellent filling of even very long, narrow features with undoped polysilicon, as shown in Fig. 4.9 [4.28].

As discussed in Sec. 1.8, when heavily doped films are deposited, the dominant species reaching the wafers may change, and the deposition process may no longer be totally surface-reaction limited. The effective sticking coefficient then increases, and filling deep features with conformal coatings becomes difficult. If gas-phase diffusion is not adequate (*ie*, if mass transport begins to limit), surface diffusion of adsorbed silicon species can also aid conformal coverage. Temperature produces two competing effects: Higher temperatures favor mass-transport-limited operation, which should degrade conformal coverage, but surface diffusion also increases at higher temperatures, improving the conformality.

If both gas-phase and surface diffusion are low, obtaining conformal coverage is difficult, and the shape of the deposited layer depends on the mean free path in the gas phase [1.32]. A long mean free path allows penetration of at least some of the gas to the bottom of deep device features. The deposited thickness is limited primarily by the angles at which molecules can travel from the gas supply above the wafer to different regions of the device feature. If the mean free path is short, molecules are likely to scatter before penetrating into narrow device features; they reach convex corners near the top of device features from many different angles and form a thick deposit there. The resulting *cusp* grows from both sides of a deep feature, finally bridging the feature and leaving a subsurface void.

Trench filling

Under typical reaction-rate-limited operating conditions, the gas-phase silicon species can penetrate into narrow spaces during deposition of undoped polysilicon. The thickness is approximately uniform along the walls of even very deep, narrow trenches [1.50]. The good filling of deep, narrow features relies on a low sticking coefficient. For undoped films, the sticking coefficient can be very low, as discussed above, and the narrow, deep trenches used for *trench isolation* between devices in integrated circuits (Sec. 6.10.3) are readily filled with undoped polysilicon.

As the walls are coated, the effective aspect ratio increases, making gas penetration to the bottom even more difficult, but good filling can be obtained if no subsurface "bulge" is caused by the trench etching process. Typically, the walls of the trench are sloped a few degrees from the vertical during etching to allow good filling. Adequate filling of trenches under mass-transport limited conditions at higher temperatures [1.51]

1.10. ENHANCED DEPOSITION TECHNIQUES

and filling of even finer features at lower temperatures [1.52] indicates that surface diffusion plays some role.

As n-type dopant is added, the adsorption of the less reactive species (with their low reactive sticking coefficients) is decreased by adsorbed dopant atoms, as discussed above in Sec. 1.8. As the dominant deposition pathway changes to more reactive species, the reactive sticking coefficient increases, and the conformal coverage degrades. Consequently, filling the deep trenches used for capacitors in dynamic random-access memories (Sec. 6.11) with doped polysilicon is difficult, even though filling the same trench with undoped polysilicon is straightforward.

Methods suggested for filling deep trenches with doped polysilicon include the following: (1) Depositing alternating layers of undoped and doped polysilicon and subsequently annealing to obtain a homogeneous dopant distribution [1.53]. The undoped layers provide good step coverage and filling, and the doped layers provide the dopant needed for electrical conduction. (2) Depositing undoped polysilicon to about half the thickness needed to fill the trench; then adding dopant from a gas-phase dopant source such as AsH_3 or PH_3; and finally completing the filling process with undoped polysilicon [1.54]. Subsequent thermal cycles distribute the dopant uniformly through the polysilicon.

Analogously, undoped, as well as doped, layers deposited using disilane are less conformal than those deposited using silane. The reactive sticking coefficients of the more reactive species involved in the deposition process can be 10-100 times higher than those of silane [1.49].

1.10 Enhanced deposition techniques

For integrated-circuit applications, polysilicon is deposited almost exclusively by thermal reactions; the low-600°C temperature commonly used for thermal deposition is compatible with the early part of the IC process, during which polysilicon is deposited. For thin-film transistors, however, polysilicon is often deposited on low-temperature glass, which usually mechanically deforms near 600°C or lower. At lower temperatures, the silicon deposition rate decreases markedly. To achieve a high deposition rate at temperatures compatible with inexpensive glass substrates, enhanced deposition techniques, which add in other ways some of the energy needed for deposition, have been investigated.

In the most common enhanced deposition technique, electrical energy is added from a plasma. Intense electric fields create energetic electrons, which can collide with neutral molecules and excite them. Less thermal energy is then needed to promote decomposition, and deposition proceeds more rapidly than if all the energy is supplied thermally. The deposition rate is also less sensitive to the wafer temperature. Polycrystalline silicon films can be deposited using a plasma [1.55]; increasing the low deposition rate of heavily phosphorus- or arsenic-doped films is especially attractive [1.56]. Plasma-enhanced deposition is often used to deposit amorphous silicon, which can be used in devices in the amorphous form or crystallized after deposition to form polysilicon.

A plasma is composed of electrons, ions, and excited neutral radicals. The probability of forming each species depends exponentially on its formation energy ΔE and the inverse of the energy or equivalent temperature $T_e = E_e/k$ of the exciting electron: $\exp(-\Delta E/kT_e)$. The energy needed to ionize a silicon molecule and create a SiH_4^+ ion and an electron is about 12.2 eV; the energy needed to create an excited neutral SiH_3 radical and a H atom is only about 3.5 eV. Because the energy to form a radical is much less than that needed to create an ion, neutral radicals are far more numerous. The excited radicals adsorb on the surface and decompose readily to form a deposited silicon layer.

The deposition process can be considered as a combination of deposition from the active silicon-containing species and etching by the hydrogen radicals [1.57]. As the power increases, the deposition rate first increases as the decomposition of SiH_4 is promoted by the plasma. At higher powers the deposition rate decreases as hydrogen radicals etch the growing film [1.58].

Although the fraction of ions is small, they can significantly affect the depositing layer. Most ions are confined to the plasma region, which is generally displaced from the wafer; however, some ions can impinge on the wafer surface and affect the structure of the depositing film, as we will discuss in Sec. 2.5.2. At the lower temperatures used for plasma-enhanced CVD, some of the active, atomic H ions from the plasma can be incorporated into the film, where they may passivate broken bonds, especially at the grain boundaries, improving the electrical properties of the layers [1.59], as discussed in Sec. 5.5.

Alternatively, amorphous silicon layers can be deposited by plasma-enhanced CVD, and the amorphous silicon can subsequently be crystallized by thermal or laser annealing. However, when the films are

1.10. ENHANCED DEPOSITION TECHNIQUES

deposited at lower temperatures, they may contain large amounts of hydrogen ($\geq 10\%$). If the films are heated rapidly during the crystallization process, the hydrogen can damage the layer as it evaporates or *evolves*.

In addition to SiH_4 and Si_2H_6, fluorine-containing precursors, such as SiF_4, have also been investigated as silicon sources for plasma-enhanced deposition [1.60, 1.61] to decrease the crystal nucleation rate and possibly increase the film deposition rate.

Energy can also be added to the deposition system by light, which can either couple to gas-phase or surface species to promote the decomposition reaction (*photolysis*) or simply heat the substrate surface to enhance the thermal reaction (*pyrolysis*). In the former case, photo-enhanced deposition is sensitive to the wavelength of the light. Silane molecules can be photodecomposed near the substrate using ultraviolet light from an ArF excimer laser [1.62]. CO_2 lasers can be used either to photoexcite silane by tuning the laser frequency to match the silane infrared absorption band near 10.6 μm [1.63] or to heat the substrate [1.64]. Photons from the visible radiation of an Ar^+ laser may also contribute to deposition [1.65]. In addition to stimulating the deposition, light may also provide additional energy to the adsorbed species to form a more ordered structure. Light-induced desorption of impurities from the surface may also occur, again allowing improved surface mobility of adsorbed species and a more ordered structure.

In photo-enhanced deposition, the light usually covers the entire substrate, but it can also be focused to a narrow beam to promote deposition locally, allowing polysilicon lines to be "written" directly on selected regions of the substrate [1.66]. Although this serial process is slow, it may be useful for personalizing a circuit at the last stages of fabrication to allow rapid prototyping or for correcting design errors on prototype circuits. To increase the writing speed, only a thin layer of polysilicon is deposited; the resistance is high, but the polysilicon may subsequently be selectively coated with a thicker layer of more conductive material [1.67].

The deposition rate may also be enhanced by partially decomposing the silane with a hot filament near the substrate [1.68] so that the deposition rate depends less sensitively on the substrate temperature. This method also eliminates the bombardment by energetic ions found in plasma-enhanced deposition and may reduce the amount of H included in the films at lower temperatures.

For large-area applications, such as the *active matrix* of transistors used for addressing liquid-crystal displays, the silicon layer is often deposited in amorphous form and crystallized after deposition. At temperatures compatible with low-cost glass substrates, the thermal deposition rate is low. During plasma-enhanced deposition at lower temperatures, significant amounts of hydrogen can be incorporated into the deposited film, as discussed above. Before crystallizing the film, the hydrogen must be removed by slowly heating it to an intermediate temperature to avoid bubble formation, increasing the process time and complexity. Therefore, deposition of thin layers by sputtering is being investigated for large-area applications [1.69].

1.11 Summary

In this chapter we investigated the deposition of polycrystalline silicon films by considering the series of steps comprising the overall deposition process. We first discussed convection in the gas phase and then introduced the concept of a boundary layer separating the forced-convection region from the stationary surfaces. We next considered the relative importance of diffusion through the boundary layer and reaction at the surface as limiting mechanisms of the polysilicon deposition process. With this background, we looked at the high-volume, hot-wall, low-pressure reactor and saw that operation in the surface-reaction-limited region is chosen for best control and uniformity, consistent with the constraints of the reactor and the structure of the layer being deposited. We considered the more flexible, cold-wall reactor and saw that its ability to operate as part of a cluster tool offers the possibility of controlling the interface between layers deposited in different chambers of the same system. We then looked at the surface processes and saw that, for deposition of undoped films under normal operating conditions, silane is usually adsorbed on the wafer surface where it decomposes. Adding a dopant-containing gas can change the deposition rate, most likely by blocking adsorption sites or by catalyzing the decomposition reaction. When thermal energy is not adequate to allow the deposition process to proceed at a practical rate, energy can be added electrically in the form of a plasma. In the next chapter, we will consider the structure of the material deposited and how it is influenced by the deposition conditions.

Chapter 2

Structure

In the discussion of polysilicon deposition in Chapter 1, we focused our attention on the processes occurring in the gas phase and the macroscopic chemical reactions occurring at or near the wafer surface. In this chapter we want to look in more detail at the microscopic processes occurring at the surface of the growing film both during and after nucleation because these processes strongly influence the structure of the material deposited. We will first discuss nucleation of polycrystalline silicon on both amorphous and single-crystal surfaces and then consider the influence on the film structure of surface diffusion of the adsorbed atoms during the continued deposition. We will next look at the techniques used to evaluate polycrystalline films and relate the information they provide to the detailed structure and other properties of the films. Finally, we will examine changes in the structure that occur by continued processing after the films are deposited.

2.1 Nucleation

We saw in Chapter 1 that deposition of polysilicon involves a complex interaction between the incoming silane molecules, the carrier gas, if any, and the substrate surface. The silicon-containing molecules — either silane, disilane, or intermediate species — reach the surface where they are adsorbed. Further decomposition occurs on the surface to produce weakly bound silicon atoms. These adsorbed atoms or their precursors migrate on the silicon surface and form nuclei, which then allow continued deposition. To simplify our discussion, we assume that the decomposition occurs before significant surface diffusion takes place so that we only need to consider the migration of adsorbed silicon atoms. Migration of the adsorbed atoms depends strongly on the temperature, the nature and cleanliness of the surface, and the presence of other gases adsorbed on the substrate surface.

2.1.1 Amorphous surfaces

When polysilicon is used as the gate electrode of an MOS transistor, it is deposited on an amorphous surface, most often SiO_2, but sometimes Si_3N_4 or an SiO_2 layer with nitrogen selectively added. In this section we discuss nucleation on such amorphous layers.

A perfectly, clean amorphous surface does not contain preferred, low-energy sites. Adsorbed silicon atoms diffuse randomly on the surface at a rate determined by their thermal energy, which is supplied primarily from the heated substrate. As the adsorbed atoms migrate, they can re-evaporate or *desorb* because they are only weakly bound to the surface, or they can encounter other diffusing atoms to form an adsorbed atom pair. Because of its larger mass, this atom pair is less likely to desorb than is an individual adsorbed atom. Additional atoms diffusing on the surface can join the atom pair, forming a larger, more stable cluster. The cluster finally reaches a critical size, beyond which it is unlikely to desorb, leading to a stable nucleus, on which continued deposition takes place. The size of the *critical cluster* is determined by the temperature, the binding energy between the individual atoms and the substrate, and the binding energy between the atoms in the cluster. To be stable, the critical cluster must be larger at higher temperatures and for species which are less strongly bound.

Because the probability of diffusing atoms encountering each other to form clusters depends strongly on the number of atoms on the surface, the concentration of critical clusters is a strong function of the arrival rate (through the partial pressure of the silicon-containing gas) and the desorption rate (through the temperature of the substrate and the binding energy of the diffusing atoms to the surface).

The number of nuclei on a surface and their size depends on several factors. For a given substrate and fixed arrival rate, the most important factor is the substrate temperature. For highly mobile nuclei in steady

2.1. NUCLEATION

state, the number of stable nuclei n_s can be expressed as [2.1]

$$\frac{n_s}{n_0} \sim \left(\frac{n_1}{n_0}\right)^{(i+1)/2} \exp\left(\frac{E_i + E_m - E_d}{kT}\right) \quad (2.1)$$

where i is the number of atoms in the critical cluster, n_1 is the adsorbed atom concentration, n_0 is total the number of surface sites per unit area, E_i is the energy of formation of the cluster containing i atoms, E_m is the activation energy for surface diffusion of mobile clusters, and E_d is the activation energy for diffusion of a single adsorbed atom. On an SiO_2 substrate, i is generally between 1 and 4, E_i between 2 and 8 eV, $E_m \approx 0.8$ eV, and $E_d \approx 0.4$ eV [2.1].

Because the desorption and surface diffusion depend on the binding energy of the atoms and clusters to the substrate, stable deposits may form on one surface while not forming on another surface under the same deposition conditions.

Figure 2.1 shows replica micrographs of the formation and growth of individual nuclei during the initial stages of deposition of a polycrystalline silicon film from silane onto SiO_2 [2.2]. The density of nuclei is shown as a function of time in Fig. 2.2 [2.3]. The initial formation of stable nuclei begins relatively slowly. At the beginning of the deposition, few diffusing atoms are present on the surface; their probability of encountering each other is low, and an *incubation period* is observed. The incubation time decreases with increasing temperature and increasing silane concentration. As the surface coverage by adsorbed species increases, adsorbed atoms more readily encounter one another, and the density of nuclei increases rapidly. As the number of nuclei increases, an incoming atom is more likely to be adsorbed within a surface diffusion length of an existing nucleus (*ie*, within the distance an adsorbed atom can diffuse before desorbing). This added atom has a higher probability of diffusing to and joining an existing nucleus than of forming a new one. Thus, the surface is rapidly covered by an array of nuclei separated from each other by approximately the surface diffusion length of an adsorbed atom, and further nucleation is unlikely. This *saturation density* of nuclei then remains relatively constant, with each nucleus increasing in size (Fig. 2.1) until the nuclei begin to coalesce to form a continuous film.

The number of nuclei generally decreases as the substrate temperature increases, as shown in Fig. 2.3 [2.4]. When the adsorbed atoms have more thermal energy, they can diffuse farther on the surface to an

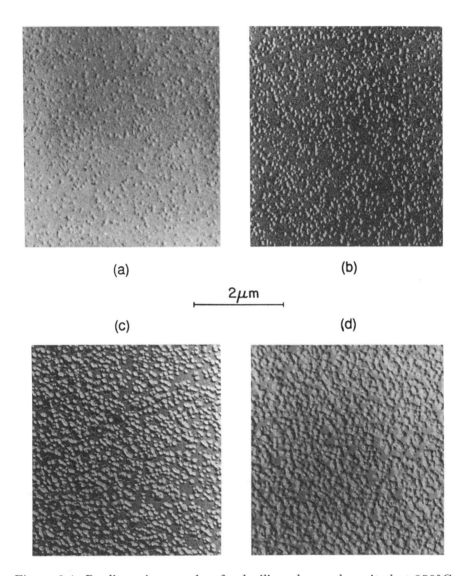

Figure 2.1: Replica micrographs of polysilicon layers deposited at 850°C in a cold-wall reactor at atmospheric-pressure to average thicknesses of (a) 1.5, (b) 5.5, (c) 14, and (d) 23 nm. (The *average thickness* is the total number of silicon atoms divided by the atomic density of silicon.) Three-dimensional nuclei form on the amorphous substrate and grow as more atoms are deposited.

2.1. NUCLEATION

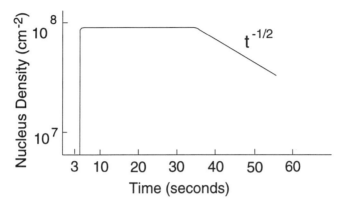

Figure 2.2: The density of stable nuclei as a function of exposure time to SiH$_4$, showing the incubation period, the rapid increase to a steady state value, and coalescence. From [2.3]. Reprinted by permission of the publisher, The Electrochemical Society, Inc.

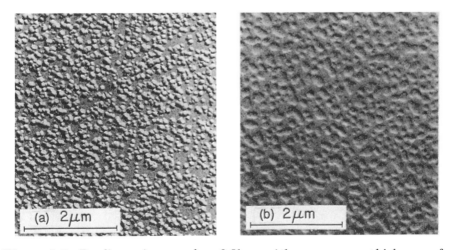

Figure 2.3: Replica micrographs of films with an average thickness of 14 nm deposited at (a) 850°C and (b) 1025°C in a cold-wall reactor at atmospheric-pressure. At higher temperatures surface diffusion is greater, and fewer, but larger, nuclei form.

existing nucleus, rather than forming a new nucleus; fewer, but larger, nuclei are then formed. In addition, small clusters, which are readily formed, are less stable at higher temperatures and more likely to desorb, again limiting the number of stable nuclei.

The nucleation of silicon on either SiO_2 or Si_3N_4 from silane in a nitrogen carrier gas can be described by the model outlined above. When a carrier gas such as hydrogen, which can also enter into the adsorption process, is used, the nucleation is somewhat more complicated. The nucleation of silicon from silane in a hydrogen carrier gas is different on SiO_2 and on Si_3N_4 [2.1, 2.5]. At lower temperatures hydrogen can adsorb on the oxide surface and block some of the surface sites where silane could otherwise be adsorbed. The saturation nucleus density then decreases, and the incubation time increases. The different behavior on SiO_2 and on Si_3N_4 at lower temperatures is attributed to the stronger bonding between hydrogen and the oxygen in SiO_2 than hydrogen and the nitrogen in Si_3N_4. The Si–O bond is only about 5-6% stronger than the H–O bond, while the Si–N bond is about 10% stronger than the H–N bond, making displacement of the hydrogen by arriving silicon atoms less likely on SiO_2 than on Si_3N_4. Adding HCl to the incoming gases also reduces the number of adsorbed atoms, leading to a lower saturation nucleus density [2.3].

At higher temperatures, the hydrogen desorbs from the surface, and the saturation nucleus density varies with temperature in approximately the same manner on SiO_2 and on Si_3N_4. (The slightly higher activation energy on SiO_2 is attributed to reaction of silicon adatoms with SiO_2 reducing the number of adatoms and, therefore, the saturation nucleus density [2.1]).

Because nucleation must occur before continuous deposition can proceed, the initial deposition rate is lower than the steady-state deposition rate. The deposition rate appears to increase approximately linearly with time during the nucleation process, before becoming constant after the amorphous substrate is covered [2.6]. Because deposition may start slowly, the steady-state deposition rate cannot be determined by dividing the film thickness by the deposition time if the incubation period is a significant fraction of the total deposition time.

During nucleation the average thickness of the silicon film deposited (the total number of silicon atoms per unit surface area divided by the atomic density of silicon) can be approximately determined by ellipsometry. Samples can be observed after removal from the reactor [2.2] or

2.1. NUCLEATION

by *in situ* ellipsometry [2.6]. If the lateral distance between nuclei is much less than the wavelength of the light used in the ellipsometer, an *effective-medium* approximation can provide useful information about the spacing between nuclei.

The joining of several adsorbed silicon species on the surface to form a stable nucleus appears consistent with the nucleation behavior observed for thermal deposition. However, alternate explanations should be considered. For thermal decomposition in the presence of hydrogen at an elevated temperature, Si can react with SiO_2 to produce volatile SiO, leaving an exposed Si atom with a broken bond. In this case, the incubation time corresponds to the time to remove a surface oxygen atom and the time for a SiH_4 molecule to reach the broken bond. This mechanism is unlikely to dominate the nucleation process for thermal deposition at moderate temperatures.

At lower temperatures with a plasma present at the start of the deposition process, ions bombarding the surface are likely to break Si–O bonds, removing surface oxygen and leaving exposed Si atoms with broken bonds. These exposed Si atoms are likely to serve as nucleation sites, decreasing the incubation time. The plasma also aids decomposition of the silicon-containing gas, forming reactive radicals, which can bind firmly to the surface. The effect of a plasma on nucleation is especially critical for the deposition of alloys of Si and Ge on an oxide surface. Without a plasma, the Ge source GeH_4 retards nucleation on oxide. When a plasma is present at the initial stages of the deposition, nucleation starts readily [2.7].

Treatment of the oxide surface before inserting it into the reactor can also affect the nucleation rate. Wet-chemical treatments, especially, can modify the nucleation sites. The enhanced nucleation caused by the wet-chemical treatment used to "clean" oxide surfaces can be seen by examining the grain size in the subsequently deposited layer [2.8]. When the density of nuclei is higher, less lateral growth of a grain can occur before it impinges on grains growing from adjacent nuclei, and the grain size is smaller. The grain sizes in polysilicon layers grown on surfaces cleaned with typical peroxide cleaning solutions used in integrated-circuit processing and in hydrofluoric acid are smaller than the grain size in layers deposited on freshly grown SiO_2 [2.8], indicating that the wet-chemical treatment increases the density of nuclei. By contrast, annealing at a high temperature after exposing the wafer to wet-chemical treatment decreases the density of nuclei and increases the grain size.

2.1.2 Single-crystal surfaces

Although nucleation phenomena are most readily observed on amorphous surfaces, they can be important even when deposition occurs on a single-crystal substrate. If a single-crystal substrate is nearly atomically clean, the diffusing silicon atoms are influenced by a regular array of low-energy sites associated with the substrate lattice. If they have enough thermal energy to migrate on the surface, they preferentially position themselves over these low-energy sites. When adequate energy is available, the ordering is complete, and a single-crystal or *epitaxial* layer is formed by layer-by-layer growth. More commonly, however, the surface is not atomically clean, and three-dimensional nuclei form, even on a single-crystal substrate [2.9, 2.10]. At high temperatures, each individual nucleus can align itself with the substrate crystal lattice; consequently, the nuclei are aligned with each other, and a single-crystal layer without grain boundaries forms when the individual nuclei grow together.

In more typical cases, a monolayer or so of oxide can be on the surface before the wafers are inserted into the reactor or form as the wafers are loaded into a hot-wall reactor. Without an *in situ* cleaning step, this thin oxide remains and prevents epitaxial alignment of the depositing silicon, and a polycrystalline structure forms. At lower temperatures, even when the surface is atomically clean, the nuclei do not have enough energy to align with the substrate lattice. Because the nuclei are randomly oriented, grain boundaries form as the isolated nuclei grow together into a continuous layer, and a polycrystalline film forms over the single-crystal substrate.

Although a single-crystal structure is desired for some applications, a polycrystalline structure is needed for optimum performance of the polysilicon emitters used in bipolar transistors. Epitaxial alignment of the depositing film is usually avoided to allow rapid dopant diffusion, as will be discussed in Sec. 6.8.2.

At even lower temperatures on either a crystalline or an amorphous substrate, the adsorbed atoms do not have enough thermal energy to diffuse significantly on the substrate surface before they are covered by subsequently arriving silicon atoms. Once they are covered, their random arrangement is locked into place. If the atoms are essentially immobilized where they first reach the substrate surface, the deposited film is *amorphous* (*ie*, it contains no long-range order). [The amorphous

2.2. SURFACE DIFFUSION AND STRUCTURE

silicon films of primary interest to us are formed by chemical vapor deposition at temperatures only slightly below the transition temperature above which a polycrystalline structure forms. These amorphous films are dense and differ markedly from the porous, hydrogen-containing, amorphous films deposited at much lower temperatures by plasma-enhanced CVD (PECVD), sputtering, or glow-discharge techniques. We will discuss both CVD and PECVD amorphous films in more detail in Sec. 2.11.4.]

2.2 Surface diffusion and structure

After a continuous film forms, the developing grain structure is strongly influenced by the amount of thermal energy available for surface migration. Deposition conditions that allow the adsorbed silicon atoms to diffuse farther on the surface before being immobilized by subsequently arriving silicon atoms lead to larger grains and a better defined structure. Surface diffusion increases when the substrate surface temperature increases because of the greater random motion associated with the increased thermal energy. As the arrival rate of silicon atoms increases, the amount of time during which the adsorbed atoms can diffuse on the surface before being immobilized by subsequently arriving atoms decreases, limiting the amount of migration possible. Surface diffusion decreases when more of the adsorption sites are filled by other (non-silicon) adsorbed atoms, as at higher total gas pressures or, especially, when strongly adsorbed contaminant atoms are present. Because temperature, rate, pressure, and impurities all modify the amount of surface migration possible, and therefore the order in the growing layer, they are interrelated, as we will discuss below.

Rate and temperature

Deposition rate and deposition temperature both affect the structure of the growing layer through surface diffusion. As the temperature increases, the atoms have more energy to diffuse. As the rate decreases, the adsorbed atoms can diffuse for a longer period of time, and a given structure should form at a lower temperature. The compensating effects of rate and temperature can be semi-quantitatively related through the surface diffusion length L:

$$L \approx \sqrt{Dt} \sim \frac{1}{\sqrt{R_D}} \exp\left(-\frac{E_a}{2kT}\right) \qquad (2.2)$$

where D is the surface diffusion coefficient of the diffusing silicon species, t is the time available for surface diffusion ($\sim$ the time to deposit one monolayer), R_D is the deposition rate, and E_a is the activation energy associated with surface diffusion.

Because the surface diffusion length varies rapidly with temperature, but relatively slowly with deposition rate, a small change in temperature can compensate for an appreciable change in deposition rate, and the structure is more sensitive to deposition temperature. The surface diffusion of adsorbed silicon atoms significantly influences the structure during the entire deposition, not only during the initial part of the deposition (*ie*, the structure is *growth-controlled*, rather than *nucleation-controlled*). Because of this continued development of the structure as the growth continues, the grains can be much larger near the top of a thick film than near the bottom.

The interrelation of deposition rate and deposition temperature on the structure of polysilicon films can be illustrated by considering the dependence on these two parameters of a film property that is sensitive to the detailed structure of the film. As we will see in Sec. 3.3, the diffusion coefficient of dopant atoms in a polysilicon film depends strongly on the structure of the deposited film. The deposition temperature needed to form a film with maximum dopant diffusion coefficient increases with increasing deposition rate [3.12]; *ie*, an increased deposition temperature allows the adsorbed silicon atoms to diffuse the same distance in the shorter length of time available at a higher deposition rate, and a similar grain structure forms.

The interrelation between temperature and rate can also be seen by considering the transition temperature between deposition of amorphous and polycrystalline silicon. Figure 2.4 shows that the boundary between the deposition of amorphous and polycrystalline layers occurs at a higher temperature when the deposition rate is higher. The slope of the boundary corresponds to the activation energy in Eq. 2.2 and is about 4 eV. (The deposition rate is a strong function of temperature at a constant SiH_4 flow rate when the overall deposition process is limited by the reaction rate. The discussion here concerns deposition rate,

2.2. SURFACE DIFFUSION AND STRUCTURE

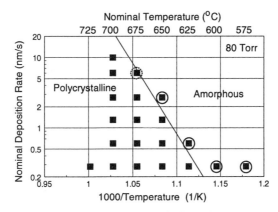

Figure 2.4: A higher temperature is needed to deposit a polycrystalline, rather than an amorphous, layer when the deposition rate is higher. These layers were deposited in a load-locked, cold-wall, single-wafer reactor at a pressure of 80 Torr using SiH_4 in a hydrogen carrier gas.

rather than SiH_4 flow rate, which can usually be adjusted to give the desired deposition rate at a given temperature.)

The trade-off between temperature and rate is limited by gas-phase impurities, which adsorb on the surface during deposition, impeding surface migration of the adsorbed silicon atoms [2.11]. If the arrival rate of impurities remains constant as the deposition rate is reduced by decreasing the silane partial pressure, the impurities can become the dominant adsorbing species. The surface diffusion length of the adsorbed silicon atoms is then limited by the impurities and does not continue increasing as the silane partial pressure is further reduced. Strongly adsorbed impurities can degrade the structure of a polycrystalline film deposited on an amorphous substrate, as we will see below. Impurities can limit the crystal structure when gas purity and system integrity are not adequately controlled.

Typical sources of impurities include the following:

- Carrier gas, which is usually obtained by vaporizing a liquid source, such as liquid hydrogen.

- Reactant gas, usually in a compressed-gas cylinder.

- Inadequate quality of reactor construction.

- Air intruding into the chamber during wafer loading in a non-load-locked system.

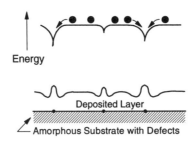

Figure 2.5: When mass transport limits the amount of incoming material, surface migration to low-energy defect sites causes a thicker film to form at these sites, while the surrounding regions are thinner.

Contamination on the wafer surface before deposition can also affect the deposited film. Randomly located contamination can cause some very-low-energy sites on the surface. As adsorbed silicon atoms diffuse randomly on the surface, they remain at these low-energy defect sites longer than at normal sites, leading to a net accumulation of silicon atoms and a thick, defective deposit there. Because some of the excess silicon near the defect is supplied by surface diffusion from the surrounding regions, the film may be thinner within a surface diffusion length of a defect than far from it, as shown in Fig. 2.5.

Effect of total system pressure

The total system pressure should also affect the crystal structure by modifying the amount of gas adsorbed on the surface of the growing layer. Adsorbed gases can block diffusion paths and retard development of the crystal structure. These adsorbed gases can be the carrier gas, the reactant gas, reaction byproducts, or impurities.

As the pressure is reduced by decreasing the carrier gas concentration, fewer hydrogen atoms are available to adsorb on the surface and impede rearrangement of the depositing silicon atoms. Surface diffusion occurs more readily, and the transition temperature between amorphous and polycrystalline deposition should decrease. However, the effect of total pressure on the structure is difficult to isolate. In the initial, cold-wall, atmospheric-pressure reactors used to deposit polycrystalline silicon from SiH_4 in a hydrogen carrier gas in the late 1960s and early 1970s, polycrystalline films were deposited above about 680°C and amorphous films were deposited below that temperatures. In the

2.2. SURFACE DIFFUSION AND STRUCTURE

low-pressure reactors first used in the late 1970s and 1980s with 100% SiH_4 at a pressure of about 0.2 Torr, the transition temperature between a polycrystalline and an amorphous deposit is about 580°C. In the extreme case of molecular-beam deposition at a pressure of 10^{-9} Torr, the transition temperature decreases further to less than 400°C [2.12].

This trend of decreasing transition temperature with decreasing pressure has been cited as evidence for the retarding effect of hydrogen on the surface diffusion. Even after compensating for the effect of the higher deposition rate in the cold-wall reactor, the transition temperature appeared to be higher at higher pressures. However, more recent observations in a load-locked, atmospheric-pressure, cold-wall reactor found a transition temperature similar to that in the low-pressure reactor. It is likely that the differences observed are dominated by the increasing purity of the deposition ambient with time. The purity improves because of improved gas quality, because of better quality construction of newer reactors, and because load locks limit exposure of deposition chambers to air.

To isolate the effect of the hydrogen pressure on the surface diffusion and structure, a more detailed study was conducted in a high-purity, load-locked, reduced-pressure, cold-wall reactor operating with a small flow of SiH_4 in a large flow of hydrogen carrier gas [2.13]. The pressure was varied from 10 to 600 Torr while keeping the deposition rate and the temperature constant. At 625°C and a rate of 20 nm/min, layers deposited at 200 Torr and above were amorphous, while those deposited at 80 Torr and below were polycrystalline. This suggests that adsorbed hydrogen does retard the surface diffusion, but much less than suggested by the earlier results. Extrapolating the very limited data to the 0.2 Torr pressure of the low-pressure reactor suggests that the difference in the transition temperature between atmospheric pressure and 0.2 Torr should be only of the order of 10–15°C, much less than thought earlier.

The effect of total system pressure on the surface diffusion length can be quantitatively considered by modifying Eq. 2.2 to read

$$L \approx \sqrt{Dt} \sim \frac{1}{\sqrt{R_D}} \exp\left(-\frac{E_a}{2kT}\right) P^{-n} \tag{2.3}$$

where n is a positive number much less than unity.

Subsurface rearrangement

Adsorbed impurities can prevent adsorbed silicon atoms from diffusing to stable, low energy positions before they are covered by subsequently arriving silicon atoms. Silicon atoms are then incorporated into the film in unstable positions, along with impurity atoms. During further heat treatment, either during or after the deposition cycle, these silicon atoms and the incorporated impurity atoms can rearrange to form a denser structure. As the silicon contracts, the resulting tensile forces deform the film and the underlying substrate wafer. Oxygen causes significant degradation when small quantities are incorporated into the film during deposition because of oxygen contamination in the gas phase [2.14]. The degradation is especially severe when the wafers are heated by energy transfer from an underlying heated support plate. As the film contracts, tensile stress causes the edges of the wafer to bend upward and lift from the support plate. The edges of the wafer cool, and adsorbed oxygen is less likely to be desorbed from the surface; more oxygen is incorporated into the film, leading to greater tensile stress and even more severe wafer distortion.

2.3 Evaluation techniques

The preceding sections discussed the primary factors influencing the nucleation and structural development of the depositing film. Before discussing details of the deposited structure, we first briefly consider the techniques used to characterize the structure. Possible techniques include the following, as well as many others:

- Optical microscopy
- Scanning electron microscopy (SEM)
- Transmission electron microscopy (TEM)
- X-ray diffraction (XRD)
- Optical reflection
- Ellipsometry
- Atomic-force microscopy (AFM)

Optical microscopy can provide some information about the structure, especially for thick films in which characteristic surface features are related to the orientation of the grains in the film [3.12]. The crystal

2.3. EVALUATION TECHNIQUES

structure can be determined from simple optical measurements in these thick films once the surface features have been correlated with the grain structure revealed by more detailed techniques. In thin films or those deposited at lower temperatures, characteristic surface features are not well enough developed to observe with an optical microscope.

The absence of characteristic surface features reduces the value of straightforward scanning electron microscopy (SEM) to provide significant information about the structure of thin films. Thin films do not appear qualitatively different even when their detailed grain structure differs significantly. The surface topography revealed by scanning electron microscopy does not directly indicate the grain size, and more time-consuming techniques must be used to evaluate thin films. Scanning electron microscopy can, however, be used to study defects in the films.

The most detailed structural information is obtained by transmission electron microscopy (TEM), in which a beam of electrons travels through a very thin section of the sample being studied. The diffraction of the electrons by crystal planes and by defects, such as grain boundaries, produces an image of the structure of the deposited film. Unfortunately, the specimen preparation needed for transmission electron microscopy is very time-consuming, so its use is generally limited to research and development, rather than being used for routine process monitoring.

Diffraction of x-rays by the crystal planes of the differently oriented grains allows quantitative comparison of films deposited under different conditions. This technique does not require extensive sample preparation, and numerous samples can be quickly analyzed and quantitatively compared.

Once the detailed structure has been determined, the macroscopic optical properties, such as reflectance, can be correlated with the structure, leading to techniques useful for obtaining a routine, nondestructive indication of the structure. In the discussion below, we will consider the different aspects of the structure revealed by each of the techniques so that a complete picture can be obtained from the complementary information.

The most extensive data about the structure of polysilicon films was obtained when polysilicon first was being used and was being intensively studied. Once the basic trends were understood, most subsequent studies were restricted to a narrow range of interest. Since the time of the

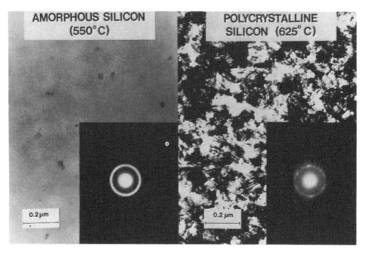

Figure 2.6: Transmission electron micrographs of 0.6 μm thick silicon films deposited onto SiO_2 at 550 and 625°C in a low-pressure reactor.

early studies, the purity of the gases and the quality of reactor construction have continuously improved. In the early studies, impurities impeded surface diffusion, requiring a higher temperature to obtain the same structure that can be obtained today at a lower temperature in a purer ambient. However, the trends generally remain the same, allowing us to use the detailed older data to gain a physical picture, which can then be applied to more modern equipment.

2.4 Grain structure

Figure 2.6 shows transmission electron micrographs of two films deposited by low-pressure CVD at about 0.2 torr (∼30 pa) using undiluted silane. The film shown in Fig. 2.6a was deposited at 550°C and is basically amorphous with a few, small crystalline inclusions. The corresponding transmission electron diffraction pattern shown in the inset exhibits the broad, diffuse rings characteristic of an amorphous film. Figure 2.6b shows a silicon film deposited at 625°C. The structure is polycrystalline with an average grain size of about 70 nm. The electron diffraction pattern exhibits the ring-and-dot pattern characteristic of a polycrystalline structure. The nearest neighbor spacing is not greatly different in amorphous and crystalline films, and the first ring appears

2.4. GRAIN STRUCTURE

at a similar location. However, the second diffuse ring of the amorphous structure splits into two rings in the polycrystalline material. This difference arises from the arrangement of the atoms in the two cases. The second ring is related to the separation of second nearest neighbors. In amorphous silicon, bond angles are flexible, and the second nearest neighbor distance has a range of values, producing the corresponding broad halo seen in Fig. 2.6a. As the silicon crystallizes, the bonds rotate to fixed angles, forming an ordered structure with two well-defined second nearest neighbor spacings. The broad halo then splits into the two rings corresponding to these spacings, as seen in Fig. 2.6b.

More detailed studies show that the transition temperature above which a polycrystalline structure forms during deposition is about 580°C in a typical, horizontal, low-pressure reactor, and perhaps slightly lower in a newer, vertical, low-pressure reactor. Near the transition temperature, a structure with elongated grains sometimes appears. At 600°C a very fine grain, but equi-axed, deposit forms. By 625°C, a well defined polycrystalline structure, which is equi-axed in the plane of the film, is obtained. The transmission electron micrograph of Fig. 2.6b shows that the grains within the polycrystalline film are often highly faulted and contain a high density of twin lamella and other, more detrimental defects. The defects within the grains, as well as the grain boundaries, can affect the characteristics of devices built in the films (Sec. 6.13). Although the grains are equi-axed in the film plane, they are often columnar perpendicular to the film surface, as shown in Fig. 2.7a [2.15]. A high density of microtwins appears in the grains, and surface asperities or hillocks also form. We will correlate this columnar structure with a particular crystal orientation in Sec. 2.5.

Grain size

The average grain size can be determined from transmission electron micrographs in a number of different ways. Some techniques rely on counting the number of grains in a given area while others measure the number of grain boundaries crossed by a straight line. Because of the grain-size distribution, as well as the shapes of the grains, these two different techniques can produce markedly different results [2.16, 2.17], accounting for some of the discrepancies between data from different studies. Using a statistically significant sample size is also important, as is the possibility that overlapping large and small grains in the micrograph lead to overemphasis of the latter.

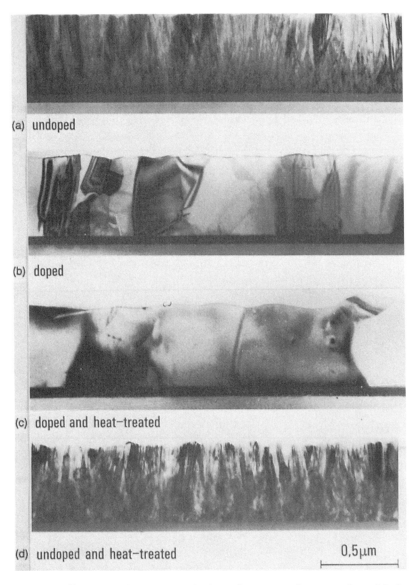

Figure 2.7: Cross section transmission electron micrographs of 0.6 μm thick films deposited in a low-pressure reactor at 625°C. The initial columnar structure of undoped polysilicon films (a) changes little when the undoped films are annealed (d), but phosphorus doping increases the grain size (b), which is further enlarged by additional annealing of the doped films (c). From [2.15]. Reprinted by permission of the publisher, The Electrochemical Society, Inc.

2.4. GRAIN STRUCTURE

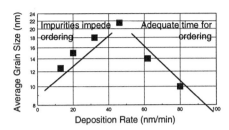

Figure 2.8: Surface diffusion and grain size first increase as the deposition rate decreases because of the longer time available for surface diffusion. When impurities dominate at low deposition rates, the grain size again decreases.

The dependence of grain-size on deposition temperature is best studied in a cold-wall reactor, in which temperature can be varied over a wide range because the forced-convection region remains much cooler than the wafer surface. This allows us to study films deposited over the range from less than 600°C to about 1100°C. Similar trends are seen in the cold-wall and in the hot-wall, low-pressure reactor over the limited range where the latter can be used.

At lower temperatures, amorphous films are deposited in both cold-wall and hot-wall reactors. Near the transition temperature, anomalous, elongated grains are sometimes obtained in both types of reactors. At higher temperatures, the more common in-plane equi-axed structure forms. As the temperature increases, the grain size also increases. In the cold-wall reactor the grain size increases from about 50 nm at 700°C to about 300 nm at 1100°C in films about 500 nm thick [2.2]. Some evidence of increasing grain size at higher deposition temperature is seen in the hot-wall, low-pressure reactor, even over the limited range that can be studied [2.18]. The grain size also increases as the polysilicon film becomes thicker.

The effect of impurities can be seen by considering the grain size as a function of deposition rate. From Eq. 2.2 we expect the grain size to increase as the deposition rate increases. Figure 2.8 [2.11] shows this expected behavior at higher deposition rates. However, as the deposition rate decreases, the ratio of impurities to depositing species impinging on the surface increases. At low deposition rates the surface diffusion

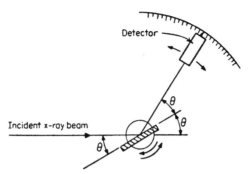

Figure 2.9: Geometrical arrangement of source, sample, and detector in the "θ-2θ" x-ray diffractometer.

decreases more because adsorbed impurities block diffusion paths than it increases by the longer time for surface diffusion. The grain size in the depositing film, consequently, decreases, as seen in Fig. 2.8 at lower deposition rates.

2.5 Grain orientation

Although transmission electron microscopy gives the most detailed information about the crystal structure, quantitative comparison of films deposited under slightly different conditions can be better obtained by x-ray diffraction. In the simplest implementation of this technique, shown in Fig. 2.9, x-rays are incident on the sample at an angle θ from the sample surface plane, and a detector is also located at an angle θ from the sample surface plane to intercept the diffracted beam. As the angle θ of the sample is varied with respect to a fixed incident beam, the detector is simultaneously moved to a new position at an angle 2θ from the incident beam. (Because of its geometrical arrangement, this type of x-ray diffractometer is often called a "θ-2θ" system.)

A maximum in the diffracted intensity is obtained when Bragg's law

$$m\lambda = 2d \sin \theta \tag{2.4}$$

is satisfied. In Eq. 2.4, d is the spacing between crystal planes, λ is the wavelength of the x-rays, and m is an integer. By varying the angle θ, a series of peaks in the diffracted intensity appears at different angles as grains with different crystal planes parallel to the film surface satisfy Bragg's law. The relative intensities of the peaks expected from

2.5. GRAIN ORIENTATION

Table 2.1: Normalization factors for x-ray diffraction measurements of silicon for Fe$_{k_\alpha}$ radiation.

Plane	Random	2θ	$\csc\theta$	G	NF
{111}	1	35.96	3.24	0.0838	1
{220}	0.6136	60.55	1.98	0.0522	0.3822
{311}	0.3617	72.48	1.69	0.0447	0.1929
{400}	0.1081	90.95	1.40	0.0372	0.0480
{331}	0.1899	101.96	1.29	0.0342	0.0775

a thick, randomly oriented sample are known [2.19] and are shown for the important measurable peaks in silicon in Table 2.1 under the column labeled "Random." Several important low-angle diffraction peaks are absent because of the crystal symmetry. The {110} peak is absent, but the amount of {110} texture can be determined from the fairly strong {220} peak. The absence of both the {100} and {200} peaks is more troublesome because the amount of {100} texture must be determined from the weak {400} peak. In addition, strong diffraction from the silicon substrate may mask the desired signal if (100)-oriented substrate wafers are used.

When using the technique for films thinner than several times the absorption length of the x-rays, the normalization factors in Table 2.1 obtained from a thick, randomly oriented sample must be corrected for the incomplete absorption in the finite film thickness. Because the effective absorption path through the film varies with the angle of incidence, a greater fraction of the incident x-rays are absorbed for grazing incidence than for near-normal incidence. The correction factor G is given by the formula [2.20]

$$G = 1 - \exp(-2\mu t \csc\theta) \quad (2.5)$$

where μ is the absorption coefficient (reciprocal of the absorption length) in silicon of the x-ray source being used (eg, Fe or Cu), t is the film thickness, and θ is the incident angle of the x-ray beam. The values of the correction factor G calculated from Eq. 2.5 for 500 nm thick films are shown in Table 2.1, along with the total normalization factor NF for each of the important peaks (ie, the product of the relative random intensity and the thickness correction factor normalized by the thickness correction factor for the {111} planes).

The absorption lengths of the common $Fe_{k\alpha}$ and $Cu_{k\alpha}$ x-rays in silicon are 38 and 71 μm, respectively. Although both absorption lengths are much greater than the thicknesses of the samples commonly deposited for integrated-circuit applications, using the more strongly absorbed iron radiation increases the strength of the diffracted signal significantly. The long absorption lengths do, however, facilitate modifying the thickness correction factor for slightly different sample thicknesses. For thin films with $t \ll 1/\mu$, the exponential in Eq. 2.5 can be approximated by $1 - 2\mu t \csc \theta$, so that

$$G \approx 2\mu t \csc \theta \tag{2.6}$$

and the correction factor varies linearly with film thickness.

To accurately determine the relative amounts of differently oriented crystals, the total area under each peak should be integrated, and these areas should be compared. However, for simplicity the peak heights are sometimes compared instead. Because small misorientations of the grains from the film plane can broaden the peak substantially in a polycrystalline sample, as can the finite size of very small grains, using the peak intensity limits detailed analysis.

After the measured values of the x-ray peaks are appropriately normalized, the relative intensities allow a useful quantitative comparison of the importance of differently oriented grains within the polycrystalline film, providing information about the dominant crystal *texture* in the film and its variation with changing deposition conditions. Determining the dominant crystal orientation in polysilicon is important for several reasons. First, the surface structure appears to depend on the crystal orientation. As we will discuss in Sec. 4.3, the quality of oxide grown on the surface of polycrystalline silicon depends on the surface structure and, consequently, on the grain orientations. Second, in polycrystalline materials, grain boundaries provide disordered regions down which dopant atoms can readily diffuse. Grain-boundary diffusion is most rapid in material with columnar grains. In polysilicon {110} texture is associated with the columnar grains seen both in transmission electron micrographs of submicrometer films (Fig. 2.7a) and in optical micrographs of thicker films (Fig. 2.10) [3.12]. Maximizing the {110} texture is important for applications requiring rapid diffusion. The influence of the structure on dopant diffusion will be discussed in Sec. 3.3.

2.5. GRAIN ORIENTATION

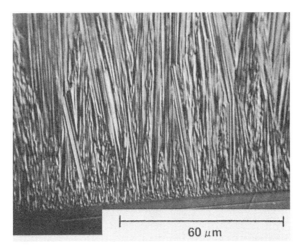

Figure 2.10: Optical micrograph of polished and etched cross section of a thick polycrystalline-silicon film deposited onto SiO_2 at 1050°C, showing the development of the columnar structure as the film becomes thicker.

2.5.1 Films formed by thermal CVD

The dominant crystal orientation (texture) depends on both energy and kinetic factors [2.21]. The energy of different crystal surfaces influences which crystal orientations are likely to grow fastest, although adsorbed species on the surface during the deposition process can modify the surface energies. Because of the finite deposition rate, kinetic factors can be equally important in controlling the preferred orientation of the grains. These kinetic limitations can restrict the ability of the arriving atoms to reach their lowest energy positions.

The kinetic factors can be described by the surface diffusion of the adsorbed silicon species before they are incorporated into the solid layer. As Eq. 2.2 and the associated discussion showed, surface diffusion increases strongly as the temperature increases, decreases moderately as the deposition rate increases, and decreases slightly as the total system pressure increases. Contamination in the deposition chamber can strongly decrease surface diffusion. When surface diffusion is very small, an amorphous layer is deposited. As the surface diffusion increases, a polycrystalline structure is obtained. Initially, a transition structure with dominant {311} texture forms; further increases in the surface dif-

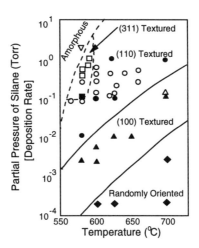

Figure 2.11: Within limits, a similar structure is obtained at a higher deposition rate by depositing at a higher temperature. Adapted from [2.22]. Used by permission of the publisher, the Electrochem. Soc. Inc.

fusion lead to a dominant {110} texture, and then a dominant {100} texture [2.22].

The tradeoff between deposition temperature and deposition rate is clearly shown in Fig. 2.11 [2.22]. A similar structure is obtained at a higher temperature and a higher rate as at a lower temperature and a lower rate. A more ordered structure is obtained by depositing at a higher temperature or a lower rate [2.22–2.24]. Because impurities impede surface migration, a given structure can be obtained at lower temperatures and higher rates in newer, better constructed reactors and in reactors with load locks and automated wafer handling, which eliminate exposure of the deposition chamber to air during wafer transfer.

As expected from our earlier discussion, the effect of the total system pressure (mainly hydrogen) on the structure is small within the polycrystalline region. At 675°C and a rate of 1.2 nm/s in a load-locked, cold-wall, single-wafer reactor, the {311} transition texture dominates over the pressure range from 600 to 80 Torr; the structure changes to the more ordered {110} texture at lower pressures (down to the 10 Torr limit of the investigation). Observing the transition texture over the wide range from 600 to 80 Torr is consistent with a very small dependence of the structure on the total system pressure.

Figure 2.12 shows the normalized x-ray signal as a function of depo-

2.5. GRAIN ORIENTATION

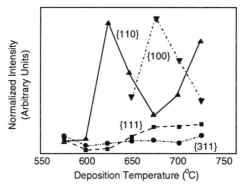

Figure 2.12: X-ray texture as a function of deposition temperature for 0.5 μm thick films deposited in a hot-wall, low-pressure CVD reactor.

sition temperature for polycrystalline silicon films about 500 nm thick deposited in a hot-wall, low-pressure reactor using undiluted silane at a pressure of about 0.2 Torr (~30 pa) [2.18]. At low deposition temperatures, the films are basically amorphous, as we saw by transmission electron microscopy, and little x-ray diffraction occurs. Above 600°C the {110} texture increases rapidly as a well defined, polycrystalline structure forms, reaching a maximum at about 625°C. Above this temperature, the {100} texture increases at the expense of the {110} texture, reaching a maximum at 675°C, where the {110} texture is minimum. At higher temperatures, the {110} texture again becomes most abundant.

The dominance of the {110} texture just above the transition temperature from an amorphous to a polycrystalline structure is not limited to films deposited at low pressures. Qualitatively similar behavior is seen in thin films deposited in an older, cold-wall, atmospheric-pressure reactor, as shown in Fig. 2.13 [2.2]. Again, a maximum and a minimum appear in the {110} texture; however, the temperatures at which these extrema occur are considerably higher for the atmospheric-pressure films than for the low-pressure films, occurring at about 775 and 900°C, respectively, in the atmospheric-pressure films, compared to 625 and 675°C in the low-pressure films. While it is likely that a similar {100} maximum is present in the atmospheric-pressure films as in the low-pressure films, the measurement equipment used in this early experiment [2.2] was not sensitive enough to detect the weak signal diffracted by the {400} planes. Although the basic trends are similar in layers deposited in the hot-wall, low-pressure reactor and in the older, cold-

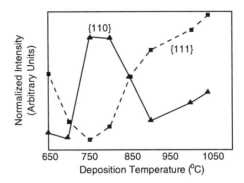

Figure 2.13: X-ray texture as a function of deposition temperature for 0.6 μm thick films deposited in an older cold-wall, atmospheric-pressure CVD reactor.

wall, atmospheric-pressure reactor, the temperatures needed to obtain similar structures are markedly different. Some of this difference is related to a faster deposition rate in the cold-wall reactor; a small part is related to the presence of a large amount of hydrogen in the cold-wall reactor; and a significant amount of the difference likely arises from the higher impurity concentration in the older reactor. Similar observations of the texture in layers deposited in a newer, load-locked, cold-wall, atmospheric-pressure reactor show the transition from amorphous to polycrystalline structure and the maximum in the {110} texture to occur at similar temperatures as in the hot-wall, low-pressure reactor.

As the films become thicker, the texture continues developing in the new material added. (To first order, the structure in the initially deposited polycrystalline part of the film does not change as more material is added, in contrast to the behavior in amorphous films to be discussed below.) Over much of the temperature range investigated, x-ray diffraction shows that {110}-oriented grains develop at the expense of differently oriented grains during the continued deposition. Figure 2.10 shows that the columnar structure associated with the {110} texture continues developing as the film becomes thicker, confirming the x-ray measurements. Because the x-ray measurements sample the structure through the entire thickness of the films, the amount of {111} texture in the upper regions is even less than indicated by the data in Fig. 2.13; much of the observed {111} texture probably is located near the lower regions of the film. The increasing structural development during the continued deposition again shows that we are dealing with a growth texture, rather than a nucleation texture.

2.5. GRAIN ORIENTATION

Although crystallites are observed with low-index planes nominally parallel to the film surface, these crystallites are not aligned precisely. The width of their distribution can be observed by tilting the sample slightly in the x-ray diffractometer so that the diffracting signal from planes perfectly parallel to the film plane is not intercepted by the detector. The observed signal decreases by three orders of magnitude when a single-crystal sample is tilted by 2°, but it only decreases by a few percent in a typical polycrystalline-silicon film. The sample must be tilted by 6° before the x-ray intensity decreases by a factor of two, indicating deviations from exact orientation by about this amount [2.25].

In addition to the deposition temperature and rate, the surface on which the film is deposited can influence the development of the film structure. For example, a thin nucleating layer with dominant {110} texture enhances the continued development of that texture even though other orientations would dominate at the deposition temperature if deposition started by random nucleation on oxide. This behavior again indicates that {110} grains, once nucleated, develop more rapidly than other orientations.

In films deposited well above the transition temperature, a polycrystalline structure develops in each layer as that layer is deposited, and the atoms are unlikely to continue rearranging after they have been covered by other silicon atoms. However, in films deposited slightly below the transition temperature, each layer of the film can be deposited in an amorphous form and crystallize during continued deposition [2.26, 2.27]. Nucleation of crystallites is most likely to occur by heterogeneous nucleation at the lower Si/SiO_2 interface. Solid-state crystallization of the amorphous silicon then continues on these initial nuclei, with the crystalline region propagating upward into the film. If the crystallization rate is less than the deposition rate, only the lower portion of the film crystallizes during the deposition, although the crystallization continues during the heat cycle after the deposition itself is terminated by stopping the silane flow. The film may be crystalline near the bottom and amorphous near the top. If the crystallization rate is greater than the film deposition rate and nucleation is rapid, the entire thickness of the film can crystallize during the deposition, possibly leading to the {311} transition structure sometimes observed over a narrow range of deposition conditions. Similar behavior has been reported for layers deposited near the transition temperature by other techniques [2.12].

Films deposited from disilane

Polysilicon films deposited from disilane are similar to those deposited from silane at similar temperatures and rates. The transition temperature between the deposition of amorphous and polycrystalline films is not significantly different from that found using SiH_4 after compensating for the higher deposition rate with Si_2H_6 [2.28, 2.29]. A similar {311} transition structure is also seen [2.29]. Undoped films deposited in a low-pressure reactor at moderate rates and a temperature of 665°C are polycrystalline [2.30]. At very high rates, however, the time for surface migration is limited; the deposited films are basically amorphous, although some solid-phase crystallization can occur while the wafers remain hot during the latter stages of the deposition cycle. Phosphorus-doped films are polycrystalline at all deposition rates investigated, consistent with phosphorus inclusion aiding rearrangement of the silicon atoms. As for boron-doped films deposited using silane, the amorphous-crystalline transition temperature is decreased by incorporating moderate concentrations of boron into the silicon film [2.31]. At very high boron concentrations ($\sim 1.6 \times 10^{21}$ cm^{-3}), however, crystallization is suppressed, possibly by the formation of a Si-B compound or the precipitation of boron.

2.5.2 Effect of plasma on structure

During plasma deposition the structure can be affected by ion bombardment, as well as by the factors already discussed. As we saw in Sec. 1.10, only a small fraction of the excited species in a plasma are ions. However, this small number of ions can have a significant effect on the depositing layer. Although most ions are confined to the plasma region, which is generally displaced from the wafer, some can impinge on the wafer surface and either promote or degrade the structure of the depositing film [2.32–2.34].

Ions may give up energy to the migrating surface species, increasing their ability to move on the surface and promoting the development of a more ordered structure. Ions may also knock weakly bound surface atoms out of the layer, again enhancing the order. More energetic ions may break well-formed bonds in the layer, degrading its structural order. Ions may also be implanted slightly below the surface and incorporated into the film, increasing the impurity content. The implanted

2.5. GRAIN ORIENTATION

ions may allow rearrangement of the subsurface atoms to form a more ordered structure [2.35]. The ions can also knock off weakly bound impurity atoms from the surface so that surface migration of the depositing species occurs more readily. At high powers, sputtering may smooth the surface of the depositing layer.

In some cases the major improvement in the structure occurs if the plasma is present before the deposition begins. This initial ion bombardment may remove impurity atoms from the substrate, leaving a cleaner surface [2.36, 2.37]. It may also break bonds in the amorphous layer on which polysilicon is to be deposited, leaving unsatisfied bonds which can act as nucleation sites during the subsequent deposition. If these mechanisms affecting nucleation dominate, the presence of the plasma is not as critical after deposition starts. The dominant mechanism depends sensitively on the energy of the ions striking the surface and, therefore, on the geometry of the reactor and the power and frequency used to generate the plasma.

2.5.3 Evaporated and sputtered films

In our discussion we saw similar changes in preferred orientation for films deposited by CVD in hot-wall and in cold-wall reactors as the deposition temperature changes. The development of the {110} texture just above the transition temperature, and the subsequent decrease of the relative importance of this texture at intermediate deposition temperatures, is not limited to CVD silicon films, but occurs in vacuum-evaporated films also. In films formed by electron-beam heating of the silicon source at a pressure of about 10^{-6} torr ($\sim 10^{-4}$ pa), {110} texture is observed just above the transition temperature, with some {111} texture visible at higher deposition temperatures [2.38, 2.39].

Similar behavior is seen in films deposited in an ultrahigh vacuum deposition system with a base pressure of 10^{-10} torr ($\sim 10^{-8}$ pa) [2.40]. The transition temperature above which polycrystalline films are deposited is only about 400°C in ultra-high vacuum, primarily because of reduced surface adsorption of impurity atoms. Above the transition temperature, {110} texture and a columnar structure are seen. At higher temperatures, the {110} texture decreases and the {111} texture develops, as for CVD films. Similar trends are also seen for germanium films.

Polycrystalline silicon films can be deposited by sputtering with the substrate at moderate temperatures in an argon atmosphere, but nucleation can be enhanced by depositing in a hydrogen ambient [2.41]. Sputtered amorphous silicon films deposited at a low temperature can be crystallized by annealing, but incorporation of oxygen during sputtering limits the properties of the films [2.42].

2.5.4 Other mechanisms controlling structure

The development of the structure has been related [2.40] to the different work functions of the differently oriented crystals and to the modified Bravais theory of crystal growth, in which the growth velocity of a particular plane is proportional to the density of surface lattice points in that plane. For silicon and germanium, the $\{110\}$ plane is the densest, followed by the $\{111\}$ and $\{100\}$. In addition, the work function of the $\{110\}$ plane is greater than that of the $\{111\}$. Both factors favor the growth of $\{110\}$-oriented grains [2.40].

2.6 Optical properties

In a number of applications the optical properties of polysilicon are critical to proper device operation. In charge-coupled-device (CCD) image sensors, light must often pass through a polysilicon gate electrode into the underlying light-sensing region. In polysilicon photoconductors or solar cells, light must be efficiently absorbed in the polysilicon to create electrons and holes, which are then collected. The optical properties of polysilicon are also important for several evaluation techniques used to determine the thickness and structure of polysilicon films deposited for more general applications.

The optical properties of polysilicon differ somewhat from those of single-crystal silicon because of grain boundaries and other structural defects. These defects can lead to states within the bandgap that act as generation-recombination centers and absorb light with energy less than the bandgap energy. The defects can also scatter light, and they distort the band structure, further changing the optical properties. Because the optical properties are sensitive to the crystal structure, they depend on the deposition conditions.

2.6. OPTICAL PROPERTIES

2.6.1 Index of refraction

The commonly used low-pressure CVD polysilicon deposited near 620°C has been studied in most detail [2.43]. In the visible region of the spectrum from about 400-800 nm wavelength, the index of refraction of undoped polysilicon is similar to that of single-crystal silicon. Adding phosphorus does not appreciably change the index of refraction in this wavelength region.

In the infrared region up to at least 1.8 μm wavelength, the index of refraction of undoped films becomes constant with a value of approximately 3.50 [2.43], close to the value of 3.42 commonly used for single-crystal silicon. In this wavelength region, the index of refraction decreases as phosphorus is added, as in single-crystal silicon. The decrease is attributed to the interaction of light with free carriers, and the index of refraction is expected to depend quadratically on the carrier concentration according to the expression [2.43]

$$n_1^2 = n_{10}^2 - \frac{q^2 N \lambda^2}{4\pi^2 \epsilon_0 c^2 m^*} \qquad (2.7)$$

where n_1 and n_{10} are the indices of refraction of the heavily doped and undoped films, respectively, q is the electronic charge, N is the active carrier concentration, λ is the wavelength, ϵ_0 is the permittivity of free space, c is the speed of light, and m^* is the effective mass of the carriers.

The index of refraction also depends on the structure of the deposited silicon film. It is much higher in amorphous films than in polycrystalline ones, with an abrupt change in the index of refraction occurring at the transition temperature separating the deposition of amorphous and polycrystalline films, as shown in Fig. 2.14 [2.44]. After annealing amorphous films to crystallize them, the index of refraction decreases and becomes similar to that of polysilicon (Fig. 2.15) [2.18].

Near the transition from amorphous silicon to polycrystalline silicon deposition, the index of refraction can provide some indication of the fraction of the material that is crystalline. In a single-wafer reactor operating at 80 Torr at a deposition rate near 60 nm/min, a transition temperature of 645°C was found for undoped silicon, with a decrease of the transition temperature by about 15°C for heavily phosphorus-doped films [2.45]. The transition temperature is much higher than observed in the low-pressure reactor mainly because of the higher deposition rate and somewhat because of the higher total pressure.

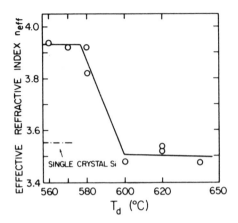

Figure 2.14: The effective refractive index measured after film deposition as a function of deposition temperature. From [2.44]. Reprinted with permission.

2.6.2 Absorption coefficient

For both undoped and doped films, the absorption coefficient α is considerably greater in polysilicon than in single-crystal silicon. The difference is about a factor of five for undoped films at 400 nm wavelength. However, the measured absorption coefficient may appear abnormally high because of scattering from the rough surface of a polysilicon film, as well as by higher absorption. Within the visible region of the optical spectrum, the absorption coefficient decreases with increasing dopant concentration, but it remains significantly greater than that of single-crystal silicon in all cases, as shown in Fig. 2.16 [2.43]. The added dopant atoms possibly compensate some of the structural defects that lead to excess light absorption. At longer wavelengths, where free-carrier absorption becomes significant ($\lambda > 700$ nm), the absorption coefficient of heavily doped films becomes greater than that of undoped films, as it does in single-crystal silicon.

The optical absorption also depends strongly on the structure of the deposited film and is higher for amorphous films than for polycrystalline ones. As we will see in Sec. 2.6.4, little light penetrates amorphous films only 0.5 μm thick for wavelengths shorter than 600 nm, while light penetrates polycrystalline films of a similar thickness at wavelengths longer than 500 nm. Although the absorption coefficient is markedly higher for amorphous silicon films than for polycrystalline films, (Fig. 2.17)

2.6. OPTICAL PROPERTIES

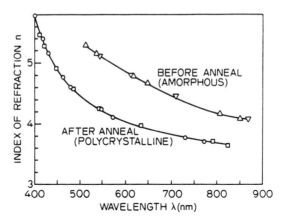

Figure 2.15: The index of refraction as a function of wavelength for amorphous silicon films (upper curve) and for the same samples after annealing to crystallize the films (lower curve).

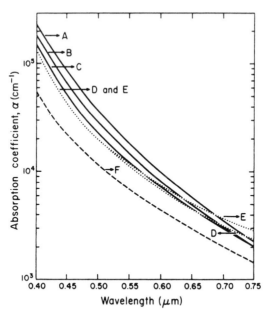

Figure 2.16: The absorption coefficient for undoped (curve A) and increasingly phosphorus doped (curves B-E) polycrystalline films. The dopant concentration increases from about 3×10^{19} cm^{-3} for curve B to about 7×10^{20} cm^{-3} for curve E. Curve F corresponds to lightly doped, single-crystal silicon. From [2.43]. Reprinted with permission.

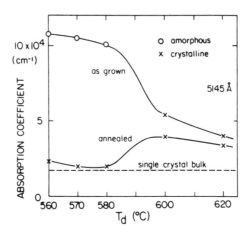

Figure 2.17: Absorption coefficient of silicon films at a wavelength of 514.5 nm as a function of deposition temperature before and after annealing at 900-1000°C. From [2.46]. Reprinted by permission of the publisher, The Electrochemical Society, Inc.

[2.46], after the initially amorphous films are crystallized their absorption coefficient becomes less than that of films which are initially polycrystalline and can approach that of single-crystal silicon [2.46]. The low absorption coefficient in the crystallized, initially amorphous films is consistent with the large grain size seen in these films after annealing, as will be discussed in Sec. 2.11.4. The absorption coefficient of the initially polycrystalline films decreases only slightly on annealing, consistent with the structural stability of these films [2.25]. At 450 nm wavelength the absorption coefficient of polysilicon films deposited at higher temperatures in an atmospheric-pressure reactor is similar to that of low-pressure films, but the absorption coefficient is somewhat higher in the atmospheric-pressure films at 600 and 750 nm [2.47].

2.6.3 Ultraviolet surface reflectance

The optical properties of polysilicon films also differ from those of single-crystal silicon in the ultraviolet region of the spectrum. In this high-energy region, the optical properties are influenced more significantly by the detailed energy-band structure of the deposited film than they are at longer wavelengths. Because silicon is highly absorbing in the ultraviolet, its optical properties are most readily observed by reflectance

2.6. OPTICAL PROPERTIES

measurements, rather than by the transmission techniques sometimes used at longer wavelengths.

In single-crystal silicon two pronounced maxima appear in the reflectance spectra at 280 nm and 365 nm. These maxima arise from optical transitions at the X point and along the Γ-L axis of the Brillouin zone, respectively, and are strongly affected by distortions in the crystal structure. The maxima are significantly less intense in polycrystalline-silicon films than in single-crystal silicon, but they are still discernible. In amorphous-silicon films, however, the long-range order which gives rise to the details of the band structure is absent, and the maxima do not appear. (In typical reflectance measurements, a single-crystal silicon wafer is often used as the reference. When comparing amorphous films to the crystalline reference, strong extrema are seen at these critical wavelengths, implying a marked difference between the structure of amorphous silicon and that of crystalline silicon. Even when polycrystalline films are being measured, some signal is seen, corresponding to the less perfect structure of the polycrystalline material.)

2.6.4 Use of optical properties for film evaluation

The dependence of the optical properties on the structure of silicon films allows some structural information to be obtained rapidly and nondestructively by measuring the reflectance over a suitable wavelength range of films deposited on oxidized silicon substrates. This measurement can serve as a rapid, nondestructive technique to distinguish between amorphous- and polycrystalline-silicon films. It also provides information about the surface roughness. The output from a typical spectrophotometer, which is determined by the light reflected from all interfaces of the $Si/SiO_2/Si$ structure, is shown in Fig. 2.18 over the wavelength range from 200 to 800 nm for an amorphous-silicon film deposited at 550°C and for polycrystalline-silicon films deposited at 625 and 700°C, all in a low-pressure reactor onto oxidized silicon wafers.

Surface roughness

At short wavelengths ($\lambda < 400$ nm) the light cannot penetrate the silicon films. The observed reflectance primarily arises from the surface structure and provides a quantitative measure of surface roughness. The rms value of surface roughness σ_0 can be calculated from the reflectance in

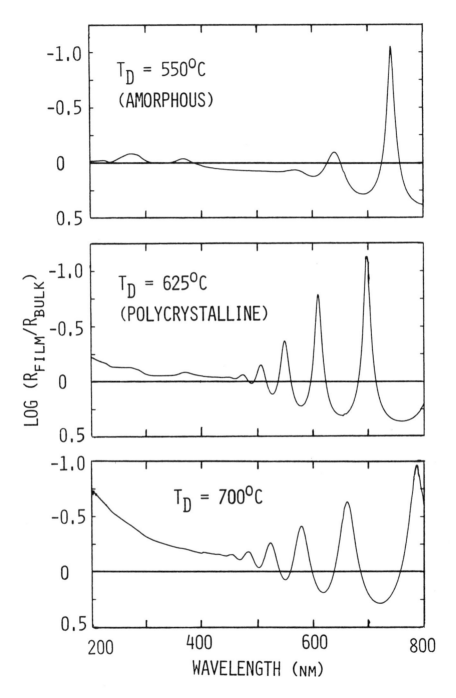

Figure 2.18: Reflectance as a function of wavelength (using a single-crystal reference wafer) for silicon films deposited in a low-pressure CVD reactor at 550, 625, and 700°C.

2.6. OPTICAL PROPERTIES

this wavelength range using the formula [2.48]

$$\sigma_0 = \frac{0.12\lambda}{\cos\phi}\sqrt{\log_{10}\left(\frac{R_0}{R}\right)} \qquad (2.8)$$

where λ is the wavelength, ϕ is the angle of incidence (measured from the normal to the sample surface), R is the reflectance of the sample, and R_0 is the reflectance of an ideally smooth surface of the same material.

The low reflectance at short wavelengths seen in Fig. 2.18 for the amorphous film indicates a very smooth surface. (As discussed above, the two small features at about 280 and 370 nm arise from the structural differences between the amorphous film and the single-crystal silicon used as a reference in the measurement.) Because of limited surface diffusion and the lack of faceted crystal growth during deposition, amorphous silicon films are markedly smoother than are polycrystalline films. The surface roughness increases abruptly at the transition temperature above which polycrystalline films are deposited (Fig. 2.18b) and increases further at higher deposition temperatures (Fig. 2.18c) [1.12].

The smoother surface of an initially amorphous film is retained even after the film is crystallized by annealing because the surface is stabilized by a thin, native surface oxide that forms when the film is exposed to air [2.49]. (Note that the film must be exposed to air before annealing to form the stabilizing oxide. Without this oxide the surface becomes extremely rough during annealing. As we will see in Sec. 2.12, the surface atoms are free to migrate if they are not constrained by the oxide layer.) The smoother surface of the initially amorphous film may improve the ability to pattern fine lines in small-geometry integrated circuits. In addition, as we will discuss in Sec. 4.3.2, the leakage current of oxide grown on a polysilicon film is lower and the breakdown voltage is higher for a smooth film deposited in an amorphous form, even though the film is crystallized before the oxide is grown. A smooth surface is also needed for optimum electrical characteristics of thin-film transistors formed with their conducting channel within the layer of polysilicon [2.50]. The smooth surface may be obtained by depositing the silicon film in an amorphous form and then crystallizing it or by using *chemical-mechanical polishing* (CMP) of an initially rougher film deposited in polycrystalline form [6.144, 6.145].

The surface roughness can also be measured over a small area by an atomic-force microscope (AFM). With this technique a small probe

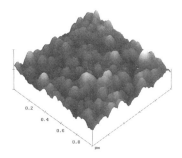

Figure 2.19: Atomic-force micrograph of polysilicon, showing surface roughness over a 1 µm × 1 µm area, with an exaggerated vertical scale of 50 nm/div; rms surface roughness of this film is 6 nm.

travels over the surface at a constant distance above the surface and the height of the probe is measured by the position of a light beam reflected from the probe onto a position-sensitive photodiode. As the probe moves across the surface, a map of the height is generated, as shown in Fig. 2.19. Atomic-force microscopy can provide a great deal of information about the structure within a small region (typically 0.1-10 µm), but sampling several areas on a wafer is time consuming. More conventional surface profilometers can provide limited information about the surface roughness. However, if the surface is rough and the irregularities are closely spaced, the stylus cannot follow the contours of the surface, and erroneously low values of surface roughness are indicated.

In addition to ultraviolet reflection and atomic-force microscopy, the coupling of light to surface plasmons at a metal surface can be used to determine the surface roughness [2.51]. Plasmons are the quanta of the longitudinal polarization field of the electron gas. Light does not couple effectively to the plasmons when the surface is ideally smooth because of the different momenta involved. A rough surface allows momentum matching so that coupling is possible, leading to a reflection minimum at the wavelength corresponding to the surface plasmon frequency. To use this technique, the silicon film must be covered with a metal which has a surface plasmon frequency at a convenient wavelength (eg, silver with $\lambda \approx 350$ nm).

Film thickness

Light of wavelength greater than 500 nm can travel through the thickness of a 500 nm thick polycrystalline film, producing constructive

2.6. OPTICAL PROPERTIES 95

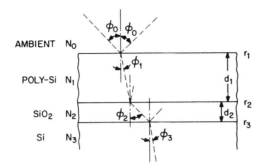

Figure 2.20: The overall reflectance from a polysilicon sample deposited on an oxidized wafer can be used to determine the film thickness from the wavelengths at which constructive and destructive interference occur. From [2.48]. Reprinted by permission of the publisher, The Electrochemical Society, Inc.

and destructive interference between light reflected from the top surface of the silicon film and that reflected from the underlying interfaces (Fig. 2.18). An amorphous film of the same thickness is still highly absorbing for light of 500 nm wavelength. Little light can pass through the film, and little interference is visible. Significant light can penetrate the amorphous film only for wavelengths well above 600 nm. However, even at longer wavelengths light transmission through the amorphous film, and consequently the strength of the interference signal, is weaker for an amorphous film than for a polycrystalline one. After it is crystallized by annealing, the initially amorphous film becomes more transparent, and stronger interference maxima and minima are seen. For thinner films, of course, less light is absorbed, and the wavelength at which interference is visible decreases. For example, interference can be seen at 400 nm wavelength for a polycrystalline film about 250 nm thick, but not for an amorphous film of the same thickness.

Once the basic structure of a film is known (ie, whether it is amorphous or polycrystalline), the reflectance can be used to determine the film thickness if the thickness of the underlying dielectric layer is known. As shown in Fig. 2.20 [2.52], incident light is reflected from the top surface of the silicon film, from the underlying silicon-dielectric interface, and from the dielectric-substrate interface. These three reflected beams combine to produce constructive and destructive interference as the wavelength varies. Using the index of refraction corresponding to either polycrystalline or amorphous silicon, the overall reflectivity can be calculated [2.52], and the wavelengths corresponding to constructive and destructive interference can be found for given dielectric and silicon film thicknesses. For a known dielectric thickness, comparison of these wavelengths with those observed allows the thickness of the deposited silicon film to be easily determined. More accurate algorithms determine the thickness by fitting the entire measured reflectivity-vs.-wavelength curve to calculations. The data reduction algorithms are built into commercially available measurement equipment, so that the thickness is provided directly for a known thickness of the underlying dielectric layer (often 100 nm of SiO_2). A more complicated, multilayer substrate can be approximated as an effective homogeneous substrate to simplify analysis, with adequate accuracy for many applications [2.53].

Because the index of refraction of amorphous silicon films is considerably higher than that of crystalline films (Figs. 2.14 and 2.15), the same values of refractive index cannot be used to reduce the data for amorphous films. The peaks in the reflected signal occur at different locations for amorphous and polycrystalline films of the same thickness. (The thickness of these dense amorphous films does not change appreciably when they are annealed to convert them to polysilicon.)

2.7 Thermal conductivity

High thermal conductivity is needed in polysilicon for proper heat dissipation when it is used as the mechanical support in the dielectrically isolated, integrated-circuit structure to be discussed in Sec. 6.10.1 and for sensors and other microelectromechanical systems (MEMS) to be discussed in Sec. 6.14. Although silicon is a good thermal conductor, with a thermal conductivity of about 1.4 W cm^{-1} K^{-1}, grain boundaries can significantly impede the heat flow. Therefore, the thermal conductivity depends on the deposition conditions.

When thin films of polysilicon are used as integrated thermal sensors, the thermal conductivity must be known for proper device calibration. In the fine-grain films normally used for this application, the thermal conductivity is about ~ 0.3 W cm^{-1} K^{-1}, 20-25% of the single-crystal value [2.54–2.56]. In the thicker films with larger grains used for the dielectric isolation structure, the thermal conductivity is 50-85% of that in single-crystal silicon [2.57, 2.58]. In an anisotropic, columnar polysilicon structure, the thermal conductivity in the direction of the columnar grains is about 60% greater than that perpendicular to the long direction of the grains [2.57, 2.58].

2.8 Mechanical properties

The mechanical properties of polycrystalline silicon are important both in conventional integrated circuits and, especially, in the structures used in the *microelectromechanical systems* (MEMS) to be discussed in Sec. 6.14, which inherently rely on polysilicon as a mechanical element [2.59].

The mechanical properties of crystalline silicon depend on the crystallographic orientation. Polysilicon contains grains with several different crystal orientations, so its mechanical properties are, at least in part, an average over the dominant grain orientations. For example Young's modulus is about 150-170 GPa in polysilicon [2.60, 2.61], comparable to the 190 GPa in single-crystal silicon. This reasonable agreement suggests that the elastic constants are dominated by the grains, rather than the grain boundaries. Because the elastic properties of crystalline silicon can vary by 50% in different crystallographic directions, the elastic properties of polysilicon are also expected to vary appreciably as the structure varies.

On the other hand, deformation and fracture may be affected more significantly by the grain boundaries. The grain boundaries may tend to block propagation of dislocations that can lead to plastic deformation; consequently, polysilicon tends to be linearly elastic until it fractures [2.62], with a fracture strength comparable to or possibly greater than that of single-crystal silicon [2.63]. On a macroscopic scale the grain boundaries may impede crack propagation, increasing the strength. On the other hand, the grain boundaries can be weak regions, along which grain boundary "sliding" can occur [4.26]. In addition, grain boundaries may cause irregular edges on polysilicon structures, providing more sites for crack initiation than in single-crystal silicon structures with smooth edges. The fracture strength of polysilicon has been measured to be 2-3 GPa [2.61]. The coefficient of friction of n^+ polysilicon appears to be lower than that of lightly doped, single-crystal silicon, probably because the rougher surface of polysilicon leads to a smaller contact area [2.64].

In integrated circuits high stress can deform wafers, preventing accurate alignment of fine-geometry masks. However, much smaller stresses in MEMS can deform the structural elements; for example, small compressive stresses can buckle polysilicon beams anchored at both ends. The stress gradient through the thickness of the film, as well as the average stress, is also important in MEMS structures; for example, nonuniform stresses can bend the cantilever beams used in polysilicon sensors [6.156].

Undoped films are generally under compressive stress [2.65–2.68]. The stress depends on the structure of the film and the preferred orientation, with the highest compressive stresses in amorphous films and in those with {110} texture [2.66]. For films without strong {110} texture, the stress appears to decrease with increasing deposition temperature, consistent with larger grains and fewer defects being formed during deposition. However, as discussed below, inclusion of impurities, such as oxygen, during deposition can affect the stress greatly [2.14]; at higher temperatures, impurities can be desorbed more readily, reducing the stress in films deposited at higher temperatures.

In films grown with compressive stress, the stress generally decreases with increasing film thickness [2.65]. If the compressive stress is related to the disordered grain boundaries, such a decrease in stress with increasing film thickness is expected. The grain size is greater and the grain-boundary density is lower in the upper portion of a film than in the first part deposited. The stress may also be reduced in thicker films by atomic rearrangement during the longer time at the deposition temperature. The development of the dominant crystal orientation as the film becomes thicker can affect the stress uniformity through the thickness of the film. The {110} texture develops as the film becomes thicker, so that this texture is more pronounced in the upper portion of the film.

Annealing tends to reduce the compressive stress in initially polycrystalline films. The large compressive stress in initially amorphous films also decreases as they crystallize during annealing, and the stress may even become slightly tensile at moderate annealing temperatures [2.66]; at high annealing temperatures, the stress relaxes and may become slightly compressive again [2.68]. Films deposited at low temperatures may contain a small amount of hydrogen. When the hydrogen evolves during subsequent heat treatment, the stress may become less compressive or even tensile [2.68].

2.9. OXYGEN CONTAMINATION

The decrease of compressive stress as polycrystalline films are doped with phosphorus after deposition is probably related to grain growth [2.65]. As will be discussed in Sec. 2.11.3, phosphorus doping greatly increases grain growth during annealing. As the grains grow, the volume initially occupied by the disordered, possibly low-density, grain boundaries is available to the growing grains, reducing the overall stress [2.69]. In addition to stress in a large area of polysilicon, the effect of stress from the surrounding materials on the polysilicon must be considered, as discussed in Sec. 6.10.

2.9 Oxygen contamination

Many of the properties of polycrystalline silicon are influenced by oxygen contamination included in the polysilicon during its deposition. As the deposition temperature decreases, the partial pressure of oxygen contamination in the deposition chamber must be reduced to preserve an oxygen-free surface during deposition. At temperatures near 800°C, oxygen is stable on the surface for oxygen partial pressures greater than 10^{-6} Torr [2.70, 2.71]. At the lower temperatures generally used to deposit polysilicon conformally, the requirements are more severe. If the incoming gas supply is the primary source of oxygen contamination, the oxygen partial pressure in the deposition chamber is lower for the same purity gas when the total deposition pressure is lower. Therefore, obtaining a low oxygen partial pressure is generally easier at lower deposition pressures.

In a recent survey [2.72] of the average oxygen concentration in polysilicon films deposited near 625°C in low-pressure reactors, the average oxygen concentration was found to be in the mid-10^{18} cm^{-3} range. In films deposited in 12 different reactors at 10 different locations of 6 different organizations, the oxygen concentration varied over a wide range from less than the detection limit of 4×10^{17} cm^{-3} to more than 1×10^{19} cm^{-3}.

If the oxygen concentration on the surface during deposition is excessive, the surface mobility of the arriving silicon atoms is reduced; they do not reach low-energy positions; and the structure is less ordered than expected for the nominal deposition conditions used. Not only is a different structure formed, but its stability can also be degraded. The silicon atoms are in unstable positions when they are covered and immobilized by subsequently arriving atoms. Further heat treatment can cause the atoms to move to low-energy positions. The resulting contraction of the film produces large tensile stresses that can deform the film and even the substrate. If the film is deposited on both sides of the wafer, as in a tube-type, low-pressure reactor, the symmetrical stress may initially prevent deformation, but subsequent processing may remove the film from the wafer back, allowing the stress to deform the wafer.

The oxygen concentration can be nonuniform through the thickness of the polysilicon layer, causing stress gradients. As the deposition proceeds, residual oxygen in the chamber is gettered into the depositing silicon, either on the wafers or on the walls of the reactor, and the partial pressure of oxygen available to enter into the upper portion of the polysilicon layer may be lower, reducing the stress there.

2.10 Etching

Wet chemical etching is usually used to remove polysilicon from large areas, such as the back of wafers. (As we saw in Sec. 1.5.1, the wafers are only supported at the edges in the low-pressure CVD reactor, and polysilicon is deposited on the back of each wafer, as well as on the front.) Dry, reactive-ion etching in a plasma is used to define small features.

Polycrystalline silicon films can be readily etched in typical wet silicon etchants composed of HF-HNO_3-HAc mixtures or in basic solutions such as KOH. Polysilicon is not readily etched by HF alone. The chemical properties of amorphous silicon films differ from those of polycrystalline silicon, and amorphous silicon is attacked by hydrofluoric acid [2.73]. After amorphous silicon is crystallized by annealing, its properties become similar to those of polycrystalline silicon, and it is no longer etched by HF-based solutions. Undoped polysilicon is not attacked by H_3PO_4, but attack of heavily phosphorus doped polysilicon sometimes occurs, perhaps because of phosphorus segregating at grain boundaries to create a Si-P phase. (See Sec. 4.2.3.)

Because wet chemical etching is generally isotropic — etching laterally as well as vertically — fine device features are usually defined by "dry" etching in a plasma [2.74–2.76]. As in plasma deposition (Sec. 1.10), excited neutral radicals and ions impinge on the wafer surface during plasma etching. When neutral radicals dominate the etch

2.10. ETCHING

process, etching is mainly isotropic. However, when the surfaces are passivated by etch products or the radicals cannot chemically attack the exposed surfaces, the etching can be highly directional (*anisotropic*). The ions are accelerated toward the surface by the electric field between the plasma region and the surface, so that they strike mainly horizontal surfaces. In these regions, they activate the surface by removing etch products and perhaps by weakening bonds, so that chemical reaction with the neutral radicals can readily occur. Ions do not reach the sidewalls as readily to remove the etch products deposited there, so these layers inhibit lateral etching.

Etching occurs by the combination of surface activation by ions and chemical etching by neutral radicals in the activated regions. By properly controlling the ion bombardment and sidewall etch-inhibiting layers, the etching process can be made very anisotropic. The entire thickness of the silicon layer can be removed in selected areas with very little lateral attack of regions protected by a mask.

However, when the physical *sputtering* action of the ions dominates the etching process, the chemical *selectivity* (which allows etching of one layer with little attack of the underlying layer of another material) decreases. Thus, a tradeoff between anisotropy and selectivity is necessary [2.77]. To maximize both anisotropy and selectivity, the etch process can include two different steps. The first step is adjusted to minimize the lateral etch rate while etching vertically rapidly (high ion contribution to the etching process and low selectivity). When most of the silicon layer has been etched, the etch conditions are changed to increase the selectivity of the etch so that etching of the underlying layer is minimum (high chemical contribution to the etching). The lateral etch rate generally is higher during this second step, so this portion of the etch process is usually kept short. The etch rate can also depend on the dopant species and their concentration [2.78, 2.79]. If the dopant distribution is not uniform through the thickness of the layer, the amount of lateral etching may vary with depth. Because of the importance of the vertical and lateral etch rates and selectivity, extensive computer simulation of the etch process has been used to reveal details of the physics and chemistry involved [2.80, 2.81].

The strong electric fields that attract the ions to the surface, the resulting ion bombardment, and the presence of charged species on the surface, can all degrade the device being fabricated. When etching the polysilicon to form the gate of an MOS transistor, charged species on the surface can migrate from surrounding regions to an isolated polysilicon region during the later stages of etching. When a large region of polysilicon collects charge (large Q) and only a small fraction of the polysilicon is over thin gate oxide (small C), the resulting voltage (V=Q/C) can exceed the breakdown voltage of the gate oxide, destroying the transistor. Because charge migrates to the polysilicon from surrounding regions, the degradation depends on the perimeter of the polysilicon line, the amount of surrounding oxide, and connection to the silicon substrate [2.82–2.85]. Charging also distorts the electric field lines and the resulting ion bombardment, changing the edge profile of the etched features [2.85].

As wafer size increases, single-wafer etchers become more attractive and are now widely used to obtain precise control of the etch process and good uniformity. To be economically acceptable, the throughput and etch rate must be high. To achieve good anisotropy, the ion bombardment must also be high, increasing the possibility of damaging the gate oxide [2.86]. The etch conditions in a such a *high-density plasma* (HDP) can rapidly degrade a photoresist mask, and an oxide *hard mask* is often used [2.87]. The walls of etch chambers are often made of anodized aluminum. The active species in the plasma (often Cl) can gradually attack the walls and transport aluminum to the wafer. The aluminum can degrade the etching uniformity and sidewall profile, as well as contaminating the wafer [2.88].

2.11 Structural stability

After a polysilicon film has been deposited, it usually undergoes significant thermal processing during subsequent steps of the device fabrication sequence. Possible changes in the grain structure during these heat cycles can affect the fabrication and performance of integrated circuits. As the minimum feature size of integrated circuits decreases, the grain size can become a significant fraction of the minimum dimension if appreciable grain growth occurs, and lithographic definition and pattern etching of fine features can be impeded by grain boundaries.

Changes in the structure of the polysilicon depend on the initial structure formed during deposition and the dopant present in the polysilicon during annealing, as well as the annealing conditions. The rapid grain growth which occurs in doped films is particularly impor-

2.11. STRUCTURAL STABILITY

tant because the polysilicon used for MOS gate electrodes is usually very heavily doped to obtain the lowest resistivity, and the dimensions of the gate electrode are most critical to transistor performance.

2.11.1 Recrystallization mechanisms

Recrystallization generally has three stages: nucleation, primary recrystallization, and secondary recrystallization. Nucleation of new grains occurs in fine-grain regions containing a high density of grain boundaries. After forming, the nuclei grow at a rate determined by the nature of the driving force causing grain growth. Three different phenomena can influence the grain growth [2.89]:

1. **Strain-induced growth.** When the main driving force for grain growth is stored strain energy (*eg*, caused by ion implantation or mechanical deformation), the grain size d increases linearly with annealing time t:
$$d = d_0 + k_1 t \qquad (2.9)$$
where d_0 is the initial grain size and k is a constant.

2. **Grain-boundary- or interface-induced growth.** The extra energy associated with the grain boundaries causes the grains to grow to minimize the grain-boundary area. This driving force is inversely proportional to the radius of curvature of the grain boundary, and the grain size increases as the square root of the annealing time:
$$d = \sqrt{d_0^2 + k_2^2 t} \qquad (2.10)$$
The lower surface or interface energy of certain crystal orientations can similarly cause growth of grains with these orientations.

3. **Impurity drag.** Although the other mechanisms discussed enhance grain growth, impurities incorporated at the grain boundary can retard grain growth, which then exhibits a cube-root time dependence:
$$d = \sqrt[3]{d_0^3 + k_3^3 t} \qquad (2.11)$$

2.11.2 Undoped or lightly doped polycrystalline films

The polycrystalline structure formed during deposition of an undoped film is relatively stable during annealing. As Fig. 2.21 shows, the pre-

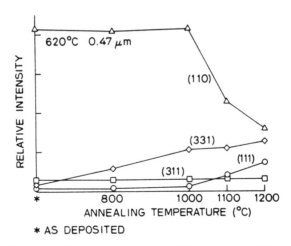

Figure 2.21: Structural changes on annealing observed by x-ray diffraction for undoped polycrystalline films deposited at 620°C in a low-pressure reactor.

dominately {110}-oriented structure changes little on annealing at temperatures up to at least 1000°C [2.25]. As seen by comparing Figs. 2.7a and 2.7d, more detailed cross-sectional transmission electron microscopy also indicates that the columnar structure formed during deposition changes little on annealing at 1000°C when the film is not doped. At higher temperatures the dominant {110} texture decreases, and the amount of other orientations increases. TEM analysis shows that the grain size increases slightly after annealing at these high temperatures. However, temperatures approaching the 1415°C melting temperature of silicon are needed to obtain very large grains [2.90–2.92].

More detailed investigation has shown that two stages of recrystallization occur in undoped films [2.90]. At very high temperatures new nuclei form, and the fine-grain structure coarsens into a large-grain structure containing grains of nearly uniform size by *primary recrystallization*. New nuclei form more rapidly in regions containing a high density of grain boundaries. The activation energy for nucleus formation appears to be about 9 eV [2.91], much greater than the ~5 eV activation energy for silicon self-diffusion. After nucleation, the new grains, which are nearly defect free, grow into the highly defective grains formed during deposition. The new grains continue growing until they impinge on other growing grains or on initially large grains. The final grain size

2.11. STRUCTURAL STABILITY

is, therefore, determined by the number of nuclei formed; if nucleation occurs uniformly within the sample, the final grain size is relatively uniform after primary recrystallization is complete. In thin films the grain growth by primary recrystallization may be limited when the growing grains reach the top and bottom surfaces of the silicon film, and the final grain size is approximately the same as the film thickness. The decreasing rate of grain growth with increasing time suggests that the growth rate is a function of the density of grain boundaries and that the energy associated with these grain boundaries provides the driving force for recrystallization. Mechanical stress can provide additional driving force for recrystallization [2.92].

At even higher temperatures *secondary recrystallization* occurs. Nucleation does not appear to play a dominant role in secondary recrystallization. Rather, some of the existing grains grow at the expense of neighboring grains so that the resulting structure contains some abnormally large grains together with the smaller grains formed during primary recrystallization.

The temperatures needed to form large grains in undoped polysilicon are markedly higher than those typically encountered in integrated-circuit fabrication processes, and undoped, polycrystalline silicon films are expected to be quite stable at moderate temperatures, as we saw in Fig. 2.21.

2.11.3 Heavily doped polycrystalline films

In contrast to the structural stability seen in undoped films on annealing (Fig. 2.7d) [2.15], significant grain growth occurs at normal integrated-circuit processing temperatures if the films are heavily doped with phosphorus or arsenic. This grain growth can take place during the gas-phase doping process (typically at 900-950°C) often used to add phosphorus to the polysilicon gate electrode of an MOS transistor. After doping, the grains are no longer columnar, but have an equi-axed structure (Fig. 2.7b), and the microtwin density within the grains decreases greatly. The maximum surface *asperity* height decreases, and the average grain size is approximately equal to the film thickness. After annealing at 1000°C (Fig. 2.7c), the average grain size increases further.

For an annealing temperature of 1000°C, phosphorus concentrations in excess of 3×10^{20} cm^{-3} enhance grain growth markedly, as shown in Fig. 2.22 [2.89]. Little enhanced grain growth occurs at 1×10^{20}

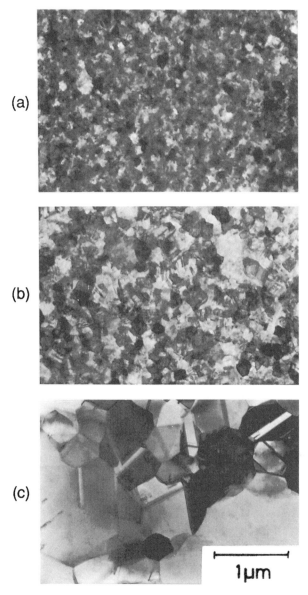

Figure 2.22: Transmission electron micrographs of 400 nm thick, phosphorus-doped polysilicon films after annealing at 1000°C. The average dopant concentrations are (a) 1×10^{20}, (b) 2.5×10^{20}, and (c) 7.5×10^{20} cm^{-3}. From [2.89]. Reprinted by permission of the publisher, The Electrochemical Society, Inc.

2.11. STRUCTURAL STABILITY

cm^{-3}, while the grain growth is greatly accelerated at 7.5×10^{20} cm^{-3}. (In this study the phosphorus was added by ion implantation, and the average concentrations cited correspond to the implanted dose divided by the film thickness. Doping by ion implantation amorphizes the top of the polysilicon layer during implantation, further complicating the rearrangement [2.93]; however, the amorphized material probably recrystallizes on the underlying fine grains at the beginning of the recrystallization process, so that the final structure only depends moderately on the amount of implantation damage.)

Adding a large quantity of n-type dopant raises the Fermi level so that the film can be extrinsic at the annealing temperature, changing the point-defect concentration. The additional point defects increase the motion of silicon atoms across grain boundaries, aiding the grain growth that minimizes the energy associated with the grain boundaries, as we will discuss below. The critical concentration for grain growth is also close to the 3.5×10^{20} cm^{-3} solid solubility of phosphorus in silicon at 1000°C [2.94, 2.95]. However, this agreement may be coincidental because not all of the dopant is expected to be located within the grains; a significant fraction can be segregated at the grain boundaries, as we will discuss in Sec. 3.6.

For the high dopant concentration of 7.5×10^{20} cm^{-3}, significant grain growth occurs at temperatures above 700°C, especially between 800 and 900°C. At 1000°C two stages of growth occur. The grain size readily increases by primary recrystallization until it is approximately equal to the film thickness. Some grains then further enlarge at the expense of others by secondary recrystallization. The activation energies for primary and secondary recrystallization are 2.4 and 1.0 eV, respectively, as shown in Fig. 2.23 [2.89]. When primary recrystallization dominates, the grain size increases as the the square root of the annealing time, indicating that grain-boundary energy drives the grain growth.

Phosphorus influences the grain growth more than do the other common dopants, arsenic and boron. Like phosphorus, arsenic increases the rate of grain growth (at concentrations above 1×10^{19} cm^{-3} [2.94]), but unlike phosphorus, arsenic enhancement of grain growth diminishes at concentrations higher than about 5×10^{20} cm^{-3} [2.94, 2.96]. When the arsenic concentration is high, arsenic clusters form; the grain boundaries apparently cannot easily move past the clusters; and the rate of grain growth decreases. Boron does not appear to enhance the grain growth

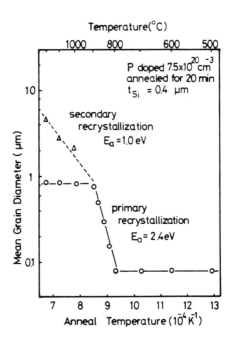

Figure 2.23: Dependence of the mean grain size on annealing temperature, showing the 2.4 eV activation energy for primary recrystallization and the 1.0 eV activation energy for secondary recrystallization. From [2.89]. Reprinted by permission of the publisher, The Electrochemical Society, Inc.

significantly. Some studies report no increase [2.96], while others cite a slight enhancement at moderate boron concentrations and higher annealing temperatures [2.97]. However, at high dopant concentrations no enhancement is seen, possibly because of boron precipitation at the grain boundaries.

Grain growth is postulated to occur by grain-boundary migration resulting from the diffusion of silicon atoms across the grain boundaries. The grain-boundary migration rate is influenced both by the driving force (ie, the decrease in the total system energy as the grains grow and reduce the grain-boundary area) and by the grain-boundary mobility. Segregation of phosphorus to the grain boundaries probably decreases the average grain-boundary energy [2.98], reducing the driving force. Therefore, the dominant effect of phosphorus on grain growth is most likely caused by an increased grain-boundary mobility (eg, an increased diffusion coefficient of silicon across the grain boundary) when large

2.11. STRUCTURAL STABILITY

amounts of phosphorus are segregated at the grain boundaries.

The diffusion coefficients of phosphorus and arsenic in silicon and also the self-diffusion coefficient of silicon all increase when high concentrations of these dopants are present. The correlation between an increased phosphorus diffusion coefficient and the enhanced grain growth in phosphorus-doped polysilicon suggests that a similar mechanism causes both [2.89]. Both effects are probably related to the presence of charged point defects, which increase as the Fermi energy moves toward the conduction-band edge with increasing dopant concentration [2.99].

Further evidence for the importance of electrical effects is seen by considering samples doped with both phosphorus and boron. We saw that phosphorus doping markedly increases grain growth while boron has little effect. However, when boron is added to a phosphorus-doped sample, the grain-growth enhancement *decreases* [2.98], although boron by itself does not decrease the grain-growth rate. Total compensation by adding equal amounts of phosphorus and boron appears to completely remove the grain-growth-enhancing effect of phosphorus. One explanation for this behavior [2.98] suggests that the limiting step in the process by which phosphorus enhances the grain growth is the ability of phosphorus to diffuse to and resegregate at grain boundaries after grain-boundary motion. Boron doping decreases the electron concentration and, therefore, the rate of phosphorus diffusion to the grain boundaries; the low phosphorus concentration at the grain boundaries, in turn, reduces the mobility of silicon atoms across the boundary. Boron may also change the amount of phosphorus segregated to grain boundaries at equilibrium.

Severe heat treatments, such as rapid thermal processing at a high temperature for a few seconds, can cause significant grain growth and roughen the polysilicon surface. Protrusions also develop at the edges of defined polysilicon patterns [2.100]. As these protrusions extend over the oxide of device structures damaged during processing (*eg*, dry etch damage of the oxide over the source or drain regions adjacent to the gate of an MOS transistor), short circuits can form between device elements. Development of these protrusions is attributed to *grain grooving*, in which silicon diffuses away from the intersection of the grain boundary with the surface and moves outward to reduce the sum of the grain-boundary and surface energies. Protrusions form more readily in inert ambients; their formation is suppressed in an oxidizing ambient or with a stabilizing oxide on the polysilicon surface.

Very thin films

In very thin films, the different surface or interface energies of differently oriented grains promote the growth of grains with the lowest surface energy. In films tens of nanometers thick, this interface or surface energy anisotropy at the top or bottom of the film can provide significant driving force for secondary recrystallization so that grains substantially larger than the film thickness can be obtained [2.98, 2.101]. In this regime the grain-size enhancement by secondary recrystallization appears to vary inversely with the thickness of the film, becoming small for films thicker than about 100 nm. Within the range where surface energy anisotropy dominates, {111}-oriented grains are often found after secondary recrystallization of a bare polysilicon surface because the energy of a free silicon surface with this orientation is minimum. On the other hand, the energy anisotropy of an Si/SiO_2 interface favors forming {100}-oriented grains [2.101].

2.11.4 Amorphous films

Although undoped films deposited in a polycrystalline form change little as they are annealed at moderate temperatures, initially amorphous films are not stable when heated even mildly [2.25]. They crystallize readily, and significant amounts of crystal texture form during annealing, with {311}- or {111}-oriented grains often dominating, as shown in Fig. 2.24 [2.25]. The diffuse rings in the transmission-electron-diffraction pattern of an initially amorphous film (Fig. 2.6a) change to the ring-dot pattern characteristic of a polycrystalline film (Fig. 2.6b).

Because larger grains improve the properties of polysilicon thin-film transistors, considerable effort has been devoted to maximizing the grain size by appropriately crystallizing films deposited in an amorphous form. The final grain size in these films can be much greater than that in films deposited in a polycrystalline form.

Crystallization is a two step process: (1) Nucleation of crystallites in the amorphous layer. (2) Crystallization (ordering) of adjacent amorphous material around these initial nuclei until all of the amorphous material is crystallized and the crystalline grains touch each other.

2.11. STRUCTURAL STABILITY

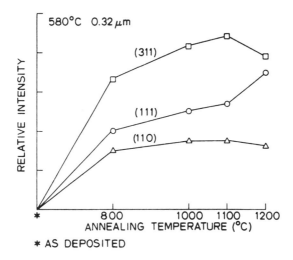

Figure 2.24: Development of the crystal structure on annealing an initially amorphous silicon film deposited in a hot-wall, low-pressure reactor at 580°C. No diffracted x-rays are observed from the initial film (denoted by *).

The detailed structure obtained after crystallizing an initially amorphous film depends on the deposition conditions, as well as the crystallization method. As we saw in Sec. 2.4 and Fig. 2.6a, a film deposited slightly below the critical temperature for polysilicon formation (~high 500°C temperature range) is basically amorphous but contains some crystallites occupying a few percent of the volume. These crystallites are generally located near the bottom of the silicon layer, and their size and number decrease as the deposition temperature decreases [1.31]. When the film is annealed, adjacent amorphous material orders around these initial crystallites, which then expand until all the amorphous material between nuclei is consumed and the growing grains impinge on adjacent growing grains. At moderate temperatures, the primary crystallization process then stops. To obtain the largest grains, the amorphous material should crystallize around a very few nuclei, allowing room for the laterally growing grains to expand before impinging on adjacent growing grains [2.102, 6.127]. The grain size L is then related to the initial density of nuclei N_n by the relation $L \approx 1/\sqrt{N_n}$.

Fewer nuclei are found in films deposited at lower temperatures. However, the deposition rate decreases rapidly as the deposition temperature decreases and becomes impractically low below the mid-500°C range for silane under typical conditions. Disilane can be used to increase the deposition rate in the low-500°C range. At even lower temperatures (~300°C), the deposition rate may be enhanced by using a plasma to decompose the silicon-containing gas.

Amorphous silicon layers deposited by thermal CVD at temperatures moderately below the amorphous/polycrystalline transition temperature contain only small amounts of hydrogen. At the low temperatures used for plasma-enhanced CVD, significant amounts of hydrogen from the silicon source gas and possibly added hydrogen diluent gas can be incorporated into the depositing silicon layer to produce *hydrogenated* amorphous silicon (a-Si:H) containing ~10-20% hydrogen. The lower temperature also produces a less compact structure than the amorphous silicon deposited at higher temperatures. Both the incorporated hydrogen and the less dense structure complicate the crystallization process for layers deposited by plasma-enhanced CVD.

The hydrogen can be incorporated with two hydrogen atoms attached to a silicon atom (SiH_2) or as the more strongly bound SiH, with one hydrogen attached to a silicon atom. The ratio of the two types of bonding depends on the hydrogen dilution during deposition and the other deposition parameters [2.103]. Hydrogen evolves readily from SiH_2 on heating to about 400°C while heating to about 600°C is required to remove the hydrogen bound as SiH. Much of the hydrogen evolves from the film at temperatures below the crystallization temperature, leaving a disordered structure with many broken bonds. At higher temperatures, the bonds rearrange, and the film becomes more compact. At high enough temperatures, the film crystallizes. When hydrogenated amorphous silicon is annealed to crystallize it, the substrate must be heated slowly to allow the hydrogen to escape from the film. Rapid heating allows the hydrogen to form bubbles, which can destroy the film as they escape.

When the density of initial nuclei formed during the deposition of the silicon layer is very low, nuclei forming during the crystallization process must also be considered. To obtain the maximum grain size, this *spontaneous nucleation* rate should be low so that few additional nuclei form during the crystallization process. Because of its high activation energy, above about 500°C the spontaneous nucleation rate increases rapidly with increasing annealing temperature [2.104, 1.31], resulting in a film with more numerous, but smaller, grains. The more rapid

2.11. STRUCTURAL STABILITY

increase in nucleation rate than grain-growth rate at higher temperatures suggests that a large-grain structure is best obtained by annealing at the lowest practical temperature. An ideal process would first form nuclei separated by a distance compatible with the time available for crystallization and then crystallize the remaining amorphous material around these nuclei. Crystallizing at a low temperature is also necessary when glass substrates are used. The inexpensive glass substrates needed for large-area applications have limited thermal stability, which often restricts the crystallization temperature to less than 600°C.

When the initially deposited layer contains widely separated nuclei, the time required for the crystallites to expand and consume all the amorphous material without excessive spontaneous nucleation and within the temperature constraints of a glass substrate may be impractically long (*eg*, ≥ 24 hours). To speed the conversion of the amorphous film to a polycrystalline layer at a low temperature, either extra nuclei can be introduced or the crystallization rate can be increased.

One method of inducing nucleation at lower temperatures is to use a bilayer structure with a thin layer of an amorphous silicon-germanium alloy above or below the amorphous silicon layer [2.105]. Because of the lower temperatures associated with crystallizing germanium, nucleation can occur in the amorphous Si-Ge layer at temperatures too low for significant nucleation in the amorphous silicon. The nuclei develop in the Si-Ge layer and the crystallization then propagates into the adjacent Si layer, thus decoupling the nucleation and grain-growth processes and providing additional flexibility. If the Si-Ge layer is above the Si layer, very few nuclei form; the temperature is too low for the usual spontaneous nucleation at the bottom of the amorphous Si layer. Large grains form, but at the expense of a long crystallization time because of the slow crystallization around the widely spaced nuclei.

In addition to controlling the nucleation, the rate of grain growth can be increased. A thin ($\sim$monolayer) metal layer, such as Cu [2.106, 2.107], Ni [2.108], Au [2.109], Al [2.110, 2.111], or Ag [2.112], formed on the amorphous silicon by evaporation or by adsorption from a solution can increase the crystallization rate at a given temperature or allow the crystallization to occur in a practical time at a lower temperature. However, this method of enhancing the grain-growth rate does not appear to increase the grain size. In addition, the effect of the metal impurities on the electrical properties of the layer is not well understood.

Surface preparation conditions that affect the condition of the underlying SiO_2 surface on which the amorphous silicon is deposited also influence the spontaneous nucleation rate, probably by varying the amount of OH groups on the surface [2.113]. Unintentionally introduced impurities in the deposited film can decrease the crystallization rate. Oxygen incorporated into the layer during deposition at concentrations above 10^{19} cm^{-3} decreases the rate of crystallization around nuclei and also degrades the electrical properties of the layer [2.114].

As the bonds rearrange during crystallization, the stress in the layer can change. Crystallization can increase the stress, but the stress decreases on further annealing [2.68]. Along with the bond rearrangement into a more ordered arrangement, the volume decreases as much as a few percent as the amorphous silicon crystallizes [2.115, 2.116].

To avoid the long time needed for solid-phase crystallization in a furnace, a pulsed or cw laser can be used as the heat source to achieve very rapid crystallization, often in the liquid phase [2.117–2.119]. The amorphous silicon film is heated to its melting point (which may be lower than the melting point of crystalline silicon) so rapidly that the underlying substrate (often a moderate-temperature glass) is not heated excessively. The low thermal conductivity of the thermal and impurity barrier layer between the amorphous silicon and the substrate allows the required large temperature difference. To obtain large grains, the entire thickness of the silicon layer must be melted. A region several millimeters across is crystallized in one laser pulse, and the laser is then moved to an adjacent area, with some (often considerable) overlap of the previously processed area [2.120]. This localized crystallization allows crystallizing selected portions of a large substrate so that the properties of the silicon layer can be optimized for a specific electrical function. [As we will see in Chapter 6, high-mobility polysilicon thin-film transistors are needed in the peripheral circuitry of active-matrix addressed liquid-crystal displays, while the switching transistor within each picture element (pixel) may be a low-leakage amorphous-silicon transistor.]

Amorphization by ion implantation

Rather than using low deposition temperatures to obtain an amorphous film with a low density of nuclei, ions (usually silicon) can be implanted into a polycrystalline silicon film to destroy the structure so that it

2.11. STRUCTURAL STABILITY

becomes amorphous. If all the crystallites are amorphized by ion implantation, spontaneous nucleation of new crystallites is needed before recrystallization can occur. This spontaneous nucleation is most likely to occur by heterogeneous nucleation at the upper and lower surfaces of the silicon film. Because of its high activation energy, spontaneous nucleation is slow at low temperatures. The few nuclei that form can grow to produce large grains before being stopped by impinging on adjacent grains. The implanted ions also produce stress, which can play a role in the recrystallization process, as we saw in Sec. 2.11.2. In addition to silicon, germanium can be implanted to amorphize a silicon film [2.121].

Completely amorphizing the polycrystalline material is critical to this technique, however. If small crystallites remain, the amorphized material can regrow on the these crystallites, rather than forming new nuclei, and small grains form again. Even though the film may appear amorphous by conventional transmission electron microscopy, small residual grains can still be present to act as nucleation centers, and higher doses than expected are needed to amorphize the silicon film completely [2.121].

As an extension of this recrystallization technique, ion channeling can be used to reduce the amount of damage created in grains with a selected orientation, preserving the structure in these grains while amorphizing grains with other orientations. The surviving grains can act as seeds for solid-phase epitaxial recrystallization, so that large grains are obtained by annealing [2.122, 2.123]. Annealing temperatures of about 500-550°C are often used because significant solid-phase epitaxial growth of existing crystallites can occur in this temperature range [2.124] with very little spontaneous nucleation [2.125–2.127].

As we saw in Sec. 2.5, under typical low-pressure deposition conditions, {110}-oriented grains dominate the crystal structure, and many are preserved by channeling along the open <110> direction when the amorphizing implant is perpendicular to the film surface. However, the orientation of these nominally {110}-oriented grains can vary by several degrees from the exact surface normal. Implanting normal to the film surface amorphizes not only the grains with other orientations, but also the nominally {110}-oriented grains misaligned by more than about 4° [2.128]. Thus, the alignment of the regrown crystals varies less than the alignment of the {110}-oriented grains in the initially deposited film.

For many electronic applications, {100}-oriented silicon is preferred because of the superior characteristics of the Si/SiO_2 interface on this orientation. As we saw in Fig. 2.12, deposition at a somewhat higher temperature than normally used can produce films with dominant {100} orientation. Starting with such {100}-oriented films obtained by depositing at 700°C, the majority of the grains in the film can be amorphized by ion implantation, retaining primarily the {100}-oriented grains [2.129]. Subsequent solid-phase regrowth at 550°C allows these grains to expand. However, because channeling is more difficult along the <100> direction and because large amounts of other orientations are present in the initial films, obtaining large-grains is more difficult for films with dominant {100}-texture than for predominantly {110}-oriented films.

To completely amorphize grains in an initially polycrystalline film requires a high dose of implanted ions and may be slow. A lower dose is adequate if the initial film is basically amorphous with a few nuclei. The implanted ions then need only amorphize the nuclei, which are often located near the bottom of the deposited film. If a non-silicon species is implanted, the implanted atoms may suppress the nucleation rate by decreasing the ability of the silicon atoms to form critical nuclei and may also decrease the diffusion rate of silicon atoms to incipient nuclei. Because the nuclei usually form first near the bottom of the layer, nucleation is suppressed most effectively when the implanted species stops near the underlying Si/insulator interface [2.130]. For example, spontaneous nucleation at the bottom of the silicon film can be suppressed by implanting fluorine there [2.130].

2.12 Hemispherical-grain (HSG) polysilicon

When polysilicon is used as the gate electrode of an MOS (metal-oxide-semiconductor) transistor, a smooth surface is generally desired to allow patterning of small device features. However, when it is used as an electrode of the capacitor of a dynamic random-access memory (DRAM), a rough surface may be useful to increase the surface area of the capacitor and, consequently, the amount of charge it can store at a given voltage.

A polysilicon structure with a very rough surface [the so-called *hemispherical-grain* (HSG) polysilicon] can be formed by depositing the polysilicon very near the transition temperature between deposition of amorphous silicon and polysilicon. The rough surface occurs only over a very narrow range of deposition conditions [2.131], requiring very good control over the temperature, pressure, and impurity content — all parameters that control the surface diffusion during deposition and immediately after deposition. The rough surface may be related to surface migration of the silicon atoms during deposition to form a rough {311}-oriented crystal texture in the polysilicon or may be caused by surface diffusion of the atoms after the deposition terminates, but before the surface is stabilized by exposure to air. Similar HSG structures are formed using either SiH_4 or Si_2H_6 [2.28].

HSG may also be formed from amorphous silicon after deposition by first depositing amorphous silicon, carefully removing the native oxide and then depositing silicon seeds on the amorphous surface. During a subsequent heat treatment in a high vacuum or other oxygen-free environment, the amorphous silicon crystallizes around the seeds. Because there is no native oxide on the surface to constrain the surface, silicon atoms migrate on the surface, creating the desired rough surface, which forms one plate of the capacitor. The rough surface of HSG polysilicon can approximately double the surface area and the capacitance [2.132].

2.13 Epitaxial realignment

Although polysilicon is usually deposited on amorphous surfaces, it is sometimes deposited directly on single-crystal silicon. The primary application for deposition on single-crystal silicon is the advanced bipolar integrated-circuit transistor to be discussed in Sec. 6.8, in which polysilicon serves as an electrical contact and often as a dopant source. When polysilicon is immediately adjacent to single-crystal silicon, epitaxial realignment of the polysilicon on the single-crystal silicon during heat treatment after deposition is possible, as shown in Fig. 2.25. As in the case of grain growth within a layer of polysilicon, high concentrations of dopant atoms enhance the epitaxial restructuring while residual oxygen at the interface retards it.

The epitaxial realignment process has an activation energy of 4.7-4.8 eV for undoped [2.133] or lightly doped [2.134] films, close to that for silicon self-diffusion. This similarity is explained by assuming that alignment involves diffusion of silicon atoms across the polycrystalline–single-crystal interface, analogous to the silicon diffusion across grain boundaries that plays a role in grain growth [2.133]. The driving force for both is the minimization of the interface or grain-boundary energy.

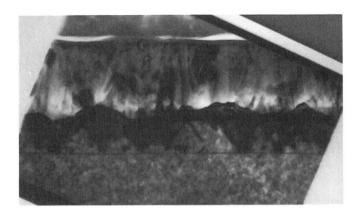

Figure 2.25: Cross-section transmission-electron micrograph showing that polysilicon deposited on a nominally bare silicon wafer can epitaxially realign on the underlying single-crystal structure during a subsequent high-temperature heat treatment. Note that the realigned region contains a significant number of defects.

The annealing ambient also affects the epitaxial realignment. At temperatures below 950°C, greater realignment occurs in an oxidizing ambient than in an inert ambient [2.135]. It is likely that a fraction of the excess silicon interstitials injected into the polysilicon during oxidation reach the polysilicon–single-crystal interface. They break some of the residual Si–O bonds there and enhance silicon self-diffusion, so that epitaxial realignment occurs more readily. (Other experiments have shown that injected interstitials can travel through a submicrometer-thick polysilicon layer to affect dopant diffusion in the underlying single-crystal silicon [3.50, 3.53].) Realignment also proceeds much more readily in a hydrogen ambient than in nitrogen [2.136]. The hydrogen is thought to diffuse rapidly through the polysilicon, reaching the interface where it can reduce the interfacial oxide and promote epitaxial realignment.

High concentrations of arsenic increase the alignment rate of polycrystalline silicon on single-crystal silicon [2.137–2.139], as well as the rate of grain growth in the polycrystalline film. Because the driving force for epitaxial realignment is related to the number of grain boundaries in the polycrystalline silicon, grain growth can make epitaxial realignment more difficult because it reduces the number of grain boundaries. The

2.13. EPITAXIAL REALIGNMENT

rate of realignment also increases when high concentrations of boron are present [2.134]. In this case, the activation energy appears to decrease to about 3.7 eV. The alignment rate is postulated to be related to silicon self-diffusion across the interface, which, in turn, depends on the number of doubly charged, positive point defects and, consequently, the doping. Following this reasoning for p-type material, observed values of the alignment rate R_a can be written in the form

$$R_a = \left[9.7 \times 10^{19} + 2.8 \times 10^{17} \left(\frac{p}{n_i}\right)^2\right] \exp\left(-\frac{4.82}{kT}\right) \text{ nm/min} \quad (2.12)$$

where p is the hole density and n_i is the intrinsic carrier density at the annealing temperature. From Eq. 2.12, we see that at high hole concentrations, the apparent activation energy of 3.7 eV arises from the difference between the 4.8 eV activation energy of silicon self-diffusion and the 1.1 eV activation energy of n_i^2.

Epitaxial realignment is also greatly affected by the surface treatment of the single-crystal silicon substrate before deposition of the polycrystalline silicon [2.140]. Wafers can be briefly etched in dilute HF before being inserted into a hot-wall, low-pressure polysilicon reactor to remove most of the native oxide, or they can be chemically oxidized in a solution such as $NH_4OH:H_2O_2:H_2O$ to form an oxide layer about 2 nm thick. Although the usual columnar grain structure is observed in both cases immediately after polysilicon deposition at 620°C, the behavior during subsequent heat treatment at 1000°C is different. During annealing, the chemically grown oxide layer becomes discontinuous over about 5% of the silicon surface area. The polysilicon in these regions makes direct contact with the single-crystal substrate and realigns epitaxially (Fig. 2.26a) [2.140]. A slightly thicker oxide should suppress epitaxial realignment completely. For samples etched in HF before polysilicon deposition, about 20-30% of the area realigns (Fig. 2.26b). The polycrystalline silicon does not realign over the entire area, probably because of the native oxide formed as the wafers are transported and loaded into the hot reactor. The interfacial oxide can be minimized by reducing the temperature of the reactor while the wafers are being inserted [2.141] or by enclosing the loading area of a vertical reactor in a nitrogen ambient [1.10].

In the realigned regions the original uniform native oxide at the interface agglomerates into small spherical oxide particles 2-3 nm in

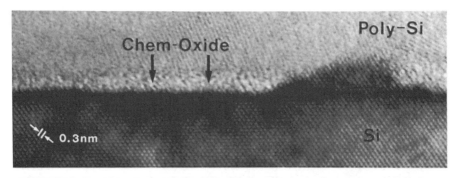

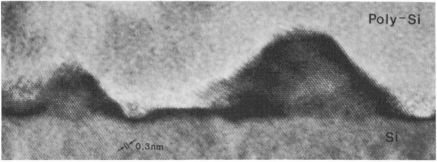

Figure 2.26: More epitaxial realignment occurs during annealing at 1000°C when the residual oxide layer at the Si/SiO_2 interface is thinner. (a) Chemically formed oxide, and (b) HF-etched surface. From [2.140]. Reprinted with permission.

diameter. As the realignment process proceeds, aligned material grows vertically above regions free of interfacial oxide, and then expands laterally into the polycrystalline regions over residual oxide. A model of the microscopic breakup of the native oxide suggests that oxide first breaks up near grain boundaries [2.142, 3.16]. Oxygen diffuses away from the native oxide along grain boundaries, and the nearby native oxide becomes thinner. A slit develops in the oxide near the grain boundary, and the surface tension of the oxide pulls it away from the slit so that the slit widens. Epitaxial realignment of the polysilicon can then occur on the exposed single-crystal silicon. As the process proceeds, the native oxide compacts into cylinders and then spheres to reduce the Si/SiO_2 interfacial area. Dopant diffusion to the interface is postulated to speed the process, but the oxide breakup occurs for undoped polysilicon, as well as for heavily doped layers. Fluorine in the polysilicon layer (typically from a BF$_2$ implant) migrates to the polysilicon–single-crystal silicon interface and aids breakup of the native oxide and, consequently, epitaxial realignment of the polysilicon [2.143, 2.144]. Fluorine may also passivate interface states [2.145, 2.146]. Using ion implantation to totally amorphize a polycrystalline-silicon film deposited on single-crystal silicon can promote solid-phase epitaxial growth of the silicon layer at temperatures in the 500°C range if the interface is adequately clean [2.147]. The ion implantation can also help disrupt any interfacial layer at the substrate-film interface, allowing the crystallization to proceed through this interface.

When polysilicon is used over the emitter of a bipolar transistor, excessive diffusion from the polysilicon into the underlying single-crystal must be reduced to avoid degrading the transistor performance (Sec. 6.8). If epitaxial realignment of the polycrystalline silicon is desired, it must occur more rapidly than does diffusion. The different activation energies of alignment and diffusion can aid this process. The apparent activation energy of realignment is greater than that of the diffusion length $L = 2\sqrt{Dt}$ in single-crystal silicon. Therefore, by using high temperatures and restricting the time of the anneal, the ratio of alignment to diffusion can be increased. The higher temperature may also reduce the effect of residual interfacial oxide on the alignment. Although the times required for these high-temperature anneals are much shorter than can be reproducibly obtained in a conventional diffusion furnace, low-mass, rapid-thermal-annealing systems allow the use of high temperatures for times of the order of tens of seconds [2.137, 2.139].

2.14 Summary

In this chapter we considered the structure formed during deposition of polysilicon by chemical vapor deposition and changes in the structure as the films are annealed. We saw that polycrystalline films form when small clusters of adsorbed atoms diffusing on the surface join into isolated nuclei, which then grow to form the grains of the polycrystalline film. Even after the surface is covered with a continuous film, surface diffusion of arriving atoms influences the structure; the amount of surface diffusion possible depends on deposition variables such as the deposition temperature, rate, and pressure and impurities in the deposition chamber. Amorphous films lacking long-range order form if appreciable surface diffusion cannot occur during deposition. A polycrystalline structure with columnar grains is often formed under typical deposition conditions. The grain structure can best be seen by transmission electron microscopy; x-ray diffraction allows quantitative comparison of different films. Optical properties can be used to obtain a rapid indication of the film structure once they have been correlated with the structure by more detailed analysis. Undoped polycrystalline films are stable when annealed at moderate temperatures while amorphous films crystallize readily. Adding phosphorus to polysilicon films allows rapid grain growth as the films are annealed.

Chapter 3

Dopant Diffusion and Segregation

3.1 Introduction

In a defect-free, single-crystal of silicon, dopant atoms diffuse into the perfect lattice by interacting with point defects, such as silicon vacancies and interstitials. Some point defects exist in equilibrium with the crystal lattice; others are introduced externally (*eg*, by injecting interstitials into the silicon lattice as the surface of the crystal is oxidized). These additional interstitials can significantly enhance the dopant diffusion rate in perfect crystals of silicon.

When extended structural defects are present, the associated weak or broken bonds also enhance the dopant diffusion rate by reducing the energy needed for a dopant atom to move from one allowed site to another. The disorder at grain boundaries and other defects in polycrystalline silicon provides high-diffusivity paths along which dopant atoms can easily move. Even though the grain boundaries occupy only a small fraction of the sample volume, dopant migration along these paths greatly increases the overall dopant diffusion in polysilicon; consequently, dopant diffusion depends strongly on size and shape of the grains in the polysilicon and on the detailed structure of the grain boundaries. The dopant atoms are probably not electrically active while they are in a grain boundary; however, after diffusing along the grain boundary, they can move back into the grain, where they influence the electrical conduction.

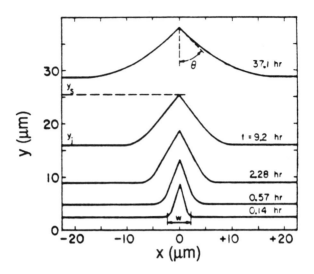

Figure 3.1: A spike-shaped diffusion front surrounds the grain boundary (located at $x = 0$), along which the dopant diffuses rapidly. The diffusion front is sharper for shorter diffusion times. From [3.3]. Reprinted with permission.

In addition to providing paths along which dopant atoms can readily diffuse, grain boundaries can also be favorable, low-energy sites for dopant atoms. This *dopant segregation* reduces the number of dopant atoms in the grains, especially for the n-type dopants, and decreases the number of active charge carriers. In this chapter we will discuss dopant diffusion both *in* and *from* polysilicon and also grain-boundary segregation. Understanding these processes is important for controllable fabrication of devices containing polysilicon layers.

3.2 Diffusion mechanism

3.2.1 Diffusion along a grain boundary

Enhanced diffusion of dopant atoms along grain boundaries in polysilicon can be understood by first considering an isolated grain boundary in an otherwise defect-free, single crystal of silicon [3.1–3.3]. Assume the grain boundary to be a thin region with high diffusivity normal to the sample surface between two grains with low diffusivity. Enhanced diffusion associated with the grain boundary can be described as a two-

3.2. DIFFUSION MECHANISM

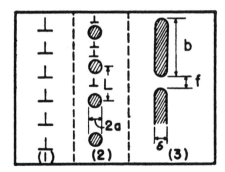

Figure 3.2: The isolated dislocations characteristic of a small-angle grain boundary (1) merge into rods (2), and then slabs (3), as the grain-boundary angle increases. From [3.1]. Reprinted with permission.

step process: The dopant atom first diffuses rapidly along the disordered structure of the grain boundary and then moves laterally into the grain, where it diffuses less rapidly. The mathematical analysis of grain boundary diffusion [3.2, 3.4] shows that the diffusion front is spike-shaped, with its deepest penetration along the grain boundary, as shown in Fig 3.1 [3.3]. The shape of the diffusion front is determined by the relative ease of diffusion along the grain boundary and in the grain; the spike is sharper as the ratio increases. The diffusion front is also more pointed for shorter diffusion times, as shown in Fig. 3.1.

The nature of the grain boundary significantly influences the amount of enhanced diffusion. When the grain orientations on the two sides of the boundary only differ slightly, the grain boundary can be described as an array of isolated dislocations, as shown in Fig. 3.2(1) [3.1]. As the misorientation increases, the dislocation density increases rapidly, and the dislocations collect into rod-like bunches. With a further increase in misorientation, the dislocation density increases, and finally the dislocations coalesce into a slab-like, high-angle grain boundary.

At small misorientation angles the average number of neighbors and the space available for the diffusing atoms in the grain boundary are not markedly different from those inside the grain, and diffusion is not greatly enhanced. As the angle increases, grain-boundary diffusion increases markedly as the volume of disordered crystal increases.

A high-angle grain boundary can be considered to be a uniform slab of high diffusivity because the distance between the dislocations comprising the grain boundary is small. With this and some further assumptions [3.1, 3.5–3.7], the dopant concentration C can be expressed as [3.2]

$$C(x,y,t) = C_0 \exp\left(-\frac{2^{1/2} D_G^{1/4} y}{\pi^{1/4} \delta^{1/2} D_{GB}^{1/2} t^{1/4}}\right) \mathrm{erfc}\left(\frac{x}{2\sqrt{D_G t}}\right) \quad (3.1)$$

where y is the distance from the surface along the grain boundary, x is the distance perpendicular to the grain boundary into the surrounding single-crystal regions, t is the diffusion time, C_0 is the surface concentration, D_{GB} and D_G are the dopant diffusion coefficients or *diffusivities* in the grain boundary and in the lattice of the surrounding grains, respectively, and δ is the width of the grain boundary.

For grain-boundary diffusion to be significant, the flux of dopant diffusing along the grain boundary must be comparable to or greater than that diffusing in the nearby lattice. This requirement can be quantitatively expressed by comparing the grain-boundary diffusivity D_{GB} times its width δ to the lattice diffusivity D_G times the diffusion length $2\sqrt{D_G t}$ in the lattice; for significant grain-boundary diffusion to occur

$$D_{GB}\delta \geq 2D_G \sqrt{D_G t} \quad (3.2)$$

Although the grain boundary is very narrow, the associated diffusivity can be high enough so that this inequality is satisfied. For example, the ratio of the diffusivity in the grain boundary to that in the grain is about 8×10^5 in cast polysilicon [3.8], while the value is about 3×10^5 in a silicon bicrystal containing a grain boundary with 7.5° misorientation [3.3]. The energy needed for a dopant atom to diffuse along the relatively open paths in a grain boundary can be much less than that required for diffusion within the dense lattice of the grain, and the observed activation energy can be much smaller for grain-boundary diffusion than for lattice diffusion. Because of their different activation energies, the difference between lattice diffusion and grain-boundary diffusion increases as the temperature decreases, leading to more irregular diffusion fronts at lower temperatures. The difference also allows additional control of a diffusion process by enhancing grain-boundary diffusion with lower temperature processing or de-emphasizing it with higher temperature processing.

More elaborate theories of grain-boundary diffusion have been developed, but the differing structures of nearby grain boundaries and

3.2. DIFFUSION MECHANISM

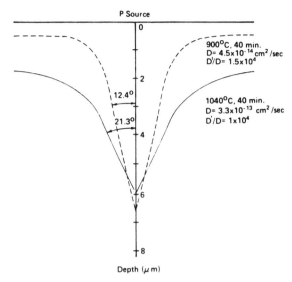

Figure 3.3: Far from the grain boundary the junction is deeper after the *higher*-temperature diffusion. Near the grain boundary the junction is deeper after the *lower*-temperature diffusion because the dopant cannot diffuse out of the grain boundary as readily at the lower temperature. From [3.8]. Reprinted with permission.

interactions between closely spaced grain boundaries limit their practical use in studying grain-boundary diffusion in polycrystalline silicon.

Fig. 3.3 shows the diffusion fronts calculated for two different heat treatments [3.8]. Far from the grain boundary, the diffusion front is controlled by vertical diffusion within the lattice, and the diffusion is deeper for the higher temperature heat treatment. However, the diffusion depth along the grain boundary can be deeper at the *lower* temperature. Less dopant is lost from the grain boundary to the surrounding grains, allowing the dopant within the grain boundary to penetrate deeper. The spike is also sharper at the lower temperature. If the amount of dopant on the surface near the grain boundary is limited, the rapid diffusion of dopant along the grain boundary can deplete the dopant on the nearby surface so that surface diffusion of the dopant to the grain boundary limits the overall diffusion process [3.9].

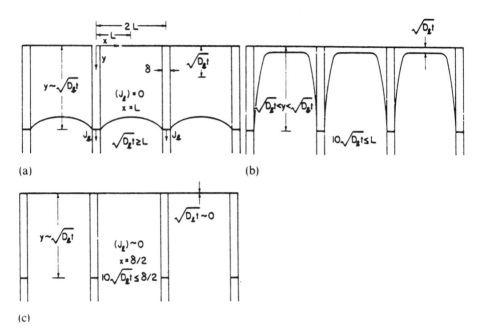

Figure 3.4: The importance of diffusion along grain boundaries (narrow vertical spaces) depends on the grain size, as well as the diffusivity within each region. The curved lines indicate the diffusion fronts, and the diffusion source is located at the top. From [3.10]. Reprinted with permission.

3.2.2 Diffusion in polycrystalline material

In polycrystalline material, many grain boundaries exist, and the spike-shaped diffusion fronts near adjacent grain boundaries can interact. Three different cases are shown in Fig. 3.4 [3.4, 3.10]. If the lateral diffusion length in the grains is comparable to or larger than the grain size, a significant amount of dopant diffuses from the grain boundaries across the grains, and the diffusion front is reasonably uniform, as shown in Fig. 3.4a (A-type kinetics). If diffusion is much more rapid along the grain boundaries than in the grains, little dopant enters the grains (Fig. 3.4c) (C-type kinetics). The more general case for polycrystalline material lies between these two extremes; grain-boundary diffusion is much greater than diffusion in the grains, but a significant amount of dopant diffuses out of the grain boundary into the surrounding grains (Fig. 3.4b) (B-type kinetics). In this case, the diffusion front is irregular, with sharp spikes near the grain boundaries. The irregularity of the diffusion front depends on the relative ease of diffusion along the grain boundary and in the grain and also on the grain size (*ie*, the distance that the dopant must diffuse laterally within the grain to dope its entire width). Because of the higher activation energy for diffusion in the grains than in the grain boundaries, movement of the diffusing dopant from a grain boundary into the surrounding grains increases with increasing temperature; the grains become more completely doped; and the diffusion front becomes less irregular.

The enhanced penetration of dopant atoms along grain boundaries is most readily observed in large-grain polysilicon. Voltage-contrast scanning electron microscopy and electron-beam-induced-current measurements can reveal the regions with electrically active dopant, indicating the spike-shaped diffusion fronts surrounding the grain boundaries [3.11]. As expected, the dopant diffuses most rapidly along the grain boundaries and then spreads into the grains, where it is electrically active and sensed by these electrical techniques.

3.3 Diffusion *in* polysilicon

In our discussion of dopant diffusion associated with polysilicon, we need to consider two different cases. In both cases, the dopant is introduced into the polysilicon by diffusion from an external gas-phase or solid source or by ion implantation. In the first case, we are concerned with the movement of the dopant *within* the polysilicon during an annealing or oxidation cycle. The dopant may diffuse vertically into the film [*eg*, doping the polysilicon gate electrode of an MOS transistor (Sec. 6.2)], or it may diffuse laterally [*eg*, the undesired lateral diffusion of dopant into the lightly doped region of a high-value polysilicon resistor from the heavily doped electrodes(Sec. 6.4)].

The second case of interest involves diffusion of dopant *from* the polysilicon into an adjacent region of single-crystal silicon. For example, arsenic can be diffused from polysilicon into an underlying single-crystal silicon region to form the emitter of a bipolar transistor. Diffusion from polysilicon can be more complex than diffusion within polysilicon. In addition to the migration of dopant within the polysilicon structure and the interaction with grain boundaries, dopant accumulation at the polycrystalline/single-crystal interface and epitaxial realignment of the

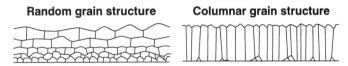

Figure 3.5: (a) The dopant atoms must diffuse both vertically and laterally along grain boundaries to move through a random polycrystalline structure. (b) Grain-boundary diffusion is primarily vertical when the structure is columnar.

polysilicon must be considered. In this section, diffusion *within* polycrystalline silicon will be discussed. Diffusion *from* polysilicon will be treated in the following section.

In the fine-grain polysilicon normally used in integrated circuits, the spacing and nature of the grain boundaries depends strongly on the deposition conditions, complicating quantitative analysis of the diffusion. In the random polycrystalline structure indicated schematically in Fig. 3.5a, the grain boundaries are not aligned in the vertical direction. The dopant atoms can diffuse rapidly down one grain boundary until they encounter an underlying grain; they are then likely to diffuse laterally along a horizontal grain boundary to find another vertical grain boundary, which can provide an easy path for vertical diffusion. Because of this combination of vertical and lateral diffusion, the total path length to the bottom of the polycrystalline film can be greater than the film thickness. However, even with the longer path length, the effective diffusivity is greater in the polycrystalline structure than in single-crystal material because of the rapid dopant diffusion along the grain boundaries. If the structure is columnar (Fig. 3.5b), continuous vertical grain boundaries extend through much of the sample thickness, and the effective diffusivity is enhanced even more than in a random polycrystalline structure. When the structure of polysilicon is anisotropic, especially with columnar grains perpendicular to the film surface, the diffusion is also anisotropic, with higher values of effective diffusivity perpendicular to the film plane.

The importance of the detailed crystal structure is seen in Fig. 3.6, which shows the vertical diffusion depth of boron and phosphorus in thick polysilicon films deposited over the temperature range 830-1170°C in a cold-wall reactor [3.12]. (Because diffused regions are more readily visible in thicker films with their larger grains, thicker films are used

DIFFUSION IN POLYSILICON

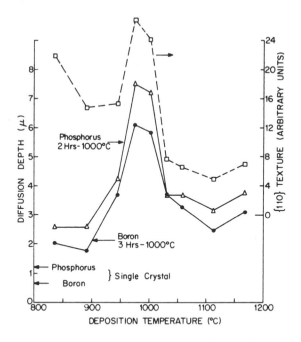

Figure 3.6: The diffusion depth in polycrystalline-silicon films depends on their structure and, therefore, the deposition conditions. These 15 μm thick films were deposited onto SiO_2 in a cold-wall, atmospheric-pressure reactor over the temperature range from 830 to 1170°C.

to illustrate the concepts being described. These concepts also apply to thinner layers with their smaller grains.) While the diffusion depth is greater in all the polycrystalline samples than in single-crystal silicon (arrows on the left axis), it depends critically on the deposition temperature. As we saw in Sec. 2.5 and Fig. 2.10, a columnar structure is related to the {110} texture. The dashed line in Fig. 3.6 shows that this structure is favored near the deposition temperature where the diffusion depth is greatest. A similar columnar structure is frequently found in thin films deposited under typical low-pressure CVD conditions (Fig. 2.7a) [2.15].

The effective diffusivity also varies with the distance from the bottom of the film because the columnar structure can become better defined toward the top of the film. The effective diffusivity is then greater near the top than near the bottom of the same film [3.12]. On the other hand, if random, equi-axed grains grow larger near the top of the film, the effective diffusivity can *decrease* because fewer grain boundaries are available for rapid diffusion. Because of the strong dependence of diffusion on the structure of the polysilicon, reported experimental data varies widely. Even within a single sample, the diffusion enhancement can differ greatly from one grain boundary or portion of a grain boundary to another. Segregation of impurities, such as oxygen, at the grain boundaries [3.13] is also likely to affect the grain-boundary diffusivity.

In many cases the diffusion depth increases with the square root of the diffusion time, allowing an *effective diffusivity* to be calculated by applying conventional diffusion analysis to the average penetration depth as if the polycrystalline material were homogeneous (Fig. 3.7) [3.12]. For example, for samples deposited at 1040°C, the effective boron diffusivity can be written as

$$D_{\text{eff}} = 7.7 \times 10^{-7} \exp\left(-\frac{1.39 \text{ eV}}{kT}\right) \text{ cm}^2/\text{s} \quad (3.3)$$

As indicated in Fig. 3.7, the activation energy depends strongly on the structure of the polycrystalline silicon. It is lowest in material with well-defined, columnar grain boundaries.

Because the diffusion occurs in two distinct regions (grains and grain boundaries), the effective diffusivity does not uniquely describe the ease of dopant diffusion in the grain or in the grain boundary and relates only to the particular polycrystalline structure being studied. It depends on the size of the grains, as well as on the ease of diffusion along the grain boundaries. More accurate analysis can determine the grain-boundary diffusivity from the effective diffusivity. From the data of Fig. 3.7, Eq. 3.3, and the grain size, an expression for the grain-boundary diffusivity can be written [3.14]:

$$D_{GB} = 6.6 \times 10^{-3} \exp\left(-\frac{1.86 \text{ eV}}{kT}\right) \text{ cm}^2/\text{s} \quad (3.4)$$

To derive the grain-boundary diffusivity, the width of the grain boundary is needed; a value of 0.5 nm is often used, as it was in the calculations leading to Eq. 3.4. Conceptually, the grain-boundary diffusivity should be independent of grain size so that it can be used, along with the grain size and the diffusivity in the grains, to provide a value of the effective diffusivity for a particular crystal structure. In practice, however, the grain-boundary diffusivity varies with the deposition conditions and the

DIFFUSION IN POLYSILICON

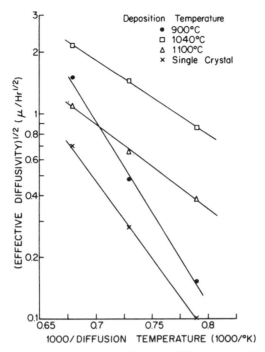

Figure 3.7: Square root of the *effective diffusivity* of boron in 15 μm thick polysilicon films deposited in an atmospheric-pressure reactor as a function of reciprocal diffusion temperature, showing the lower activation energy for diffusion in polysilicon than for diffusion in single-crystal silicon.

prior thermal history of the sample, so that a unique value cannot be used to characterize all grain boundaries in polycrystalline silicon.

In the submicrometer films conventionally used in integrated-circuit fabrication, the grain size is small and determining the actual diffusivity in the grain boundaries is difficult. Analysis is further complicated by segregation of dopant at grain boundaries (to be discussed in Sec. 3.6) and by the dependence of the diffusivity on the dopant concentration in the grains and possibly in the grain boundaries. Grain growth during annealing must also be considered; in particular, the extensive dopant enhancement of grain growth discussed in Sec. 2.11.3 makes stabilizing the grain structure prior to dopant addition difficult. When diffusion is limited by diffusion within the grains (*eg*, $2\sqrt{D_G t} \ll L_G$, where D_G is the dopant diffusivity in the grain and L_G is the grain size), diffusion is normally slow. However, if a grain boundary moves through the film during the diffusion process, dopant can be transferred from the grain boundary to the recrystallized grain its motion leaves behind, increasing the net dopant in the grain [3.15, 3.16]. Efforts to model some of these effects to allow numerical prediction of the dopant movement are discussed in Sec. 3.7.

Different methods of dopant introduction also complicate comparison between different studies. Deposition of the dopant from a gas-phase source generally produces a surface concentration in the grains equal to the solid solubility of the dopant in silicon at the doping temperature. However, additional dopant can be incorporated into the disordered structure of the grain boundaries, possibly in the form of precipitates. When the dopant is introduced by ion implantation, the quantity of dopant added is accurately known, but crystal damage from the implantation and annealing of the damage during diffusion can influence the diffusion process. In addition, the dopant atoms implanted near the edges of a grain can easily migrate to the nearby grain boundaries, where they rapidly diffuse. The dopant atoms implanted near the center of a grain must diffuse a significant distance within the grain before reaching a grain boundary, and diffusion within the crystalline grains, rather than grain-boundary diffusion, may dominate the overall diffusion process. Diffusion of the dopant atoms from a deposited oxide or from an additional polysilicon layer allows introducing a controlled amount of dopant without crystal damage, but is less frequently used in modern integrated-circuit fabrication processes.

In spite of these uncertainties, numerically predicting the concentration and location of dopant atoms in polysilicon and their movement between polysilicon and nearby single-crystal silicon is increasingly important in finer-geometry integrated circuits. Numerical simulators have been developed to predict dopant diffusion in polysilicon, as discussed in Sec. 3.7. However, experimental values of the effective diffusivity are usually reported, although the terms *effective diffusivity* and *grain-boundary diffusivity* are often used interchangeably. Because of the strong dependence of diffusion on the experimental conditions, widely varying values of effective diffusivity are reported.

3.3.1 Arsenic diffusion

Vertical diffusion of arsenic in polysilicon layers has been studied in most detail [3.17–3.20], both in polysilicon deposited on oxide [3.19] and in

DIFFUSION IN POLYSILICON

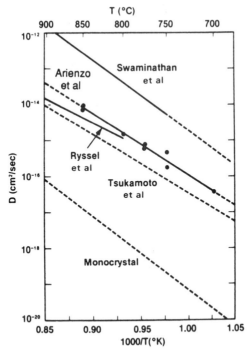

Figure 3.8: Effective diffusivity of arsenic in thin films of polysilicon found in different studies. From [3.20]. Reprinted with permission.

that deposited on single-crystal silicon [3.17, 3.20]. Observing diffusion in films deposited on either a clean, etched surface or on a native oxide in the same experiment allows direct comparison of diffusivity in the structures formed on different underlying materials [3.18]. Lateral diffusion has also been investigated [3.21, 3.22]. The widely varying experimental results are summarized in Fig. 3.8 [3.20] and Table 3.1.

Unlike diffusion of arsenic in single-crystal silicon, arsenic diffusion in polysilicon does not appear to depend on the dopant concentration. When the diffusion is dominated by the grain boundaries, concentration enhancement of the diffusivity is less important. However, segregation of dopant to the grain boundaries [3.23, 3.24] (Sec. 3.6) may also lower the dopant concentration in the grains below the value at which concentration-dependent diffusion is expected. The data also suggest that a structure with more high-diffusivity grain boundaries is formed on oxide than on bare silicon surfaces [3.18].

Table 3.1: Effective diffusivity of arsenic in polysilicon

Reference	Substrate	D (900°C) (cm^2/sec)	E_A (eV)	D_0 (cm^2/sec)
3.17	Si	9x10^{-15}	3.2	0.5
3.20	Si	3.7x10^{-14}	3.36	10
3.18	Si	1.5x10^{-14}	2.74	8.5x10^{-3}
3.18	Oxide	2.5x10^{-14}	3.22	1.66
3.19	Oxide	1x10^{-12}	3.9	8.6x10^4
3.22	Oxide	2x10^{-13}	3.1	4.1
3.21	Oxide	1.3x10^{-12}		
Single-crystal (intrinsic)		7x10^{-17}	4.1	28.5
Single-crystal (high-conc.)		8x10^{-15}	at $N_D = 2 \times 10^{20}$ cm^{-3}	

Several factors can lead to the different values of diffusivity obtained in different studies. First, the structure of the polysilicon deposited probably varied considerably because of the different film deposition temperatures and pressures. In one study [3.19] the films were annealed at a high temperature before introducing the arsenic in an attempt to stabilize the grain structure. The method of dopant introduction and the dopant concentration also varied [3.22, 3.25]. In many of the studies grain growth, especially during the initial stages of diffusion, may have occurred along with diffusion, possibly affecting the effective diffusivity [3.21]. Finally, the different methods of measurement and data analysis can also cause apparent discrepancies in the data.

Lateral dopant diffusion is also of great concern. As polysilicon resistors are scaled to smaller geometries, lateral diffusion of dopant atoms from the heavily doped polysilicon electrodes into the lightly doped resistor can reduce the effective length of the resistor and possibly cause a short circuit. In addition to its practical importance, study of lateral diffusion allows greater diffusion lengths to be observed, increasing the accuracy of experimental data. At 1000°C the reported values of lateral diffusivity vary from the low 10^{-12} to the low 10^{-11} cm^2/s range [3.21, 3.26], a narrower range than reported for vertical diffusivity [3.17–3.19]. Because columnar grains separated by well-defined vertical grain boundaries allow rapid vertical dopant diffusion, the vertical diffusivity should depend very sensitively on the structure. In the lateral direction,

DIFFUSION IN POLYSILICON   137

however, the diffusion path is not expected to vary as significantly as the structure changes, consistent with the smaller variations of lateral diffusivity reported. Anisotropic diffusivity has been observed in thick films with a well-defined columnar structure [3.12]; however, in thin polysilicon films the diffusivity is often less anisotropic [3.22].

Because dopant diffusion depends sensitively on the grain structure, it also depends on the deposition conditions. If the polysilicon is deposited in a single-wafer reactor, the high deposition rate required for adequate throughput creates a very different grain structure than in films deposited slowly in the hot-wall, low-pressure batch reactor. Using films with different grain structure requires considerable modification of the processes used to dope the polysilicon. Alternatively, other deposition parameters, in addition to the deposition rate, can be changed to achieve a similar grain structure in the single-wafer reactor as in the low-pressure, hot-wall batch reactor.

Initial studies of dopant diffusion in polycrystalline silicon employed conventional diffusion furnaces. However, rapid thermal processing, in which the samples are heated to a high temperature for a few seconds with an incoherent light source, is now widely used to provide the process control needed for fabrication of advanced devices. Diffusion of dopants in polycrystalline silicon during rapid thermal annealing has been investigated [3.27–3.29]. Even in short times of about 10 s, significant diffusion occurs in polysilicon, and appreciable dopant can also be lost to the ambient if the surface is unprotected. Arsenic is lost more readily than the other dopants during rapid thermal annealing, and its loss from polysilicon is much greater than from similarly treated single-crystal silicon. A thin SiO_2 cap on the polysilicon surface can substantially eliminate the loss of arsenic. Annealing in an oxygen ambient also decreases the arsenic loss. For short times at a peak wafer temperature of 1050°C, little grain growth occurs, and a diffusion coefficient of 4×10^{-12} cm^2/s is obtained for arsenic diffusion [3.27]. At higher temperatures, significant grain growth is seen.

3.3.2 Phosphorus diffusion

Diffusion of phosphorus is also significantly faster in polysilicon than in single-crystal silicon. This enhancement is especially marked in thicker films with a columnar grain structure, as shown in Fig. 3.6 [3.12], but the diffusivity is also greater in thin polysilicon films than in single-crystal silicon. At 900°C the diffusivity in thin films is about 1×10^{-12} cm^2/s for phosphorus and is approximately the same in the lateral and vertical directions [3.22], possibly reflecting phosphorus-enhanced grain growth. The 3.4 eV activation energy for the diffusion of phosphorus in these fine-grain polysilicon films is only slightly less than that for diffusion in single-crystal silicon [3.30]. Diffusion of phosphorus in thin, initially amorphous silicon films deposited at 580°C in a low-pressure CVD reactor has also been investigated [3.31]. A lower diffusivity was reported in this material than in films which were deposited in a polycrystalline form; however, the phosphorus was implanted before the films were crystallized so that the crystallization and diffusion occurred during the same heat cycle. In addition, as we saw in Sec. 2.11.4, the grain size in initially amorphous silicon after it is crystallized can be larger than that in material deposited in polycrystalline form. For thick, large-grain polysilicon films deposited at high temperatures, a single, well-defined activation energy is not seen [3.12]. The slope of an Arrhenius plot appears to decrease with increasing diffusion temperature. Although the diffusivity of arsenic in polysilicon was reported to be independent of arsenic concentration, the diffusivity of phosphorus in polysilicon appears to depend somewhat on the phosphorus concentration [3.31].

3.3.3 Antimony diffusion

Antimony also diffuses more rapidly in polysilicon than in single-crystal silicon. At 900°C, the diffusivity of antimony in low-pressure CVD silicon films is approximately 10^{-14} cm^2/s [3.32], about two orders of magnitude higher than the diffusivity in single-crystal silicon at the same temperature.

In large-grain polysilicon formed by CVD, the activation energy for diffusion is 2.9 eV [3.33], compared to 3.9 eV in single-crystal silicon. The actual value of the grain-boundary diffusivity depends on assumptions about the width of the grain boundary, and extrapolated values at 900°C vary from about 1×10^{-12} cm^2/s [3.33] to 1×10^{-10} cm^2/s [3.9] in large-grain material.

3.3.4 Boron diffusion

Both vertical and lateral diffusion of boron have been studied in polycrystalline silicon [3.34, 3.35], with more consistent values reported than for arsenic. For diffusion at 1000°C, reported values of the diffusivity include 3×10^{-13} cm^2/s in submicrometer-thick films [3.34], 7×10^{-13} cm^2/s in somewhat thicker films deposited at higher temperatures [3.35], and 2×10^{-12} cm^2/s in thick, columnar films deposited at high temperatures [3.12]. The diffusivity is at least an order of magnitude greater than the diffusivity in single-crystal silicon in all cases. For both thin and thick films, the diffusivity varies with the deposition temperature of the film, again showing the dependence of the diffusivity on the film structure. In thinner films, the differences between lateral and vertical diffusion are smaller.

The reported activation energies for boron diffusion vary appreciably. For thin polysilicon films deposited at lower temperatures, the activation energy is about 3.3 eV [3.34], only slightly lower than the 3.4 eV observed in single-crystal silicon. For intermediate-thickness films deposited at a higher temperature, the activation energy for boron diffusion is about 2.4–2.5 eV [3.35], markedly lower than the value for single-crystal silicon, but greater than the 1.4 eV found in the thick, large-grained films deposited at high temperatures [3.12]. Both higher deposition temperatures and thicker films favor the development of the well-defined grain boundaries apparently needed for low activation energies.

3.3.5 Limits of applicability

In this section we have seen that the *effective diffusivity* of dopant atoms in polysilicon depends very sensitively on the film structure, the method of dopant introduction, and other experimental details. Impurities, such as oxygen, at the grain boundaries may have a large effect on the movement of dopant along the grain boundaries. As the purity of the deposition ambient has improved over time, the structure of the grain boundaries that allows rapid dopant diffusion has probably changed also. No unique expression can be given to express diffusivity in polysilicon because of the great structural differences between different types of polysilicon. The data presented above can only serve as a guide, and more exact values must be determined for the particular material and experimental conditions being used.

3.3.6 Heavy doping

Heavy doping of polysilicon is especially critical for the electrodes of dynamic random-access memories. The non-planar structure of such electrodes makes the doping even harder. We saw in Sec. 1.9 that conformal deposition of *in situ* doped films is difficult. The active silicon species needed to compete for adsorption sites on the surface of the growing film increases the sticking coefficient of arriving molecules and degrades the conformality. However, heavily doped polysilicon can be formed on non-planar structures by depositing alternating layers of undoped and doped polysilicon [1.53] or by adding dopant from the gas phase after a portion of the polysilicon thickness has been deposited and then completing the deposition [1.54]. In either case, moderately annealing the polysilicon distributes the dopant uniformly through its thickness.

3.3.7 Nitrogen

Nitrogen is sometimes introduced into polysilicon during deposition or by implantation after deposition to retard boron diffusion through polysilicon and the underlying gate insulator of an MOS transistor [3.36] (Sec. 6.2). Because of the low solid solubility of nitrogen in silicon, much of the nitrogen is likely to form precipitates, and it is not free to diffuse [3.37]. Only the low concentration below solid solubility is mobile. As the mobile nitrogen diffuses away from the high-concentration regions, some of the immobile nitrogen becomes mobile and is able to diffuse. Much of the diffusing nitrogen segregates at the underlying polysilicon/SiO_2 interface. Although the nitrogen retards boron diffusion, concentrations greater than $\sim 1\%$ also reduce the electrical activity of boron. The resistance of the polysilicon then increases [3.36], decreasing the coupling between the gate electrode and the transistor channel and degrading the transistor behavior.

3.3.8 Implant channeling

Penetration of implanted dopant atoms through the large grains that form in heavily doped films can seriously degrade the electrical properties of MOS transistors. In the silicon-gate MOS transistor to be discussed in Sec. 6.2, polysilicon serves as an implant mask to prevent the source and drain dopant atoms from entering the channel region of the transistor. Fine-grain polysilicon appears to mask the implant effectively because the ions must penetrate through several differently oriented grains before reaching the gate dielectric and the underlying single-crystal silicon [3.38]. However, as we saw in Sec. 2.11.3 and Fig. 2.7, large grains can form in heavily doped polysilicon so that a single-grain extends through the entire thickness of the film. Because the orientation of the grains is broadly distributed, some are likely to be aligned with an open *channeling* direction along the incident ion beam. Channeling through the polysilicon can lead to unexpected and undesired doping in the underlying single-crystal silicon, with resulting changes in transistor threshold voltage. Channeling is found for phosphorus [3.39], arsenic [3.40], and boron [3.41]. Forming an amorphous layer over the polysilicon reduces channeling to a limited extent [3.39, 3.40], but tilting the sample is not effective because of the varying crystal orientations in the polysilicon layer. Because of the random location of grains correctly aligned for channeling, threshold voltage variability between transistors increases when significant channeling occurs.

3.4 Diffusion *from* polysilicon

In the previous section, we considered the vertical and lateral movement of dopant atoms *within* a polysilicon layer. Understanding and characterizing this diffusion is important for devices such as MOS transistors and high-value polysilicon resistors. However, in other applications, the polysilicon serves as a source of dopant atoms for the adjacent regions of single-crystal silicon that contain active device elements. Implanting the dopant into a polysilicon layer and then diffusing it from the polysilicon into the adjacent single-crystal silicon can form shallow junctions and avoid implant damage in the critical device regions [3.42–3.45]. In addition to serving as a dopant source, the polysilicon can also be used to make efficient contact to the single-crystal silicon. In some devices, the polysilicon plays an important role in determining the device characteristics, as we shall see when we consider the polysilicon-emitter bipolar transistor in Sec. 6.8.2.

Because dopant diffusion is much more rapid in polysilicon than in single-crystal silicon, the dopant concentration is often assumed to be uniform throughout the thickness of the polysilicon film; that is, the

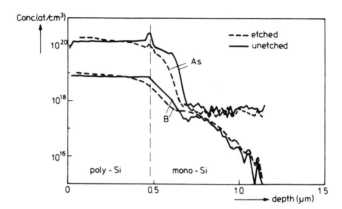

Figure 3.9: Boron and arsenic concentration profiles for samples on which the polysilicon was deposited on HF-etched or chemically oxidized single-crystal silicon; annealed for 95 min at 1000°C after dopant implantation. From [3.47]. Reprinted with permission.

dopant is assumed to redistribute through the thickness of the polysilicon in a time short compared to that needed for the desired dopant diffusion into the adjacent single-crystal silicon. In this case, a simple analysis predicts the dopant profile in the single-crystal silicon by assuming that the concentration at the surface of the single-crystal silicon remains constant at a value equal to that in the grains of the polysilicon (but limited to the solid solubility of the dopant in silicon) [3.46].

However, a more detailed treatment may be needed. When polysilicon is deposited on single-crystal silicon, the substrate surface is not atomically clean; the thin oxide often present between the single-crystal silicon substrate and the polysilicon film can significantly affect the diffusion of dopant. Arsenic, and possibly other dopant species, tend to accumulate at the polysilicon/single-crystal silicon interface (Fig. 3.9) [3.47]; the amount of dopant segregated near this interface depends on the thickness of the interfacial oxide.

In addition, structural changes within the polysilicon can alter the amount of dopant diffusing into the underlying single-crystal silicon. As we saw in Sec. 2.13, if the interface is sufficiently clean, the atoms within the polysilicon layer just above the interface can align epitaxially to the underlying single-crystal silicon during a moderate heat cycle. Dopant atoms can enhance this epitaxial restructuring. The amount of epitaxial realignment and its effect on dopant diffusion depend strongly on the amount of interfacial oxide between the polysilicon and the single-crystal silicon, as seen in Fig. 3.9. In samples from which the native oxide is removed, epitaxial realignment can occur. Dopant diffusion in the epitaxially realigned silicon proceeds slowly because of the absence of grain boundaries. The dopant in the realigned regions diffuses into the underlying substrate and is replenished slowly from the remaining polysilicon. Because the dopant atoms have to diffuse a longer distance from the remaining polysilicon, the amount of dopant entering the substrate under the realigned regions decreases. Thus, the amount of dopant diffusing from the polysilicon into the substrate can be *greater* when a thin oxide layer is present to prevent the epitaxial realignment. Of course, as the oxide thickness increases further, it begins to block the diffusion, and the amount of dopant entering the substrate decreases. In addition to blocking diffusion, a thicker oxide also increases the electrical resistance from the polysilicon into the single-crystal silicon. An oxide sufficiently thick to retard epitaxial realignment, but not thick enough to block diffusion maximizes the amount of dopant entering the substrate. A thickness ≤ 1 nm is optimum, while 2 nm is considerably too thick [3.47, 3.48]; however, controlling an oxide of this thickness in a manufacturing environment is difficult. In many cases, the residual oxide is made as thin as possible and the remainder of the device process appropriately designed to accommodate this very thin oxide.

The lower activation energy for diffusion compared to epitaxial realignment suggests that the dopant can first be redistributed throughout the thickness of the polysilicon and perhaps slightly into the underlying single-crystal silicon at a lower temperature ($\sim 850°C$). The structure can then be heated to a much higher temperature ($\sim 1000°C$) for a short time in a rapid thermal processor to break up the native oxide and epitaxially realign a portion of the thickness of the polysilicon. Breaking up the native oxide reduces the contact resistance of the polysilicon emitter bipolar transistor, as will be discussed in Sec. 6.8, but may also decrease the gain of the transistor (Sec. 6.8.2). This brief, high-temperature treatment also diffuses more of the dopant from the polysilicon into the underlying single-crystal silicon.

The annealing ambient influences the amount of epitaxial alignment (Sec. 2.13). It also affects dopant diffusion in the substrate. For example, some of the excess silicon interstitials injected into the polysilicon in an oxidizing ambient can travel through a submicrometer-thick polysilicon layer. The excess interstitials that reach the substrate can affect dopant diffusion there. The ability of polysilicon to absorb silicon interstitials depends on its grain structure. The absorption increases as the grain size decreases, suggesting that the interstitials are trapped by interacting with broken bonds at the grain boundaries. The absorption in polysilicon is always much greater than that in a single-crystal film of the same thickness [3.49], with an effective diffusion length of the order of 100 nm [3.50]. As the grains in the polysilicon grow during annealing, the ability to absorb interstitials decreases [3.51].

Some of the excess interstitials that diffuse through the polysilicon are absorbed at the interface, further reducing the amount of oxidation-enhanced diffusion in the underlying single-crystal substrate [3.50]. However, even with an oxide of moderate thickness (*eg*, 10 nm) between the polysilicon and single-crystal silicon, an appreciable number of interstitials can enter the underlying single-crystal silicon and affect dopant diffusion there [3.49]. When different or similar dopant species are separately introduced into single-crystal silicon under polysilicon, excess interstitials injected during the second doping process can markedly affect the previously introduced dopant [3.52, 3.53]. Interstitials from ion implantation damage generated in the polysilicon can have a marked effect on diffusion in the underlying single-crystal silicon.

3.5 Interaction with metals

Although diffusion of dopant atoms has received the most attention, interaction of other materials with polysilicon must also be considered. In integrated-circuit structures metal usually contacts polysilicon, as well as single-crystal silicon. After depositing the metal, the structure is usually annealed at a moderate temperature in the 400°C range to remove damage from the metal deposition process and to obtain intimate contact between the silicon and the metal. The time and temperature of the annealing operation are usually chosen to promote the correct amount of interaction between the silicon and the metal needed to provide good contact. However, polysilicon interacts with metals much more rapidly than does single-crystal silicon because of the weakly bonded silicon atoms at the grain boundaries. The resulting extensive interaction can degrade device properties, and the annealing process must be designed to optimize the amount of interaction of the metal with both single-crystal silicon and polysilicon.

3.5. INTERACTION WITH METALS

The interaction of polysilicon with aluminum, the primary integrated-circuit metallization material, is especially important. Other metals are also used in integrated-circuit metalization systems, either in metallic form or as silicides; the simultaneous formation of silicide on single-crystal silicon and on polysilicon by reaction with a deposited metal layer is often advantageous and again requires careful process design.

3.5.1 Aluminum

Because of its practical importance, the interaction between aluminum and polysilicon has been studied in most detail. The solubility of silicon in aluminum even below the 577°C eutectic temperature is high (about 1% at 500°C [3.54, 3.55]), and silicon can be transported from single-crystal device regions into the aluminum metallization of an integrated circuit [3.56]. The void left in the silicon can, in turn, be filled with aluminum; if this aluminum reaches sensitive regions of the integrated circuit, such as p-n junctions, it can degrade the devices. Because silicon atoms near grain boundaries in polysilicon are weakly bound, they can move into the aluminum even more readily. In turn, aluminum can penetrate the polysilicon film and reach the underlying polysilicon/SiO$_2$ interface at temperatures as low as 350°C, and some interdiffusion occurs even at 310°C [3.55]. The activation energy of this interaction is only about 2.6 eV [3.57]. To prevent possible device damage from this migration, silicon can be co-deposited with the aluminum at a concentration somewhat higher than its solid solubility at the annealing temperature. However, the loose structure of polysilicon may still allow excessive interaction, requiring use of a barrier material between the metal and the polysilicon.

The interaction occurs primarily along the grain boundaries, but a high concentration of arsenic (greater than about 10^{19} cm^{-3}) at the grain boundaries or at the top of the polysilicon layer can retard the interdiffusion [3.58]. The rapid interdiffusion and the effect of arsenic on the interdiffusion is postulated to be related to changes in the size of a silicon atom as it interacts with aluminum or arsenic [3.58]. Aluminum and silicon have about the same atomic radius (0.143 and 0.132 nm, respectively). Because aluminum is a p-type dopant in silicon, the silicon atom can lose one electron to the aluminum and become positively ionized with a radius of 0.041 nm, enhancing its probability of diffusing, especially along the open structure of the polysilicon grain boundaries. Arsenic, an n-type dopant, contributes an electron to a silicon atom, increasing its radius to 0.271 nm and slowing its diffusion along grain boundaries.

Titanium and vanadium at the interface also retard the interdiffusion, in this case by forming $TiAl_3$ and VAl_3, which do not react with the polysilicon during further moderate heat treatments [3.59]. The microstructure of the top region of the polysilicon film also influences the interdiffusion. Increasing the grain size there reduces interdiffusion [3.60], while amorphizing the top region enhances the interdiffusion greatly.

In silicon-gate MOS integrated circuits, placing an aluminum contact directly over the polysilicon gate electrode is often desirable to increase density, but anomalous results are often observed. Changes in the capacitance-voltage characteristics indicate that the polysilicon/SiO_2/silicon structure changes to to an aluminum/SiO_2/silicon structure during annealing [3.55]. After rapidly diffusing through the thickness of the polysilicon layer along grain boundaries, the aluminum atoms accumulate and spread along the polysilicon/SiO_2 interface. The capacitance is then the same as if aluminum were deposited directly on SiO_2. Interactions between the aluminum and the gate insulator can also adversely affect device yield and reliability, as will be discussed in Sec. 6.2.5. In addition to vertical interdiffusion, measurable quantities of aluminum can be incorporated into polysilicon films substantial distances from the aluminum electrodes by lateral diffusion, which can extend as far as 10 μm during annealing at 450°C for 2 hours [3.61]. The diffusion, which is postulated to be stress related, is reduced when the grain size is larger.

Interaction of aluminum and polysilicon structures can also occur during extended operation under severe conditions. During electrical stressing at an effective temperature of approximately 400°C, significant electromigration of aluminum occurs along the polysilicon/silicide interface of a $TaSi_2$/polysilicon structure [3.62]. The aluminum can travel long distances from the bonding wires to form blisters and extrusions.

3.5.2 Other metals

Similar metal-polysilicon interactions are observed when silver and gold layers are in contact with polysilicon [3.63]. The silicon-silver system

3.5. INTERACTION WITH METALS

has a eutectic temperature of 830°C, and interaction is observed at 550°C. For the silicon-gold system, the eutectic temperature is 370°C, and interaction is observed at 200°C. Gold strongly segregates to the grain boundaries (See Sec. 3.6), where it diffuses only slowly [3.64]. At high temperatures the effective diffusivity approaches the lattice diffusivity. However, the preferential segregation of gold at grain boundaries increases at lower temperatures, causing the effective diffusivity to deviate increasingly from the lattice diffusivity and to depend on the grain structure.

3.5.3 Silicides

Metals which form silicides, such as titanium, cobalt, tungsten, palladium, nickel, and chromium, behave differently. The silicide-forming reaction appears to dominate for polysilicon, as it does for single-crystal silicon, with the same diffusion-limited behavior (*ie*, a square-root time dependence) and similar critical temperatures for forming the silicide. After all the metal is consumed to form silicide, the structure is stable during further moderate heat treatments. The temperature required to form silicide by reaction with tungsten is similar for polysilicon and single-crystal silicon. Silicide formation begins at approximately 625°C with lightly doped polysilicon, as it does with single-crystal silicon [3.65]; high n-type dopant concentrations retard silicide formation with both forms of silicon [3.66]. When silicon is deposited onto nickel substrates, silicide formation is observed at temperatures as low as 300°C [3.67].

The grain structure of the polysilicon can change during silicide formation. Enhanced grain growth of phosphorus-doped polysilicon films is observed during formation of titanium silicide $TiSi_2$ by reaction of a deposited titanium film with polysilicon [3.68]. The combination of silicon motion to form $TiSi_2$ and simultaneous interaction with phosphorus facilitates migration of grain boundaries, leading to significant grain growth in the remaining polysilicon.

In complementary MOS (CMOS) circuits, the gate of the n-channel transistor is usually doped n-type, and the gate of the p-channel transistor is usually doped p-type at the same time as the corresponding source and drain junctions are formed. The amount of dopant introduced into the polysilicon is limited to that needed to form the shallow source and drain junctions. The sheet resistance of the polysilicon is higher than desired, and a low-resistance silicide layer is usually placed over the polysilicon at the same time it is formed on the source and drain regions.

The silicide layer is often formed by evaporating metal over the entire circuit, allowing it to react with exposed single-crystal and polycrystalline silicon regions, selectively etching the unreacted metal from the regions over insulators, and possibly annealing at a higher temperature to lower the silicide resistivity. This self-aligned silicide (*salicide*) process allows silicide to be formed without adding masking steps. However, the reaction is impeded by high n-type dopant concentrations and also becomes harder as the polysilicon gate becomes narrower [3.69].

Titanium silicide is most commonly used for the salicide process. $TiSi_2$ has two dominant phases of interest for integrated-circuit fabrication: The phase initially formed at moderate temperatures ($\sim$550-650°C) is the higher-resistance C49 phase. At higher temperatures ($\geq$650°C) the C49 structure changes to the lower-resistance C54 structure [3.70]. The resulting silicide is stable to 850°C, but tends to agglomerate at about 900°C [3.71].

The transition temperature between the C49 and C54 phases increases by as much as 180°C as the polysilicon line width decreases from 1.0 μm to 0.1 μm [3.72], impeding development of the low-resistivity C54 phase on narrow lines. Silicide formation at a grain boundary is poor [3.73]; in a very narrow line, a grain boundary may extend completely across line, increasing the resistance of the line significantly. During the heat treatments during and after silicide formation dopant atoms can move from the polysilicon into the silicide. This segregation of dopant into the silicide can reduce the effective dopant concentration in the polysilicon significantly, possibly lowering the active dopant concentration near the critical polysilicon/gate-oxide interface. In addition, dopant atoms can rapidly travel laterally within the silicide to other regions connected to the same silicided line [3.74]. This lateral dopant movement can transport dopant between the oppositely doped gates of n-channel and p-channel transistors, possibly counterdoping the gates and changing the threshold voltage.

Because of the limitations of titanium silicide, other silicides are being studied. Nickel silicide avoids the limitation of high resistance at grain boundaries [3.73], while cobalt silicide forms at lower temperatures and is less sensitive to line width [3.75–3.77].

3.6 Dopant segregation at grain boundaries

In addition to affecting dopant diffusion in polysilicon, grain boundaries are also low-energy sites at which dopant atoms can be preferentially located. The disorder at grain boundaries allows dopant atoms to readily fit into the structure, lowering the total energy of the system. At a given temperature thermodynamic equilibrium governs the distribution of dopant atoms between the low-energy positions at grain boundaries, where they are usually electrically inactive, and substitutional positions within the grains, where they can contribute to the conduction process. This *grain-boundary segregation* occurs especially for the n-type dopants phosphorus and arsenic [3.23, 3.24], and some evidence indicates that antimony also segregates [3.32]. Grain-boundary segregation does not appear to be important for boron.

3.6.1 Theory of segregation

Two factors influence the distribution of the dopant atoms between the grains and the grain boundaries [3.78]. First, a *chemical effect* makes segregation more likely if adding the dopant atom lowers the melting temperature of silicon. Of the common dopant atoms, arsenic and phosphorus tend to lower the melting temperature, while boron raises it. The heat of sublimation of the dopant element also correlates with the chemical effect; segregation is favored when the heat of sublimation is less for the dopant element than for silicon. Second, if the size of the dopant atom differs from that of silicon, a *size effect* favors segregation of the dopant atom to the grain boundary, both when the dopant atom is larger than silicon and when it is smaller. Segregation lowers the energy by reducing the lattice strain, and the tendency to segregate is favored by a greater difference in size. The covalent radii and heats of sublimation for the common dopant elements and silicon are given in Table 3.2 [3.78].

The chemical effect favors segregation of the n-type dopants phosphorus and arsenic; little effect is expected for antimony; and boron should be rejected from the grain boundaries. The size effect should have little influence on segregation of phosphorus and limited effect for arsenic. The sizes of antimony and boron are markedly different from that of silicon, favoring segregation of these elements. Considering both effects, we see that segregation of phosphorus, arsenic, and antimony is

Table 3.2: Estimate of probability of dopant segregation

Element	Covalent Radius (0.1 nm)	Heat of Sublimation (Kcal/mole)	Probability of Segregation		
			Size Effect	Chemical Effect	Total
Si	1.1	40			
B	0.82	130	+	−	?
P	1.06	3	~0	+	+
As	1.2	8	+	+	+
Sb	1.4	47	++	~0	+

likely. The two effects are in opposite directions for boron, so the net effect cannot readily be predicted.

For an infinite sample with one grain boundary, the atom fraction of dopant atoms X_{GB} on the grain boundary can be related to the atom fraction of dopant atoms X_G in the bulk of the grain by equating the chemical potentials of the two regions. In the simplest case in which only enthalpy contributes to the entropy, equating the chemical potentials leads to the equation

$$E_{GB} + kT \ln\left(\frac{X_{GB}}{1 - X_{GB}}\right) = E_G + kT \ln\left(\frac{X_G}{1 - X_G}\right) \quad (3.5)$$

where E_{GB} and E_G are the enthalpies at the grain boundary and in the grain, respectively, and T is the temperature of the heat treatment during which equilibrium is established. Equation 3.5 can be rewritten as

$$\frac{X_{GB}}{1 - X_{GB}} = \frac{X_G}{1 - X_G} \exp\left(\frac{Q_0}{kT}\right) \quad (3.6)$$

where the heat of segregation $Q_0 = E_G - E_{GB}$ is the difference in enthalpy between an atom in the grain and one at the grain boundary. If the free energy is reduced by placing the dopant atom at the grain boundary, Q_0 is positive, and segregation is favored. Considering vibrational entropy, as well as enthalpy, modifies Eq. 3.6 by adding a pre-exponential *entropy factor A*. With this entropy factor, which is related to the difference in vibrational entropy between a dopant atom

3.6. DOPANT SEGREGATION AT GRAIN BOUNDARIES

on the grain boundary and one in the bulk, Eq. 3.6 becomes

$$\frac{X_{GB}}{1-X_{GB}} = \frac{AX_G}{1-X_G}\exp\left(\frac{Q_0}{kT}\right) \qquad (3.7)$$

The derivation of Eqs. 3.6 and 3.7 assumes that the total number of sites N_S (per unit volume) at the grain boundaries is negligible compared to the total number of sites N_{Si} (per unit volume) in the system. If the grains are small, as they usually are in thin films, the number of sites at the grain boundaries may be a significant fraction of the total number of sites in the system, and Eq. 3.7 must also be modified to account for the finite size of the grains. The occupancy of grain-boundary sites by dopant atoms can then be found from the equation

$$\frac{X_{GB}}{1-X_{GB}} \approx A\left(X_0 - \frac{N_S}{N_{Si}}X_{GB}\right)\exp\left(\frac{Q_0}{kT}\right) \qquad (3.8)$$

where $X_0 = N/N_{Si}$ is the average dopant occupancy of all sites. [N is the total number of dopant atoms, which are divided between the grains (N_G) and the grain boundaries (N_{GB}).] If the grain boundaries are far from saturated with dopant atoms, $X_{GB} \ll 1$, and the fraction of sites in the grain boundary occupied by dopant atoms is

$$X_{GB} = AX_0\exp\left(\frac{Q_0}{kT}\right)\left[1 + \frac{AN_S}{N_{Si}}\exp\left(\frac{Q_0}{kT}\right)\right]^{-1} \qquad (3.9)$$

The corresponding expression for the fraction of occupied sites in the grain is

$$X_G = X_0\left[1 + \frac{AN_S}{N_{Si}}\exp\left(\frac{Q_0}{kT}\right)\right]^{-1} \qquad (3.10)$$

For typical values of grain size (100-200 nm) and Q_0 (~0.5 eV) and assuming $A = 1$, about 25-75% of the dopant atoms are likely to segregate to the grain boundaries.

If a dopant atom is located at the grain boundary, it is unlikely to contribute to the conduction process. Because the grain boundary is a disordered region with many dangling bonds, the five bonds of the n-type dopants or the three bonds of the p-type dopant can be readily satisfied. Consequently, few weak bonds should be associated with the dopant atoms at the grain boundaries, and the dopant ionization energies should be too large for significant carrier excitation to the conduction band or valence band.

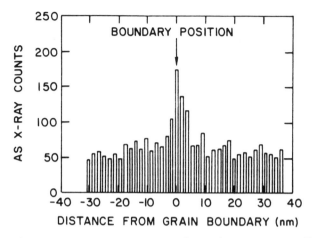

Figure 3.10: Arsenic segregates at grain boundaries, especially after annealing at lower temperatures. From [3.24]. Reprinted with permission.

3.6.2 Experimental data

Analytical measurements

Although much of the evidence for grain-boundary segregation is inferred from electrical measurements, advanced analytical techniques allow direct observation of dopant accumulation at grain boundaries. Using energy-dispersive x-ray analysis in a scanning transmission electron microscope (STEM) has permitted the arsenic concentrations to be detected near grain boundaries and in the bulk of grains. As shown in Fig 3.10 [3.24, 3.79], higher concentrations of arsenic are found at grain boundaries than within the grains, providing direct evidence for dopant segregation there.

Arsenic can segregate significantly during deposition. In samples doped with arsenic during deposition at approximately 670°C in a cold-wall, atmospheric-pressure reactor (grain size ≈ 20 nm), about 20% of the arsenic segregates to the grain boundaries even without any additional heat treatment. Changes in the amount of segregated dopant during annealing can provide information about the heat of segregation of the dopant to the grain boundaries. However, in a polysilicon sample, observing changes in dopant segregation from different heat treatments is complicated by grain growth changing the number of grain-boundary sites during the heat treatment. In addition, dopant may be redistributed by the moving grain boundaries [3.68]. To minimize

3.6. DOPANT SEGREGATION AT GRAIN BOUNDARIES

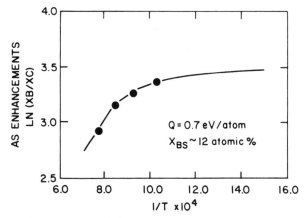

Figure 3.11: Enhancement of the arsenic concentration at a grain boundary indicated by the ratio of the dopant concentration at the grain boundary to that in the grain. The dopant concentration at the grain boundary probably saturates at the lowest temperatures investigated. From [3.24]. Reprinted with permission.

grain growth during dopant segregation experiments, the films can first be annealed at a high temperature to promote grain growth. Subsequent annealing at lower temperatures changes the structure less than without high-temperature annealing; however, some grain growth does occur, making detailed quantitative measurements difficult.

In the STEM study of arsenic segregation, after heating to 1000°C to stabilize the grain size, annealing at lower temperatures increases the amount of dopant segregating to the grain boundaries, as shown in Fig. 3.11 [3.24]. At low annealing temperatures (eg, 700 and 800°C), the grain-boundary sites can become saturated with dopant atoms.

The amount of segregation can vary considerably from one grain boundary to another, even in the same sample, complicating quantitative investigation. It is likely that this nonuniform segregation corresponds to the nonuniform electrical properties of grain boundaries seen within one sample [3.80]. For a single grain boundary of known orientation, the density of segregation sites can be correlated with the number of dangling bonds predicted by a structural model of the grain boundary [3.81]. For a polycrystalline sample the presence of different types of grain boundaries prevents directly relating segregation to grain-boundary angle or detailed structure. However, the similar binding energy seen for different grain boundaries at which moderate or high concentrations of arsenic segregate suggests that the primary difference between different grain boundaries is the concentration of segregation sites, rather than their nature. Analogous segregation of arsenic to a single-crystal/polycrystalline-silicon interface has also been seen by ion backscattering [3.82]. However, unlike grain boundaries, twin boundaries do not appear to be significant segregation sites.

In addition to arsenic, phosphorus also segregates to grain boundaries, with more segregation at lower annealing temperatures [3.83]. Other impurities also segregate to grain boundaries in polysilicon [3.13, 3.84]. Auger analysis and secondary-ion mass spectroscopy (SIMS) of large-grain polysilicon reveal the presence of C, O, and several metallic impurities. Again, the amount of segregation differs significantly from one grain boundary to another and even at different positions along the same grain boundary. Although oxygen is sometimes seen at grain boundaries, it does not appear to significantly affect either the number of grain-boundary sites available for phosphorus segregation or the heat of segregation [3.85].

Electrical measurements

Although sensitive analytical measurements can detect dopant segregated at grain boundaries in polysilicon, electrical measurements are more useful for quantitatively determining the amount of dopant segregated and changes during processing. The effect of grain-boundary segregation is best separated from that of the carrier trapping to be discussed in Sec. 5.4.1 by considering higher dopant concentrations, at which the carrier traps are saturated and do not affect the resistivity significantly. In this range, the carrier concentration approximately equals the dopant concentration remaining in the grains and can be found from Hall measurements. However, the dopant concentration should be below the solid solubility of the dopant in crystalline silicon.

Figure 3.12 [3.23] shows the measured carrier concentration as a function of annealing temperature for arsenic- and phosphorus-doped polycrystalline-silicon films. The active carrier concentration decreases and the resistivity increases with decreasing annealing temperature, implying that more dopant atoms segregate to the grain boundaries during annealing at lower temperatures, consistent with the analytical measurements. The lower active dopant concentration for arsenic-doped polysilicon suggests that more dopant segregates to the grain bound-

3.6. DOPANT SEGREGATION AT GRAIN BOUNDARIES

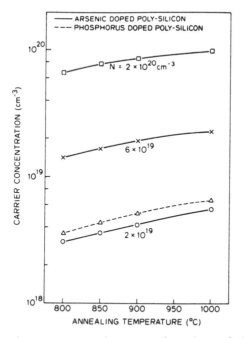

Figure 3.12: Carrier concentration as a function of the final annealing temperature for arsenic- and phosphorus-doped polysilicon with the average dopant concentrations indicated.

aries for arsenic than for phosphorus. However, as we saw in Sec. 2.11.3, grain growth is less pronounced for arsenic than for phosphorus, so the difference in the carrier concentration may indicate a larger number of grain-boundary sites for arsenic, rather than a greater tendency to segregate at a grain-boundary site. Consistent with decreasing segregation at higher annealing temperatures, little segregation of phosphorus is observed in large-grain polysilicon samples annealed at a very high temperature (about 1225°C) [3.86].

The change in resistance on annealing for antimony-implanted polysilicon films suggests that antimony also segregates to the grain boundaries [3.32]. Antimony segregation to the two surfaces of a polysilicon film has also been observed, although it has been tentatively attributed to a different mechanism (formation of a stable oxide complex). Resistivity and Hall data for boron show little change in the carrier concentration for different annealing temperatures, implying that little boron segregates to the grain boundaries at moderate concentrations

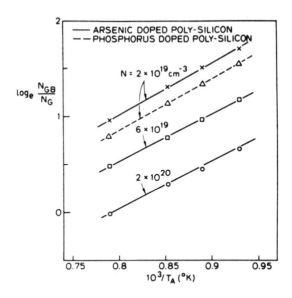

Figure 3.13: Ratio of dopant at the grain boundary to that in the grain as a function of the final annealing temperature.

[3.23]; however, boron may cluster or precipitate at grain boundaries when high concentrations are present [3.87].

For the dopants that do segregate, plotting the ratio of the dopant segregated at the grain boundary to that in the grain as a function of reciprocal temperature (Fig. 3.13 [3.23]) allows the parameters describing the grain-boundary segregation to be determined. From Eqs. 3.9 and 3.10, we can write

$$\ln\left(\frac{N_{GB}}{N_G}\right) = \ln\left(\frac{AN_S}{N_{Si}}\right) + \frac{Q_0}{kT} \qquad (3.11)$$

where T is the annealing temperature. The linear behavior shown in Fig. 3.13 lends support to the model. From this analysis, the heat of segregation is about 0.41–0.44 eV for both arsenic and phosphorus, and the factor A is in the range 2–3 [3.88]. The value of A appears to decrease slightly with increasing dopant concentration, suggesting that increasing the number of dopant atoms at the grain boundaries leads to a more ordered system, a greater reduction in vibrational entropy, and a lower entropy factor A. Somewhat higher values of the heat of segregation (0.54–0.65 eV) are found in other studies [3.24, 3.89].

3.6. DOPANT SEGREGATION AT GRAIN BOUNDARIES

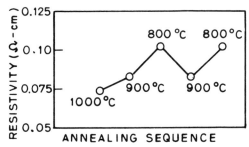

Figure 3.14: The resistivity of a polysilicon film implanted with 10^{15} cm^{-2} arsenic atoms changes reversibly on annealing at lower and higher temperatures as the dopant moves to and away from the grain boundaries.

Reversibility

Because the relation between the dopant at the grain boundaries and that in the grains is controlled by thermodynamics, the process is reversible as a sample is sequentially annealed at different temperatures [3.23]. Figure 3.14 shows that, after annealing at a higher temperature to stabilize the film structure, the resistivity increases on annealing at lower temperatures as the dopant moves to the grain boundaries where it is inactive. After further annealing at a higher temperature, the resistivity again decreases as the dopant moves from the grain boundaries into the grains. If the structure does not change during these annealing treatments, the resistivity can be repeatedly increased and decreased, so that the final resistivity depends on the last annealing treatment.

However, the amount of dopant segregation, and consequently the film resistivity, is determined not only by the final processing temperature, but also by the time at this temperature. The time required to reach equilibrium between the dopant in the grains and that at the grain boundaries is attributed to the time needed for the dopant to diffuse from the interior of the grain to the grain boundary. It consequently, increases at lower annealing temperatures and with increasing grain size [3.90, 3.91]. At lower temperatures, typical processing times may not be adequate for dopant segregation to reach equilibrium. Below about 700°C dopant loss to grain boundaries during normal heat cycles is less severe than expected from equilibrium considerations. In particular, little dopant loss to grain boundaries is expected during the final, short, low-temperature metallization anneal. At intermediate temperatures, determining the amount of dopant segregation can become complicated because it depends on the annealing time, as well as the annealing temperature [2.18].

The distribution of dopant between grains and grain boundaries in a completed device is determined by the last high-temperature processing steps at which the distribution can approach equilibrium. As device processing temperatures decrease to the 800-900°C range, grain-boundary segregation becomes of more concern. The resistivity of polysilicon films may increase unacceptably at these temperatures. Rapid thermal processing, which employs high temperatures for short times, may be used to avoid severe dopant segregation at grain boundaries.

In addition to the nominal processing temperature, temperature transitions can also also affect segregation. Wafers are cooled slowly when they are removed from a furnace to avoid crystal damage from nonuniform cooling. The slow cooling allows the wafers to remain at intermediate temperatures for enough time for appreciable dopant segregation to occur. The loss of dopant to grain boundaries becomes more severe with the slower cooling rates needed for larger wafers. A 25% increase in sheet resistance has been seen when wafers are removed from a furnace at a rate of 10 cm/min [3.92]. With very slow cooling, the sheet resistance can increase by as much as a factor of three. Rapid thermal annealing is especially beneficial in avoiding segregation during cooling because the wafer is quickly cooled through the intermediate temperatures where dopant segregation is most severe.

Dopant range and activity

Dopant segregation can only be studied unambiguously over a limited range of dopant concentrations: The concentration should be high enough that the carrier traps at the grain boundary (Sec. 5.4.1) are saturated by a small fraction of the charge carriers; it should be low enough that the active carrier concentration is not limited by the solid solubility of the dopant [3.89]. The concentration range over which arsenic segregation can be investigated is limited by its low solid solubility. Study of arsenic is also complicated by its tendency to form clusters at high concentrations; clustering is also reversible with subsequent heat treatments and increases after annealing at lower temperatures. It can be important at dopant concentrations above about 10^{20} cm^{-3} [3.23].

3.7. COMPUTER MODELING OF DIFFUSION

In polysilicon, unlike single-crystal silicon, the clusters can form at the grain boundaries, as well as within the grains.

3.7 Computer modeling of diffusion

As we have seen, dopant diffusion in polysilicon is complicated by the different dopant diffusivities in the grains and at the grain boundaries, anisotropy of the grain boundaries, movement of grain boundaries during the diffusion process, and segregation of dopant at grain boundaries. When the polysilicon communicates with the underlying single-crystal silicon, dopant segregation at the polysilicon/single-crystal silicon interface and epitaxial realignment also need to be considered.

Accurate analytic treatment of these processes is difficult. However, with the ready availability of powerful workstations and large computers, computer modeling of dopant diffusion in polysilicon has become practical [3.15, 3.16, 3.93–3.99]. The detail of the modeling requires a trade-off between accuracy and computer time, with the optimum trade-off depending on the application. When polysilicon is only used as a diffusion source, less detail may be required. When the location of the dopant within the polysilicon is critical, more details must be considered.

A detailed model can examine a few grains and the surrounding grain boundaries, surfaces, and interfaces. For example, understanding the dopant behavior while annealing an arsenic implant may require considering segregation to grain boundaries, the surface and interfaces. For a short anneal, segregation to grain boundaries and interfaces may dominate, with a significant portion of the interior of the grain not being completely doped. For a longer anneal grain growth (*ie,* movement of the grain boundaries) may be important, with the rate of grain growth depending on the dopant concentration. The diffusion rate within the grain depends on the dopant concentration also. All the relevant processes can be considered for a region containing a few grains, but the computation capability needed makes application to large regions difficult.

More practical simulation considers two-dimensional effects on the scale of device features, rather than for each individual grain, to reduce computation time. A typical approach involves a *two-stream* model. Dopant diffusion within the grains is calculated separately from the dopant diffusion within the grain boundaries. The grains and grain boundaries are then coupled by allowing dopant to move between the grains and the grain boundaries. The separate numerical treatment of grains and grain boundaries allows structural information about the grain boundaries, such as their dominant direction, to be included. Variations of the grain size (distance between grain boundaries) can also be included, as can grain growth during high-temperature processing, interfacial-oxide break-up, and epitaxial realignment (by decreasing the grain-boundary density as a function of time). Transfer of dopant from grains to grain boundaries by the movement of grain boundaries can also be included [3.15, 3.96, 3.99].

Within the grain interiors, the diffusion is influenced by dopant concentrations and their gradients, by the electric fields arising from electrically active dopant, and by point defects. Diffusion within the interior of the grain can be written as[1]

$$\frac{dC_g}{dt} = \nabla(D_g)\nabla C_g - G \quad (3.12)$$

where $D_g = D_{g0}\exp(-E_d/kT)$ is the dopant diffusivity within the grain and G describes the net transfer of dopant from the grain to the grain boundary.

Within the grain boundaries, the dopant concentration per unit area is calculated, as is the grain-boundary area per unit volume, and the two quantities are combined to obtain the dopant concentration in the grain boundaries per unit volume of polysilicon. The models for diffusion of dopant within the grain boundaries may include the concentration dependence of dopant diffusion and the effect of electric field. Diffusion of the grain-boundary component of the dopant can be written as

$$\frac{dC_{gb}}{dt} = \nabla_i(F_{ij}D_{gb}\nabla_j C_{gb}) - G \quad (3.13)$$

where F_{ij} describes the effect of the anisotropy of the grain boundaries.

After expressing the diffusion within the grains and within the grain boundaries, the two components are coupled through the term G, which depends on the segregation coefficient p_{seg} between grains and grain boundaries. This segregation coefficient, in turn, depends on the number of available sites at the grain boundaries and in the grains and on the

---

[1]Adapted from [3.100] with permission of Silvaco International.

3.7. COMPUTER MODELING OF DIFFUSION

temperature-dependent entropy and can be written in a form based on Eq. 3.11

$$p_{seg} = \frac{Q_0}{L_G N_{Si}} A \exp \frac{Q_0}{kT} \quad (3.14)$$

where Q_0 is the heat of segregation, L_G is the grain size, N_{Si} is the density of silicon atoms, and A is the entropy factor. The transfer G of dopant between the grains and the grain boundaries can then be written as

$$G = \frac{1}{\tau_0}\left(\frac{C_g}{p_{seg}} - C_{gb}\right) \quad (3.15)$$

where τ_0 is the temperature-dependent time constant for equilibration of dopant between grains and grain boundaries. Interfaces and surfaces can be handled analogously.

To include models for the physical phenomena, some assumptions must be made. A few examples are considered here:

1. The directional character (anisotropy) of grain-boundary diffusion can be treated by assuming that the grain boundaries extend through the thickness of the film in a specific direction; the grain size in the film plane is also specified.

2. Grain growth during processing after deposition is considered by writing the grain size at time t as

$$L_G = \sqrt{L_0^2 + cD_{Si}t} \quad (3.16)$$

where $c = 6 \times (\text{lattice constant})^2 \times (\text{grain} - \text{boundary energy})/kT$. D_{Si} is the silicon self diffusivity near the grain boundary and may be increased or decreased by the dopant. (For example, arsenic segregation at the grain boundaries can impede grain-boundary motion.)

3. Oxide breakup at the polysilicon/silicon interface can be considered by assuming a characteristic time for forming voids in the oxide, calculating the fraction of the oxide broken up at any time, and considering epitaxial regrowth around the broken-up oxide.

4. For an amorphous film that crystallizes during subsequent processing, the grain size after crystallization can be assumed to be a fraction of the thickness of the initially deposited amorphous film.

3.8 Summary

In this chapter we looked at the interaction of polysilicon with dopant atoms and, to a limited extent, with adjacent metal films. We saw that the disordered structure at grain boundaries provides paths along which dopant atoms can readily diffuse. After diffusing along the grain boundaries, the dopant atoms move into the grains, where they can be electrically active. This rapid grain-boundary diffusion leads to more rapid penetration of dopant, both laterally and vertically, in polysilicon than in single-crystal silicon. The magnitude of the enhanced diffusion depends strongly on the detailed grain structure; vertical diffusion is especially rapid in the columnar structure associated with $\{110\}$-oriented grains. Diffusion from polysilicon into adjacent single-crystal silicon is influenced by segregation at the polysilicon/single-crystal-silicon interface and by possible epitaxial realignment on the underlying substrate, which reduces the number of grain boundaries available to act as rapid diffusion paths. Residual oxide at the interface can also retard diffusion. Metals can interdiffuse with polysilicon more rapidly than with single-crystal silicon, and annealing temperatures and times must be restricted.

The disordered structure of grain boundaries also provides low-energy sites at which dopant atoms, especially the n-type dopants, can be preferentially located. Dopant segregation at the grain boundaries is more pronounced after annealing at lower temperatures if the segregation process has adequate time to reach equilibrium. Because segregation is a thermodynamic process, it is reversible; the amount of dopant segregated, and consequently the film resistivity, can depend on the last steps in the device-fabrication process.

Quantitatively modeling the effects described in this chapter allows more rapid and less expensive process development. With the ready availability of computers of increasing power, modeling becomes practical and models are being refined to include more effects of the polycrystalline structure.

Chapter 4

Oxidation

4.1 Introduction

In most integrated circuits, the polysilicon device features are electrically isolated from overlying conductors (either metal or additional layers of polysilicon) by silicon dioxide, which can be either thermally grown on the polysilicon or deposited. For many applications the oxide must simply be highly insulating. However, specialized devices, such as *electrically erasable, programmable, read-only memories* (EEPROMs or E^2PROMs) used with increasing frequency in VLSI circuits require a thin oxide with well-controlled conductivity above the polysilicon. Numerous studies have shown that the oxidation rate of polysilicon can differ substantially from that of single-crystal silicon and also that the electrical properties of the oxide grown on polysilicon can be inferior to those of a similar thickness of oxide grown on single-crystal silicon.

In this chapter, we first examine the oxidation rate of polysilicon and its relation to the polysilicon grain structure. We see that in some applications the different oxidation rates of polysilicon and single-crystal silicon can be used constructively, while in other applications the difference complicates integrated-circuit fabrication. Then, we consider conduction through oxide grown on polysilicon, showing that irregularities on the polysilicon surface lead to unexpectedly high conduction through thermally grown oxides. As an extreme case, we consider insulators formed on the hemispherical-grain (HSG) polysilicon discussed in Sec. 2.12.

4.2 Oxide growth on polysilicon

4.2.1 Oxidation of undoped films

Two mechanisms control the growth of moderate-thickness oxides on single-crystal silicon [4.1]. Initially, the growth is limited by reaction of the oxidizing species with silicon at the Si/SiO_2 interface. In this surface-reaction-limited regime, the oxidation rate is constant, and the oxide thickness is linearly proportional to the oxidation time. Because the oxidation process is limited by a chemical reaction with the silicon surface atoms, the rate depends on the surface orientation of the silicon. As the oxide grows, diffusion of the oxidizing species through the already formed oxide becomes more difficult, and the oxidation rate decreases. Eventually, diffusion limits the oxidation rate. In this diffusion-limited regime, the oxide thickness increases as the square root of the oxidation time.

As discussed in Sec. 2.5, polycrystalline silicon is composed of grains with various orientations. Because the oxidation rate varies for differently oriented silicon surfaces in the surface-reaction-controlled regime, the oxide thickness grown on polysilicon is expected to depend on the dominant crystal orientations, with {100}-oriented grains oxidizing most slowly, {111}-oriented grains oxidizing most rapidly, and {110}-oriented grains oxidizing at an intermediate rate [4.2, 4.3]. As the oxide becomes thicker and diffusion of the oxidizing species becomes rate limiting, the dependence of the oxidation rate on the preferred crystal orientation should decrease.

This behavior can be illustrated most clearly with thick layers of polysilicon polished so that the resulting smooth surface contains large grains with different crystal orientations [4.4]. When oxidized under surface-reaction-limited conditions, each differently oriented grain oxidizes at a rate characteristic of that particular orientation of silicon. Grains with a specific crystal orientation exhibit the same oxide interference color as similarly oriented, single-crystal wafers. The oxide thickness measured macroscopically is an average of the thicknesses on the differently oriented grains and is between those on (100)- and (111)-oriented, single-crystal silicon. Replica micrographs made after removing the oxide show the different heights of the differently oriented grains, confirming that a different amount of silicon is consumed by oxidation of grains with different orientations. As expected, the differences in ox-

4.2. OXIDE GROWTH ON POLYSILICON

Table 4.1: Oxide thickness grown on 0.5 μm thick polysilicon films and differently oriented single-crystal wafers during a 75 min, 850°C, pyrogenic steam oxidation.

Type of silicon	Lightly doped (nm)	n^+ (Phosphorus doped) (nm)
(100) single-crystal	108	350
AP 960°C polysilicon	125	220
LP 625°C polysilicon	145	199
(111) single-crystal	147	366

ide thicknesses on the differently oriented grains are smaller when the oxidation is performed under diffusion-limited conditions.

The macroscopically measured oxide thickness grown on the fine-grain polysilicon typically used in integrated circuits is also an average of the oxide thicknesses grown on the differently oriented grains. The oxide thicknesses grown on polysilicon films deposited either in an atmospheric-pressure reactor at 960°C or in a low-pressure reactor at 625°C are between those of oxide grown on (100)- and (111)-oriented single-crystal silicon, as indicated in Table 4.1 [4.5]. Similarly, the oxidation rate of initially amorphous films deposited at 575°C in a low-pressure reactor is between that of the two orientations of single-crystal silicon [4.6]; although the films are initially amorphous, they crystallize as they heat prior to oxidation or during the initial stages of oxidation. Because the oxide thickness grown depends on the dominant crystal orientations in the film being oxidized, it can differ significantly for polysilicon deposited under different conditions. We saw in Sec. 2.5 that the crystalline texture of polysilicon films depends strongly on the deposition conditions; the oxide thickness should vary correspondingly. If the differences are appreciable, modifying the oxidation process may be necessary when the polysilicon deposition conditions are changed. For unpolished samples the faceted surface structure causes the exposed surface planes to differ from the grain orientations normal to the film plane, further complicating the behavior.

Replica micrographs of the thick, undoped, large-grain samples do not show enhanced oxidation at grain boundaries [4.4]. Similarly, transmission electron microscopy of thinner, fine-grain polysilicon layers does not reveal any rapid grain-boundary oxidation in undoped films [4.7].

The well-characterized, linear-parabolic oxidation model does not apply to the growth of the very thin (<10 nm) oxides used for gate insulators of transistors and for capacitor dielectrics of memory devices. Oxidation of both single-crystal silicon and polysilicon in this thin-oxide region is attributed to Mott-Cabrera, field-assisted oxidation [4.8], which is characterized by an inverse logarithmic growth law:

$$\frac{dx_{\text{OX}}}{dt} = K_0 \exp\left(-\frac{E_a}{kT}\right) \exp\left(\frac{x_1}{x_{\text{OX}}}\right) \quad (4.1)$$

where x_{OX} is the oxide thickness and x_1 is a characteristic length. The initial oxidation rate of polysilicon at low temperatures (*eg*, 850°C) in an oxygen ambient diluted with nitrogen is lower than that of (100)-oriented, single-crystal silicon, but the rate coincides with that of single-crystal silicon at 950°C and above [4.8]. The 2.8 eV activation energy found at higher temperatures is not far from the 2.2 eV activation energy of single-crystal silicon, but a higher value (>5.2 eV) is observed in the low-temperature region. This higher activation energy is attributed to enhanced oxidation at grain boundaries or to effects of impurities or structural defects. (Detailed study of the initial stages of oxidation of undoped, large-grain polysilicon at extremely low temperatures by Auger electron spectroscopy and angle-resolved, x-ray photoelectron spectroscopy shows some enhanced oxidation at the grain boundaries [4.9]. After oxidizing at 140°C, an oxide thickness of about 4 nm is observed near the grain boundaries, while only about 1.5 nm of oxide is grown on the grains away from the grain boundaries. As shown above, similar enhanced oxidation is not seen in the temperature ranges typical of integrated-circuit processing, but further investigation of thinner oxides is needed.)

4.2.2 Oxidation of doped films

The oxidation of heavily doped polysilicon films has been studied extensively because of the need to oxidize the heavily doped polysilicon used in integrated circuits. Phosphorus has been studied most thoroughly because of the widespread use of phosphorus-doped polysilicon in storage elements of dynamic memories and in some silicon-gate MOS transistors. The oxidation of arsenic-doped polysilicon has also been

4.2. OXIDE GROWTH ON POLYSILICON

investigated in some detail, with limited consideration given to boron-doped films. Obtaining the largest difference in the oxidation rates of heavily doped polysilicon and adjacent regions of lightly doped, single-crystal silicon is important in the fabrication of some device structures, and conditions which maximize the difference have been widely studied. The difference is largest at lower oxidation temperatures, at which dopant-enhanced oxidation is greater.

The difference in the oxidation rates of heavily doped polysilicon and lightly doped single-crystal silicon is especially important in larger-geometry, silicon-gate MOS transistors, which use n-type polysilicon as the gate electrode for both n- and p-channel transistors (Sec. 6.2). In these devices the heavily doped, n-type, polysilicon gate electrode is adjacent to the lightly doped, single-crystal silicon regions that subsequently become the source and drain. After definition of the polysilicon, these single-crystal regions are still lightly doped. Oxidation under suitable conditions produces a thick oxide on the polysilicon while only a thin oxide grows on the single-crystal silicon. The thick oxide on the polysilicon can serve as an implant mask during processing, and it can reduce capacitance in the finished circuit. The differential oxidation rate can also be used to avoid a critical mask alignment; the thin oxide on the single-crystal silicon can be removed while not exposing the heavily doped polysilicon, which is covered by a much thicker oxide.

In other structures the minimum ratio of oxide thicknesses on polysilicon and on single-crystal silicon is desired (for example, to maximize the capacitance between two polysilicon electrodes when the lower, heavily doped, polysilicon layer is oxidized at the same time that the gate oxide is grown for transistors on nearby single-crystal silicon). Higher oxidation temperatures and thicker oxides favor a lower ratio [4.10]. Although most devices require a highly insulating oxide, thin oxides through which carriers can tunnel must be grown reproducibly for specialized applications (*eg*, EEPROMs), as will be discussed in Sec. 4.3.

Oxidation of doped films is complex because of the interaction of the dopant and the grain boundaries. Oxidation is affected not only by the dopant species and its concentration in the grains, but also by possible dopant precipitation at the grain boundaries, and by the different dopant diffusion rates in single-crystal silicon and in polysilicon, as will be discussed below. The grain boundaries may serve as a conduit for rapid diffusion of dopant atoms away from the surface of the polysilicon, reducing the oxidation rate. Excess dopant at the grain

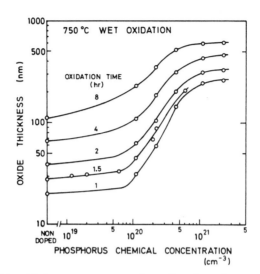

Figure 4.1: Oxide thickness grown on heavily phosphorus-doped polysilicon during a 750°C, wet-oxygen oxidation. From [4.12]. Reprinted by permission of the publisher, The Electrochemical Society, Inc.

boundaries near the surface may cause these regions to oxidize rapidly, forming a doped oxide with properties markedly inferior to those of pure, thermally grown SiO_2. The oxidation rate may also depend on the method of adding the dopant; excess dopant may be incorporated at grain boundaries if the dopant source is unlimited, while the grain boundaries may deplete dopant atoms from the grains if the source is limited.

Excess point defects in heavily doped, single-crystal silicon increase its oxidation rate [4.11]. The oxidation rate of polycrystalline silicon also increases as the dopant concentration increases, as shown in Fig. 4.1 [4.12]. However, the enhanced oxidation of heavily doped polysilicon can be less than that of similarly doped single-crystal silicon, as shown in Table 4.1 [4.5]. Figure 4.2 [4.5] shows the oxide thickness grown under surface-reaction-limited conditions on two different types of polysilicon and two orientations of single-crystal silicon for varying phosphorus doping added by gaseous diffusion from a $POCl_3$ source. For the lightly doped films (high sheet resistance) the oxide thickness grown on the polysilicon samples is between that on the two orientations of single-crystal silicon, as expected. As the dopant concentration increases, however, the oxide thickness grown on polysilicon increases *less* rapidly

4.2. OXIDE GROWTH ON POLYSILICON

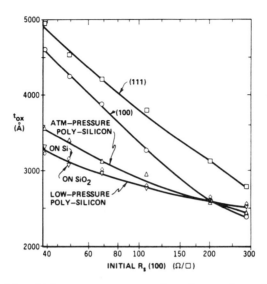

Figure 4.2: Oxide thickness grown on phosphorus-doped single-crystal and polycrystalline silicon during a 150 min, 850°C, pyrogenic steam oxidation as a function of the sheet resistance R_S measured on the (100)-oriented, single-crystal silicon wafer in each set.

than does that on single-crystal silicon, even though the initial phosphorus concentration in the polysilicon should be at least as great as that in the single-crystal silicon. This behavior is caused by the interaction of diffusion and oxidation. In Sec. 3.3 we saw that impurities diffuse much more rapidly in polysilicon than in single-crystal silicon. During doping and oxidation the dopant can diffuse away from the surface more readily in polysilicon than in single-crystal silicon, as indicated in Fig. 4.3a. Consequently, the surface concentration is lower, and the oxide grown on the polysilicon is thinner. The difference between the oxide thicknesses grown on different types of polysilicon depends on the ease of dopant diffusion in each sample, which, in turn, is governed by its detailed crystal structure.

Very thin films of heavily doped polysilicon sometimes oxidize *more* rapidly than thicker layers or even more rapidly than similarly doped, single-crystal silicon, as shown in Table 4.2. In thick films, the dopant can readily diffuse away from the polysilicon surface, reducing the surface dopant concentration. In thin polysilicon films, however, the finite film thickness confines the dopant atoms, as shown in Fig. 4.3b. As

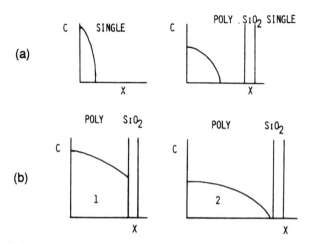

Figure 4.3: Schematic representation of dopant profiles in single-crystal silicon and in polysilicon. (a) In a thick polysilicon film, the dopant diffuses away from the surface more rapidly than in single-crystal silicon, lowering the surface concentration and the oxidation rate. (b) In a very thin film (1) the dopant is confined by the limited thickness of the polysilicon; the surface concentration and oxidation rate remain high while those in a thick film (2) decrease.

dopant atoms approach the back surface of the thin film, the increasing dopant concentration there decreases the concentration gradient driving the diffusion. Consequently, the surface concentration remains high, and the oxide grown is thicker. When the polysilicon layer is thicker than the diffusion depth, the oxide thickness grown should not depend on the polysilicon thickness (assuming the grain structure does not depend on the film thickness).

Dopant segregation and precipitation at the grain boundaries complicates the oxidation of heavily doped layers of polysilicon [4.13], as will be discussed in Sec. 4.2.3. As oxidation proceeds, the ratio of dopant atoms in the grains to that in the grain boundaries changes if the oxidation temperature differs from the temperature of the previous heat cycle. As dopant segregates and precipitates at the grain boundaries, the oxidation rate there can increase significantly, influencing the macroscopically observed oxidation rate. The situation is further complicated by grain growth during oxidation changing the amount of dopant segregated at the grain boundaries. All the effects related to the grain

4.2. OXIDE GROWTH ON POLYSILICON

Table 4.2: Oxide thicknesses grown on n^+ polysilicon films of different thicknesses during a 75 min, 850°C, pyrogenic steam oxidation.

Polysilicon thickness (μm)	Oxide thickness (nm)	
	LP	AP
0.5	393	378
1.0	280	301
1.5	276	310
n^+ (100) single-crystal	515	

boundaries depend on the grain size which, in turn, depends on the deposition conditions, the doping process, and the high-temperature heat cycles during device fabrication.

Phosphorus-doped films

The oxide thickness depends on the phosphorus concentration when the polysilicon is extrinsic (carrier concentration greater than the intrinsic carrier concentration n_i) at the oxidation temperature but the dopant concentration is still less than the solid solubility of the dopant in silicon at the oxidation temperature. At low oxidation temperatures dopant-enhanced oxidation is seen for average phosphorus concentrations from about 10^{20} cm^{-3} to about 10^{21} cm^{-3} (Fig. 4.1) [4.12]. Although the upper limit of this range is higher than the solid solubility of phosphorus at the oxidation temperature, a portion of the phosphorus segregates or precipitates at the grain boundaries so that the dopant concentration in the grains is lower. As shown in Fig. 4.1, the presence of the dopant can increase the oxide thickness by as much as an order of magnitude. Marked dopant enhancement of the oxidation rate is also seen for phosphorus-doped films oxidized in dry oxygen [4.14].

The observed behavior can be modeled by the linear-parabolic oxidation kinetics widely used for single-crystal silicon. The linear rate constant B/A is most significantly affected because it is strongly influenced by the reaction kinetics at the Si/SiO$_2$ interface [4.12]. The parabolic rate constant B also increases because incorporation of phosphorus into the oxide increases the diffusivity of the oxidizing species through the already formed oxide. As in single-crystal silicon, phosphorus is rejected from the growing oxide into the silicon during oxidation. The segregation coefficient (the ratio of phosphorus in the polysilicon to that in the oxide) is consistent with the value of 10 used for single-crystal silicon [4.15].

Phosphorus also segregates at the Si/SiO_2 interface in single-crystal silicon, producing a phosphorus-rich layer about 10 nm thick, containing about one monolayer of phosphorus [4.16]. Qualitatively similar behavior is observed in polycrystalline silicon [4.15, 4.17] although the amount of phosphorus segregated is about an order of magnitude less for polysilicon [4.17]. The amount of phosphorus included in the interface layer increases linearly with increasing bulk phosphorus concentration in single-crystal silicon [4.18], but in polysilicon it saturates for phosphorus concentrations above about 5×10^{19} cm^{-3} [4.17]. Interface segregation is attributed to locally strained Si–Si and/or Si–O bonds. Phosphorus atoms tend to pile-up near the interface to relax the strained bonds. The large number of lattice defects near the polysilicon/SiO_2 interface may possibly relieve some of the strain, decreasing interface segregation. Consistent with this interpretation, interface segregation is greater for polysilicon with larger grains. Similar segregation is also seen at the Si/SiO_2 interface at the bottom of a polysilicon layer.

Oxidation of polysilicon doped from a gas-phase source leads to high concentrations of phosphorus in the oxide, especially for wet oxidation [4.15]. During doping from the gas-phase, the polysilicon surface is in direct contact with a phosphorus-doped glass, and a phosphorus-rich skin forms on the polysilicon. Subsequent oxidation of this layer produces an oxide with a high phosphorus concentration. Oxidizing the polysilicon and removing the initial oxide layer lowers the phosphorus concentration in subsequently grown oxides. Although phosphorus inclusion in the oxide increases the oxidation rate by enhancing diffusion of the oxidizing species, excessive phosphorus in the oxide can degrade photoresist adhesion and pattern definition. The electrical properties of the oxide also degrade if the phosphorus concentration is too high.

Arsenic-doped films

The use of arsenic to dope gate electrodes of advanced, n-channel, silicon-gate MOS transistors has lead to investigation of the oxidation of arsenic-doped polysilicon [4.19, 4.20]. As for phosphorus, the oxide thickness increases when arsenic is added to the polysilicon. In the

4.2. OXIDE GROWTH ON POLYSILICON

limited range where comparable data are available for phosphorus and arsenic, the enhancement of the oxidation rate for the two dopants appears to be similar. However, lack of similarly processed samples doped with the two different elements in same study makes detailed comparison difficult. In addition, adding arsenic by implantation amorphizes the top layer of the polysilicon, complicating analysis of the oxidation behavior if the samples are not annealed to re-form the crystalline structure before oxidation.

As arsenic-doped polysilicon films are oxidized, the arsenic tends to segregate into the polysilicon, rather than having a similar concentration in the oxide and in the polysilicon. The segregation coefficient

$$m = C_{\text{As-poly}}/C_{\text{As-oxide}} \tag{4.2}$$

between the polysilicon and the oxide depends on the grain structure of the polysilicon. After the grain structure is stabilized by annealing at a high temperature, the segregation coefficient is given by [4.21]

$$m_{\text{As}} = 3.23 \times 10^9 \exp[-1.97(eV)/kT] \tag{4.3}$$

As for phosphorus, dopant enhancement of the oxidation rate saturates for arsenic concentrations above solid solubility at the oxidation temperature. For arsenic, however, formation of arsenic complexes or compounds can impede the oxidation so that the rate may decrease at high arsenic concentrations [4.20]. As a polysilicon film is oxidized, the thickness of the remaining polysilicon decreases, increasing the arsenic concentration per unit volume as arsenic is rejected from the growing oxide into the polysilicon. The increased arsenic concentration may decrease the oxidation rate as the polysilicon film is consumed. After the polysilicon film is completely oxidized, most of the arsenic remains at the original polysilicon/SiO_2 interface, although some can diffuse through a thin underlying oxide into the silicon substrate. The arsenic embedded in the oxide at the original interface is distributed laterally at distances comparable to the polysilicon grain size [4.20], suggesting that this arsenic was located at the polysilicon grain boundaries before oxidation. This behavior is consistent with the severe segregation of arsenic to grain boundaries in polysilicon.

Boron-doped films

Boron is used to dope the gate electrode of advanced p-channel, silicon-gate MOS transistors so that the desired threshold voltage can be

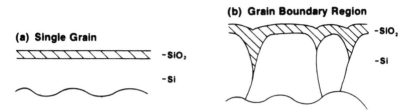

Figure 4.4: Oxidation may proceed more rapidly along a grain boundary (b) than over the grain (a). From [4.9]. Reprinted with permission.

achieved without counterdoping the channel. This important application has led to examination of the oxidation of heavily boron-implanted silicon [4.6]. In addition to its practical importance, investigating oxidation of boron-doped polysilicon allows effects related to high concentrations of dopant at the grain boundaries to be excluded because boron does not segregate to the grain boundaries, as do the common n-type dopants. Unlike the behavior of heavily doped, n-type material, boron enhancement of the oxidation rate is less than a factor of two [4.22]; under typical oxidation conditions the oxidation rate is increased only ~20% or less, and the enhancement is similar for both polysilicon and single-crystal silicon. Unlike arsenic, boron tends to segregate into the growing oxide, rather than into the polysilicon, during oxidation. The segregation coefficient appears to be independent of temperature, with a value $m = 0.36$ [4.21].

4.2.3 Effect of grain boundaries

The oxidation rate depends not only on the dominant crystal orientations present in the film, but can also be influenced by the grain boundaries themselves. Grain boundaries are disordered structures with weak bonds, which might be expected to oxidize more rapidly than the center of the grains, as suggested in Fig. 4.4 [4.9]. The interaction of grain boundaries with dopant atoms can also influence the oxidation process, and the possibility of enhanced oxidation near grain boundaries in polysilicon has been investigated both indirectly and directly.

Changes in stress during oxidation have been examined in both undoped and heavily phosphorus-doped polysilicon to detect possible rapid oxidation at the grain boundaries [4.23]. If enhanced oxidation occurs there, the additional oxide formed should cause compressive stress. On

4.2. OXIDE GROWTH ON POLYSILICON

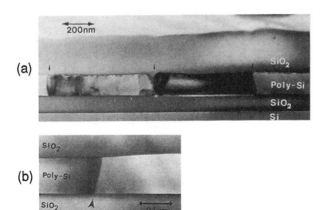

Figure 4.5: High phosphorus concentrations enhance oxide growth at grain boundaries during low-temperature oxidation (a) (800°C), but are less likely to affect the oxidation during high-temperature oxidation (b) (1100°C). From [4.7]. Reprinted with permission.

the other hand, sintering and grain growth during oxidation should decrease the volume by forming a more compact structure from the less dense, grain-boundary material, resulting in tensile stress. After oxidation, the stress in undoped films becomes more tensile, indicating that volume reduction by grain growth is more important than is enhanced oxidation at grain boundaries. The absence of enhanced oxidation at grain boundaries in undoped material is consistent with the earlier study using large-grained polysilicon [4.4] discussed in Sec. 4.2.1. On the other hand, the stress in doped films becomes more compressive during oxidation, consistent with enhanced grain-boundary oxidation.

High-resolution, cross-sectional, transmission electron micrographs of doped films also show that oxidation can occur preferentially at grain boundaries in doped films [4.7]. Although phosphorus doping from a gas-phase source at 950°C can increase the surface roughness, no enhanced oxidation occurs at the grain boundaries during the doping process, which forms an oxide [4.7]. After a subsequent oxidation at a low temperature (eg, 800°C), however, significant enhanced oxidation is visible at the grain boundaries in the shape of a "V"-groove (Fig. 4.5a) [4.7]. Longer oxidation at low temperatures increases the size of the V-grooves until they extend through a significant fraction of the polysilicon thickness. By contrast, when a doped film is oxidized at a high temperature

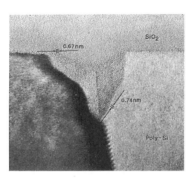

Figure 4.6: High concentrations of phosphorus or a phosphorus-containing phase precipitated at grain boundaries cause rapid oxidation of the grain boundaries, forming a phosphorus-containing oxide. From [4.7]. Reprinted with permission.

(eg, 1100°C), no enhanced oxidation is observed at the grain boundaries (Fig. 4.5b). The different behavior seen during oxidation below or above the doping temperature suggests that excess dopant leads to enhanced oxidation at the grain boundaries. After doping from a gas-phase source, the dopant concentration is expected to be close to its solid solubility in silicon at the doping temperature. During a subsequent heat treatment at a lower temperature (eg, during low-temperature oxidation), the dopant concentration is above its solid solubility at the heat-treatment temperature, and the dopant can move to the grain boundaries and precipitate as a silicon-phosphorus phase, possibly monoclinic silicon phosphide (Fig. 4.6) [4.7]. These precipitates oxidize rapidly, forming a thick phosphorus-containing oxide near the grain boundaries. On the other hand, at the high oxidation temperature of 1100°C, dopant precipitation at the grain boundaries is not expected because the average dopant concentration in the grains is less than the solid solubility at the high temperature. Therefore, few Si-P precipitates form at the grain boundaries, and enhanced grain-boundary oxidation does not occur.

The limited dopant solubility, along with related precipitation and rapid oxidation at grain boundaries, can cause severe yield loss in MOS integrated circuits unless the interrelation of these processes is well understood. The lowest possible polysilicon resistivity is needed for high-speed circuit performance. As technology progresses, the limited conductivity of polysilicon affects circuit performance more seriously, and

4.2. OXIDE GROWTH ON POLYSILICON

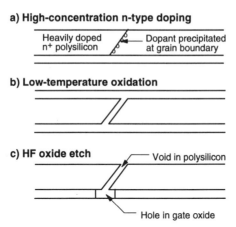

Figure 4.7: High concentrations of phosphorus at the grain boundaries can cause rapid oxidation there (a), possibly through the entire thickness of the polysilicon layer (b). Subsequent etching can rapidly remove the phosphorus-containing oxide and degrade the underlying gate oxide (c).

more emphasis is placed on maximizing the conductivity by increasing the dopant concentration to or beyond solid solubility. However, dopant concentrations above solid solubility at the processing temperature are not electrically active (Sec. 5.6). The excess dopant does not remain in the lattice, but is likely to precipitate, usually at grain boundaries (Fig. 4.7a). As we saw above, the resulting silicon-phosphorus phase oxidizes rapidly, forming a phosphorus-doped oxide, which can easily extend along the grain boundary through the entire thickness of the polysilicon [4.7] (Fig. 4.7b). This phosphorus-doped oxide etches extremely rapidly in any subsequent HF etch, leaving a void in the polysilicon (Fig. 4.7c). If the polysilicon is a gate electrode over thin gate oxide, the HF etches the underlying gate oxide also, leaving a hole which can cause an electrical short circuit between the gate electrode and the channel of the transistor.

4.2.4 Effect of device geometry

The previous discussion considered oxidation of large areas of polysilicon. In MOS integrated-circuit fabrication, however, the polysilicon is often defined before it is oxidized, so that lateral oxidation can also be important. Anomalous lateral oxidation is often observed when a polysilicon gate electrode is oxidized before a second level of polysili-

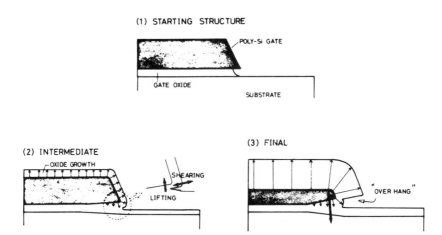

Figure 4.8: Lateral oxidation under the exposed polysilicon gate electrode forms an oxide wedge and overhang, which are more significant when the amount of oxide undercutting is greater. From [4.24]. Reprinted by permission of the publisher, The Electrochemical Society, Inc.

con is deposited. In this structure lateral penetration of the oxidizing species can cause additional oxide to grow between the polysilicon gate electrode and the underlying single-crystal silicon, as shown in Fig. 4.8 [4.24]. This lateral diffusion of the oxidizing species along the gate oxide is analogous to the formation of the *bird's beak* in a locally oxidized (LOCOS) isolation structure [4.25] by lateral oxygen diffusion along a thin stress-relief oxide beneath a silicon-nitride layer. In both cases, lateral oxidation creates a force which tends to bend the overlying layer. In the polysilicon structure, the force exerted by the oxide wedge can be relieved by plastic deformation of the polysilicon, and sliding along vertical grain boundaries [4.26]. This effect is especially severe at low oxidation temperatures, where little of the stress can be relieved by viscous flow of the oxide. Lateral oxidation is especially important in short-channel transistors where it can increase the effective gate oxide thickness over a substantial fraction of the channel length, degrading transistor performance [4.27].

The sidewall of a heavily doped polysilicon gate electrode can oxidize rapidly. Because the oxide grown is about twice as thick as the silicon consumed, an overhang and re-entrant corner are formed below

4.2. OXIDE GROWTH ON POLYSILICON

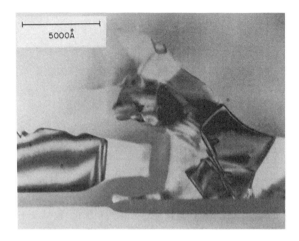

Figure 4.9: A second layer of polysilicon can fill the region beneath the overhang leading to a small separation between the two layers of polysilicon and low breakdown voltages. From [4.28]. Reprinted by permission of the publisher, The Electrochemical Society, Inc.

the edge of the polysilicon gate electrode, as shown in Fig. 4.8. When a second layer of polysilicon is deposited, it fills the region under this overhang (Fig. 4.9) [4.28] (Sec. 1.9), so that the two levels of polysilicon are separated by only a thin oxide, leading to possible dielectric breakdown at low voltages.

In some processes the exposed gate oxide is etched after polysilicon definition so that the source and drain regions adjacent to the gate electrode can be doped before the polysilicon and the source and drain regions are oxidized. Undercutting of the gate oxide beneath the polysilicon electrode exposes a small portion of the lower surface of the polysilicon. During the subsequent oxidation, the lower surface of the polysilicon is oxidized, as is the corresponding single-crystal region immediately below it, accentuating the oxide wedge and the overhang.

During oxidation of polysilicon, surface irregularities may be accentuated and may even become detached from the surface of the polysilicon, leading to silicon inclusions in the oxide [4.29]. When the polysilicon is patterned before oxidation, such inclusions are likely to form above the edges of the polysilicon lines. The upper edges may also become more pointed during oxidation, forming sharp points or *horns,* which can degrade the dielectric strength.

4.2.5 Oxide-thickness evaluation

The oxide thickness grown on polysilicon can be determined by etching the oxide from a portion of the sample and measuring the height of the resulting step with a surface profilometer or an atomic-force microscope (AFM). Although this technique is straightforward, it is destructive and time consuming; its resolution is limited; and it cannot readily be used on device wafers. In addition, the finite size of the profilometer stylus or AFM tip can prevent it from following the contours of the polysilicon surface; different surface roughness on the two sides of the polysilicon oxide makes accurate measurements difficult.

Although measuring the oxide thickness on single-crystal silicon is straightforward using optical reflection techniques with an ellipsometer or a spectrophotometer, measuring the oxide thickness grown on polysilicon is more complex because of the multilayer structure on which the oxide is grown. At the wavelengths typically used for oxide-thickness measurements (eg, $\lambda = 628$ nm for ellipsometry and $\lambda = 400\text{-}800$ nm for spectrophotometry), polysilicon is transparent. The overall reflected intensity is determined by interference of light reflected from the interfaces beneath the polysilicon, as well as from the top and bottom of the oxide layer grown on the polysilicon. In theory, the reflectance of the total multilayer structure can be analyzed, and the oxide thickness of the polysilicon can be extracted. In practice, however, variations in the thicknesses of the underlying layers introduce too much uncertainty into the indicated thickness of the oxide grown on the polysilicon to allow this technique to be used routinely.

Optical techniques can, however, be adapted to the ultraviolet wavelength range, in which silicon is opaque. In this region the reflected signal is dominated by interference at the top and bottom of the oxide grown on the polysilicon. The wavelength range from 200 to 400 nm is suitable for spectrophotometer measurements. Near the lower end of this wavelength range, the surface roughness of the polysilicon also affects the reflected signal, and at $\lambda = 280$ and 370 nm, absorption in structural bands of crystalline silicon can influence the reflected signal (see Sec. 2.6.3). The interference signal from light reflected at the two surfaces of the top oxide is strong, however, and the technique is a useful, nondestructive method of measuring the oxide thickness grown on polysilicon. Thick oxides on polysilicon can be measured by reflection techniques. Figure 4.10 [4.5] shows the interference signal obtained over

4.3. CONDUCTION THROUGH OXIDE ON POLYSILICON

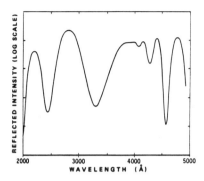

Figure 4.10: The reflectance in the wavelength range from 200 to 400 nm arises from interference in the oxide above the polysilicon, while the signal at longer wavelengths is influenced by reflection at interfaces beneath the polysilicon as well.

the wavelength range from 200 to 500 nm, illustrating interference in the top polysilicon oxide below about 400 nm, and interference throughout the multilayer structure at longer wavelengths. Ellipsometry using ultraviolet light can be used to measure thinner oxides, again taking advantage of the fact that silicon is opaque to ultraviolet light. If the oxide under the polysilicon is thin (<10 nm) and the polysilicon thickness is suitable, visible ellipsometry may provide an approximate value of the oxide grown on polysilicon, with an error of the order of the underlying oxide thickness [4.30, 4.31]. Visible ellipsometry may also be used with specially fabricated test structures [4.32].

4.3 Conduction through oxide on polysilicon

In most integrated circuits, the oxide layer above a polysilicon device element separating it from an overlying conductor must be highly insulating. However, specialized devices, such as EEPROMs, require a thin oxide with well-controlled conductivity above the polysilicon so that charge can tunnel through the oxide to erase information from the memory cell. EEPROMs will be discussed in more detail in Sec. 6.3. To obtain either high-resistivity or controlled-conductivity oxides on polysilicon, conduction through the oxide must be understood.

When oxide is thermally grown on polysilicon, current conduction through the oxide is typically several orders of magnitude higher than that through oxide of similar thickness grown on single-crystal silicon [4.33], and the breakdown field is substantially lower. In addition, the conduction can depend quite sensitively on the structure of the polysilicon and on the oxidation conditions [4.34]. The inferior properties of oxide grown on polysilicon have been correlated with its rough surface.

Current flow through thermally grown oxide is generally limited by injection of electrons from the silicon into the oxide by the Fowler-Nordheim mechanism and is given by the relation

$$I = AK_1(\mu E)^2 \exp\left(-\frac{K_2}{\mu E}\right) \qquad (4.4)$$

where A is the area, K_1 is the carrier distribution factor, K_2 is the emission factor related to the energy barrier ($\phi=3.2$ eV) and the effective mass ($m^* = 0.4\ m_0$), and μ is the ratio of the local electric field to the average field E. More complete expressions also include the field dependence of the effective emission area [4.35] and barrier height [4.36]. The potential barrier for charge injection is similar for polysilicon and for single-crystal silicon; the difference in the injected current arises from the different local electric fields. The strong dependence of the current on electric field makes the conduction very sensitive to small changes in the field.

We can understand the enhanced current flow through oxide grown on polysilicon by comparing its surface with the surface of single-crystal silicon. When a voltage is applied across the oxide grown on the smooth surface of single-crystal silicon, the electric field is uniform and equal to the applied voltage divided by the oxide thickness. The surface of polysilicon is generally not smooth, however, and the local electric field near surface irregularities (*asperities*) can be appreciably greater than the average electric field, as indicated in Fig. 4.11. Because the conduction in SiO_2 increases rapidly with increasing electric field, the observed current corresponding to the high local electric field is much greater than that expected from the average electric field. The conduction is greatly increased for both aluminum and polysilicon counterelectrodes above the oxide. For conventional polysilicon the enhancement is greater for negative voltages on the polysilicon than for negative voltages on the counterelectrode, consistent with a rougher surface at the bottom of the oxide than at the top creating higher electric fields and increased electron injection at the bottom of the oxide [4.37]. The higher local electric

4.3. CONDUCTION THROUGH OXIDE ON POLYSILICON

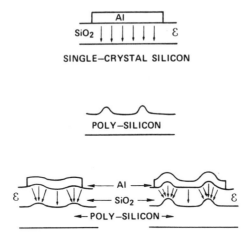

Figure 4.11: The local electric field is equal to the average electric field for oxide grown on the smooth surface of single-crystal silicon, but for oxide grown on polysilicon, the local field near the surface asperities can be much greater than the average field.

fields also decrease the average breakdown strength of the oxide [4.38]; the breakdown strength does, however, increase as the oxide thickness decreases, just as for oxides grown on single-crystal silicon [4.38].

The field enhancement from the irregular surface of polysilicon has been calculated assuming that the irregularities are bell-shaped [4.39] and found to be about a factor of three or four, in reasonable agreement with the behavior seen. The calculations are consistent with the observed 20 nm height and 70 nm width of typical surface irregularities. The field enhancement has also been calculated for isolated hemispherical irregularities [4.40], with about a factor of three field enhancement again observed. Even this moderate field enhancement can increase the Fowler-Nordheim injection current by about 50 times.

For the extremely rough surface of the *hemispherical-grain polysilicon* discussed in Sec. 2.12, the oxide roughness can be greater at the *top* of the oxide (between grains) than at the bottom (over the center of the grain) [4.41, 4.42]. The resulting high local electric field at the top of the oxide leads to greater Fowler-Nordheim injection when the counterelectrode is biased negatively.

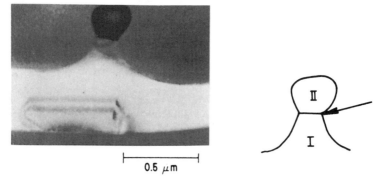

Figure 4.12: Protuberances form from slowly oxidizing crystallites. Lateral oxidation along a grain boundary (arrow) can detach the protuberance from the polysilicon, causing an isolated inclusion. From [4.43]. Reprinted by permission of the publisher, The Electrochemical Society, Inc.

4.3.1 Interface features

The dielectric properties of oxide grown on polysilicon depend sensitively on the microstructure of the polysilicon and its changes during oxidation. After oxidation four different types of interfacial features can be seen [4.43]:

1. Interface roughness corresponding to the initial *asperities* or general surface roughness of the polysilicon.
2. Anomalously high *protuberances* that develop during oxidation.
3. Inclusions of unoxidized silicon within the oxide.
4. Bumps caused by anomalous polysilicon nucleation at local contamination.

These features occur more frequently at lower oxidation temperatures, at which the oxidation process is dominated by surface reaction, rather than by supply of the oxidizing species.

Protuberances (Fig. 4.12) [4.43] form during oxidation from crystallites which appear to resist oxidation. This oxidation inhibition does not appear to be caused by surface contamination, but may be related to the different oxidation rates of differently oriented grains. In the first stage of protuberance growth, some crystallites rise above the Si/SiO_2 interface. After they begin emerging, stresses acting on the exposed

4.3. CONDUCTION THROUGH OXIDE ON POLYSILICON

crystallites are postulated to further inhibit oxidation locally, causing the protuberances to grow as the oxidation continues. When oxidation occurs laterally along a grain boundary below a protuberance (arrow in Fig. 4.12), the connection between the protuberance and the remainder of the polysilicon can be cut off, leaving an isolated inclusion. Enhanced oxidation from different directions makes formation of inclusions at the upper edges of defined patterns especially likely. The thin oxide above a protuberance degrades the observed average breakdown strength of the oxide. For thick oxides, the development of protuberances can make the upper and lower surfaces of the oxide less conformal and increase the polarity dependence of the current [4.40, 4.43]. Unlike protuberances, bumps of polysilicon formed at contaminated sites during deposition do not lead to severe thinning of the subsequently grown oxide, although the irregular surface increases the local electric field. For improved oxide quality they must be eliminated, generally by improved surface preparation before polysilicon deposition.

In most cases, the general surface roughness of the polysilicon film appears to lead to the enhanced conduction [4.44]. The current density is independent of the electrode area, and electron-beam-induced-current (EBIC) observations show a high density of active sites separated by approximately the grain size. While the more numerous asperities related to the general surface roughness appear to cause the enhanced electric field that leads to increased conduction through the oxide, isolated protuberances can markedly lower the average breakdown field.

Controlled conduction through oxide grown on polysilicon is needed for the operation of EEPROMs. In this case, field enhancement at surface irregularities is used to advantage, but control of the exact surface topography is essential to achieve reproducible device performance. Therefore, for these devices the structure of the polysilicon must be carefully controlled during its deposition and subsequent processing [4.45].

4.3.2 Deposition conditions

We saw in Sec. 2.5, that the crystal structure and surface roughness of the polysilicon, and consequently conduction through subsequently grown oxide, depend strongly on the deposition conditions of the polysilicon and its subsequent processing, with the greatest sensitivity to the deposition temperature [2.2, 2.18]. In particular, the surface roughness is related to the {110} texture [4.46], which depends critically on the deposition temperature near the 625°C normally used for polysilicon deposition in a low-pressure CVD reactor [1.9].

In conventional, low-pressure reactors used to deposit polysilicon, a temperature gradient is sometimes applied along the length of the reaction chamber to compensate for gas depletion (Sec. 1.5.1). This temperature gradient causes the structure of the deposited film to vary from one wafer to another, with corresponding variations in conduction through subsequently grown oxide. The surface roughness, and therefore the conduction, needed for a particular device may only be obtained over a limited portion of the deposition chamber, and the yield may be low. Eliminating the temperature gradient by using multiple gas injectors or by other means is often critical, as discussed in Sec. 1.5.1.

In the majority of integrated-circuit applications, the oxide above the polysilicon is simply used to isolate the polysilicon from overlying conductors, and the lowest possible conduction is desired. In these cases the enhanced conduction can be reduced by obtaining polysilicon films with smoother surfaces. We saw in Sec. 2.6.4 (Fig. 2.18) that a film in which a polycrystalline structure forms during deposition has a rough surface while one that is deposited in an amorphous form has a smooth surface. This smooth surface is retained when the film is subsequently crystallized by annealing. The smooth surface is stabilized by the thin native oxide that grows on the silicon film when it is exposed to air after deposition [2.49] or as it heats in the oxidation furnace. The stabilizing oxide must form before the temperature is high enough to allow significant rearrangement of the silicon atoms on the surface.

Because the surface of the initially amorphous film remains smooth after it crystallizes, the local electric field in subsequently grown oxide approaches the average field. Therefore, the current through oxide grown on initially amorphous films can be many orders of magnitude lower than that through oxide grown on silicon layers deposited in a polycrystalline form, as shown in Fig. 4.13 [4.38]. Similarly, the average electric field at which breakdown occurs is higher for oxide grown on initially amorphous films. The surface remains smooth and the breakdown field remains high even if the silicon film is doped by ion implantation or from a gas-phase source before oxidation [4.46–4.48]. However, these improved characteristics are obtained at the expense of a reduced deposition rate at the lower temperature needed to deposit amorphous films. Polishing a film deposited in polycrystalline form also reduces the surface roughness and improves the breakdown strength [4.49]. Advanced

4.3. CONDUCTION THROUGH OXIDE ON POLYSILICON

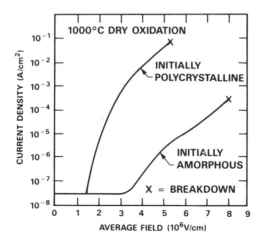

Figure 4.13: Current density through oxide grown on polysilicon films deposited (a) in an initially amorphous form and (b) as polycrystalline silicon, showing the lower conduction and higher breakdown field for the smoother, initially amorphous film. From [4.38]. Reprinted by permission of the publisher, The Electrochemical Society, Inc.

chemical-mechanical polishing (CMP) techniques can reduce the surface roughness to 1 nm or less, but can also reduce the thickness of the polysilicon layer. Polishing can, of course, only be used if the polysilicon is deposited on a plane surface, with steps much smaller than the thickness of the polysilicon layer. Obtaining such a plane surface may require substantial modification of the earlier portions of the IC process.

High-quality oxides on polysilicon are especially important when the polysilicon is used for the channel of a thin-film transistor, as will be discussed in Sec. 6.13. Although thin-film transistors on low-temperature substrates use deposited dielectrics, thin-film transistors formed on single-crystal silicon or fused silica can use thermally grown oxide. The oxide must have low conductivity and be thin to maximize the coupling of the gate voltage to the transistor channel. To achieve large grain size, the silicon layer is often deposited in amorphous form and subsequently crystallized. As discussed above, the native oxide formed when the amorphous film is exposed to air stabilizes the smooth surface of the amorphous film during crystallization. Because of the smooth surface, the oxide subsequently grown has good electrical properties.

The current through the oxide is higher for electron injection from the bottom surface of the oxide than from the top even when the polysilicon is deposited in an initially amorphous form and subsequently crystallized [4.37]. A typical rms surface roughness is about 1.6 nm at the lower interface and only 1.2 nm on the upper surface of a 100 nm thick oxide, and the interface roughness increases with increasing oxide thickness. When this initial oxide is removed, the surface of the underlying silicon film is rougher than the surface before oxidation, and the conduction through a second oxide grown on this roughened surface is greater than that through the initial oxide. Although this increased oxide conduction is not desired in most devices, oxidizing, removing the oxide, and reoxidizing can be used to controllably increase the local field and, therefore, the current through the oxide in devices such as EEPROMs [4.50]. Creating surface roughness in this manner also makes the conduction less sensitive to the polarity of the applied bias.

When surface roughness must be reduced, but the slow deposition of an amorphous silicon layer is not practical, the surface roughness formed by oxidizing a single layer of polysilicon can be reduced by (1) growing a thin intermediate oxide on the polysilicon, (2) depositing a second thinner layer of polysilicon, and (3) oxidizing the entire thickness of this second layer [4.51]. The oxide formed from the second layer of polysilicon joins the thin intermediate oxide grown on the first layer. The grains in the second layer are smaller than those in the first layer and the grain boundaries in the two layers are not aligned. Consequently, the resulting surface is smoother than that of a oxide grown directly on a single layer of polysilicon, although at the expense of more complex processing.

4.3.3 Oxidation conditions

Not only does conduction through oxide grown on polysilicon films depend on the deposition conditions of the polysilicon, but it also depends on the oxidation conditions. The surface roughness, and consequently the local electric fields and the conduction through the oxide, can change significantly during oxidation. For example, as shown in Fig. 4.14 [4.52], oxidation can accentuate the irregular surface of an asperity, leading to increased current flow through the oxide. In general, high-temperature oxidation produces a less-conductive oxide than does low-temperature oxidation [4.34], consistent with the smoother surface obtained after

4.3. CONDUCTION THROUGH OXIDE ON POLYSILICON

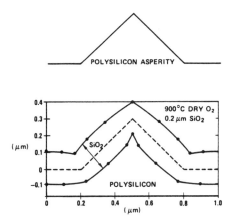

Figure 4.14: The shape of an asperity can change during oxidation, modifying the local electric field. [4.52]

high-temperature oxidation. Oxidizing in an HCl ambient produces oxide with inferior electrical characteristics; this decrease in oxide quality is postulated to be caused by water formation during HCl oxidation [4.10]. On the other hand, adding fluorine after oxidation appears to increase the quality, possibly by allowing the stress in the oxide to relax [4.53].

Because of the rougher surface obtained after oxidizing at lower temperatures, enhanced conduction through oxides grown on polysilicon is becoming of more concern as lower oxidation temperatures are increasingly used in advanced device processing. Growing the oxide at a high temperature for a short time in a rapid thermal processor keeps the "thermal budget" low, while forming a higher-quality oxide [4.54].

4.3.4 Dopant concentration and annealing

The dopant concentration also strongly affects the conduction through the oxide and the breakdown strength. We saw in Sec. 2.11.3 that phosphorus enhances grain growth. As the grain size increases with increasing phosphorus concentration, the surface of the polysilicon also becomes smoother [4.55], and the breakdown field increases (Fig. 4.15) [4.55]. However, above a critical phosphorus concentration, the breakdown strength begins to decrease. In this region, the breakdown field is less sensitive to the polarity of the voltage applied across the oxide, suggesting that the breakdown field is not determined by the surface

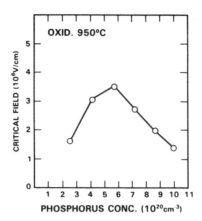

Figure 4.15: Increasing dopant concentrations initially lead to smoother surfaces and higher oxide breakdown fields. At higher concentrations, oxidation of precipitates at grain boundaries degrades the breakdown strength. From [4.55]. Reprinted by permission of the publisher, The Electrochemical Society, Inc.

roughness in this range of dopant concentration. Instead, the degradation in breakdown strength is attributed to precipitation of phosphorus atoms or a silicon-phosphide compound at the grain boundaries when the dopant concentration exceeds its solid solubility [4.7, 4.55], as discussed in Sec. 4.2.3. The quality of the phosphorus-rich oxide subsequently grown on this precipitate is low. Because the grain-boundary regions oxidize more rapidly than the grains, an indentation is left at the grain boundaries when the oxide is removed, increasing the surface roughness and the local fields in subsequently grown oxide. Therefore, after repeated oxidation and removal of the oxide, the degradation of the oxide quality by precipitates increases [4.10, 4.56].

Annealing polysilicon before oxidation causes grain growth and a smoother surface, which increases the breakdown strength in oxide grown afterwards [4.55]. The breakdown field increases with increasing grain size in the dopant-concentration range in which surface roughness controls the breakdown strength [4.55]. Breakdown fields of about 7 MV/cm — approaching that of oxide grown on single-crystal silicon — can be obtained with optimum phosphorus concentration and proper annealing.

Control of the polysilicon doping and annealing conditions is often effectively used to improve the dielectric properties of oxide grown on

4.3. CONDUCTION THROUGH OXIDE ON POLYSILICON

polysilicon. However, smoothing the surface by melting the top of the polysilicon layer has also been demonstrated [4.57–4.59]. Melting the surface with a pulsed laser before oxidation allows the silicon to flow, diminishing the asperities. Conduction through oxide subsequently grown on the smoothed surface decreases, and the breakdown field increases.

4.3.5 Carrier trapping

When a high electric field is applied across the oxide and carriers are injected, not all of these carriers travel through the oxide to the counterelectrode. Some are trapped in the oxide, either at pre-existing traps or at traps which are formed during the current flow. Trapped electrons can terminate some of the electric field lines from a positively biased counterelectrode and reduce the local electric field which caused the charge injection. Consequently, the current through the oxide can decrease with time. Because of the rough polysilicon/SiO_2 interface, the trapped charge is much higher (~ 10 times [4.60]) than that at a single-crystal Si/SiO_2 interface. Carrier injection and trapping are very nonuniform. Because carrier injection is greatest where the local field is highest, the maximum local electric fields are preferentially reduced, and the local fields can approach the average field. The oxide conduction then becomes more like that of oxide grown on single-crystal silicon.

The decrease in current with time resulting from carrier trapping and the subsequent electric-field reduction also depend on the oxidation conditions [4.61]. As discussed above, a higher oxidation temperature leads to a smoother surface and less carrier injection, changing both the initial current flowing through the oxide and (by modifying carrier injection and trapping) its time dependence. The electrical quality of the oxide is also affected by the method of doping the polysilicon. In an oxide/nitride/oxide dielectric, the time-dependent dielectric breakdown of oxide grown on *in situ* doped polysilicon is higher than that of oxide grown on polysilicon doped by diffusion from $POCl_3$ after deposition [4.62].

Nitrogen incorporation at a Si/SiO_2 interface provides a barrier to charge injection from single-crystal silicon into SiO_2. Oxidizing polysilicon in a N_2O ambient also reduced the charge injection, probably both from incorporated nitrogen and from a somewhat smoother interface [4.63]. The reduced roughness at the bottom interface also may change the polarity at which the highest current flows.

Some of the trapped charge can be detrapped by irradiating the sample with ultraviolet light, but the traps remain and are quickly filled when the field is again applied [4.61]. Similarly, annealing at 250°C removes the trapped carriers, but not the traps [4.64]. On the other hand, annealing at 450°C in hydrogen appears to annihilate the traps also [4.61].

Because carrier trapping can change the local electric fields during a measurement, obtaining the current-voltage characteristic is not always straightforward. Carrier trapping can cause the local electric fields to increase less than linearly with applied voltage, and the current flowing during a second measurement can be lower than that during an initial test with the same applied voltage. A floating-gate structure allows the current-voltage relation to be obtained without carrier trapping distorting the results [4.65].

Charge trapping in the oxide separating the two polysilicon layers of a nonvolatile memory is especially detrimental. As electrons are trapped, the electric field near the injecting electrode decreases, reducing the current markedly. Because the current through the polysilicon oxide is used to erase the cell, the useful life of the device is decreased [4.66].

4.3.6 CVD dielectrics

Chemically vapor deposited dielectrics generally do not accentuate the irregularities of the polysilicon surface as does thermal oxidation. They tend to be conformal or even smooth the surface irregularities; they also shield weak spots in underlying dielectric layers. Similarly, good coverage of the corners and edges by CVD dielectrics improves the dielectric properties compared to thermally grown oxide alone; in thermally grown oxide local electric fields are accentuated at edges and corners as they become sharper during oxidation. Therefore, CVD oxide or nitride is often deposited above a thermal oxide when low conductance is needed.

CVD dielectrics can also be used to increase device reliability by reducing time-dependent dielectric breakdown. The high electric fields found in modern ULSI structures lead to concern about long-term device reliability. Charge injection and trapping in oxide grown on polysilicon, as in oxide grown on single-crystal silicon, can increase the local fields, leading to destructive oxide breakdown. The irregular topography found in advanced structures compounds this problem. Trapping near the edges of device features, where the fields are especially high, increases time-dependent dielectric breakdown when thermal oxide is used alone. Improved reliability can be obtained by using a composite $SiO_2/Si_3N_4/SiO_2$ dielectric structure, in which the Si_3N_4 layer is formed by chemical vapor deposition [4.67]. This CVD layer tends to smooth the surface, reducing electron injection from the polysilicon, but an overlying oxide layer is needed to prevent carrier injection from the upper electrode into the nitride.

A similar $SiO_2/Si_3N_4/SiO_2$ combination is frequently used as the dielectric of the charge-storage capacitor in dynamic, random-access memories. This triple-layer dielectric reduces conduction through the dielectric compared to a single-layer oxide with the same thickness. Equally importantly, the higher relative permittivity of nitride ($\epsilon_r \approx 7$) increases the capacitance and stored charge per unit area, reducing the area needed to store a given amount of charge.

4.4 Summary

In this chapter we looked at the thermal oxidation of polysilicon, considering both growth of the oxide and conduction through it. We saw that the oxidation rate of undoped films depends on the exposed surface planes. Dopant-enhanced oxidation occurs for high concentrations of the n-type dopants, but dopant diffusion away from the oxidizing surface can cause the oxidation rate to differ from that of similarly doped, single-crystal silicon. Oxidation of dopant precipitated at grain boundaries can form low-quality oxides and cause local thinning of polysilicon. Two-dimensional oxidation can increase the irregularity of patterned polysilicon device features and complicate subsequent processing.

Conduction through oxide grown on polysilicon is greater than that through oxide grown on single-crystal silicon because the rough surface increases the local electric fields. This enhanced conduction depends sensitively on the structure of the polysilicon and is influenced by the deposition conditions, the dopant concentration, and the annealing treatments. The oxide conduction needed for EEPROMs, especially, requires careful control of the polysilicon deposition and processing for reproducible device performance. The surface is markedly smoother for films deposited in an amorphous form, and the surface remains smoother even after the films are crystallized and doped. The conduction in oxides grown on these smooth silicon surfaces is consequently lower, and the breakdown field is higher.

Chapter 5

Electrical Properties

5.1 Introduction

Polycrystalline-silicon films formed by chemical vapor deposition are used in a wide variety of ULSI applications requiring very different electrical properties. High-value load resistors for static random-access-memory (RAM) cells utilize the high resistance of lightly doped polysilicon to provide a convenient and stable resistor that limits the current flowing in the cell. At the other extreme, the excellent technological compatibility of polysilicon with high-temperature, integrated-circuit processing allows straightforward fabrication of self-aligned gates and convenient interconnections in ULSI circuits. Although a resistivity of less than about 10^{-3} Ω-cm — eight orders of magnitude less than for static RAM load resistors — is routinely achieved, the lower bound on the resistivity of polysilicon can limit the performance of silicon-gate integrated circuits that use polysilicon interconnections to conduct signals long distances across a chip [5.1]. As feature sizes become smaller and intrinsic transistor delays decrease on chips of increasing overall dimensions, the resistance of polysilicon interconnections is becoming a more serious limitation on integrated-circuit performance.

In this chapter we will examine the wide range of electrical properties that can be achieved in CVD polysilicon films and show that the high resistance of lightly and moderately doped polysilicon is related to the grain boundaries. Methods of modifying the electrical properties by controlling the number of grain boundaries or their electrical activity will be examined. For highly doped films, we will see that the resistivity is limited by the solid solubility of the dopant; *ie,* by the amount of dopant that can be incorporated in substitutional sites within the grains.

5.2 Undoped polysilicon

The range of deposition parameters accessible in a low-pressure reactor is limited, and the resistivity can be studied over a much wider range of deposition temperatures in films deposited in cold-wall reactors. For thin polysilicon films deposited in a polycrystalline form over the temperature range from 650 to 1050°C in an atmospheric-pressure, cold-wall reactor, the resistivity parallel to the surface is in the mid-10^5 Ω-cm range and is not sensitive to the deposition temperature [5.2]. The value of mid-10^5 Ω-cm is close to the resistivity of intrinsic silicon.

Although the crystal structure varies widely over the range in which polysilicon films are deposited, the resistivity of undoped films is quite insensitive to the structure. For example, the grain size increases with increasing thickness, but the resistivity varies only slightly with film thickness, changing from about 5×10^5 Ω-cm for films 0.5 μm thick to about 3×10^5 Ω-cm for films 25 μm thick. Similar values of resistivity are also observed in layers deposited in different types of reactors.

Undoped films deposited in the amorphous form at lower temperatures and subsequently crystallized by annealing at a higher temperature have a resistivity about 2 or 3 times higher than films deposited in the polycrystalline form. In films deposited at a higher rate, the discontinuity in resistivity occurs at a higher temperature, consistent with the increased transition temperature from an amorphous deposition to a polycrystalline deposition at a higher deposition rate.

At higher temperatures, epitaxial films can be formed on single-crystal wafers at the same time that polycrystalline films are deposited on oxide-covered wafers. Measurement of the conductivity type (using a *hot-point-probe*) shows that the single-crystal films are n-type while the polycrystalline films are p-type. The n-type behavior of single-crystal samples is attributed to impurities in the silane used to deposit the films. The different conductivity type of the polycrystalline films implies that structural defects influence the resistivity enough to compensate for the small amount of n-type dopant in the silane.

In thick films, the resistivity can be measured perpendicular to the film plane, as well as parallel to it, and is somewhat lower in the per-

5.2. UNDOPED POLYSILICON

pendicular direction. In films about 13 μm thick, the resistivity is about 3×10^5 Ω-cm in the film plane, but only 1.5×10^5 Ω-cm in the perpendicular direction. This anisotropy is probably related to the columnar crystal structure normal to the film plane (Fig. 2.10) [3.12]. Similar anisotropic conductivity is seen in thicker samples with a columnar grain structure deposited from a silicon halide at a high temperature (1100°C) [2.57]; the resistivity is 5×10^5 Ω-cm parallel to the film plane, but only 5×10^4 Ω-cm perpendicular to it.

For moderate electric fields, the current flow is ohmic, although nonlinear current-voltage behavior is observed for fields above 10^4 V/cm in the vertical samples, in which high fields can be readily applied. Measurements with different electrode materials and sample geometries suggest that the nonlinearity is not dominated by the electrodes.

Over the temperature range from −25 to 200°C, the conductivity of undoped films varies with the measurement temperature T according to the relation

$$\sigma = \sigma_0 \exp\left(-\frac{E_a}{kT}\right) \tag{5.1}$$

with an apparent activation energy E_a above room temperature of approximately half the bandgap of silicon. However, a weaker temperature dependence is sometimes observed below room temperature, indicating a different conductivity-limiting mechanism at low temperatures. The conduction may arise from hopping between localized states [5.3] within the grain boundaries [2.34, 5.4, 5.5], with a temperature dependence of the conductivity of the form

$$\sigma\sqrt{T} = \sigma_0 \exp\left[-\left(\frac{T_0}{T}\right)^{1/4}\right] \tag{5.2}$$

Irreproducibly low values of resistivity are sometimes observed on films with unprotected surfaces, probably because of surface conduction from adsorbed moisture or contamination. Thin films deposited in the amorphous form and subsequently crystallized are less likely to exhibit anomalously low resistivity, even without surface protection. The resistivity of all the films is stable if a thermally grown oxide is used to passivate the surface.

5.3 Amorphous silicon

The electrical properties of amorphous silicon that are not crystallized by subsequent heat treatment can be very different from those of polysilicon because of their greater structural disorder. In polysilicon, the basic conduction process within the grains is similar to that in crystalline silicon, with conduction of electrons and holes in well-defined conduction and valence bands. At all energies above the lower edge of the conduction band, electrons can readily move through the crystal. The properties of polysilicon differ from those of single-crystal silicon primarily because of the grain boundaries.

In amorphous silicon, on the other hand, the basic crystalline structure is absent, and the concept of a band structure becomes somewhat tenuous. Instead of a well-defined bandgap, there is a range of energies with few allowed states. The edges of this region are not sharply separated from regions with allowed states in which the carriers can readily move through the material. Instead, near the edges of the "bandgap," a high density of *localized* states is present. Carriers in these states cannot readily move through the material, and they do not contribute significantly to the conduction. At higher energies, carriers can move through the material, but the absence of a well-defined crystal structure greatly reduces the ability of carriers to move (*ie,* reduces their mobility).

In amorphous material the energy corresponding to the lower edge of the conduction band in a crystalline material is called the *lowest unoccupied molecular orbital* (LUMO), and the energy corresponding to the upper edge of the valence band is called the *highest occupied molecular orbital* (HOMO). Although the structure of amorphous material is very different from that of crystalline material, applying the familiar concepts of bandgap and mobility to amorphous material is useful.

The effective bandgap of amorphous silicon can be much greater than that of polycrystalline silicon; values near 1.8 eV are often found, although the value depends strongly on the deposition conditions, the bonding, and the amount of hydrogen incorporated. As discused in Sec. 2.11.4, the amorphous silicon used in very low-temperature processes can be deposited in the 300°C temperature range by plasma-enhanced CVD and contains a large amount of hydrogen. The relative amounts of hydrogen bound as Si-H and Si-H_2 can affect the electrical properties, as well as the stability of the material.

5.4. MODERATELY DOPED POLYSILICON

Conduction occurs primarily by hopping between localized states, with the temperature dependence given by Eq. 5.2, rather than by conduction in well-defined bands. The "bandgap" is important because it influences carrier generation and, therefore, the leakage current and the energy dependence of the photosensitivity. Because the bandgap is greater in amorphous silicon than in crystalline silicon, the leakage current may be lower because fewer free carriers are generated; however, the larger bandgap may lead to a greater temperature dependence of the leakage current than in polysilicon with its smaller bandgap.

5.4 Moderately doped polysilicon

For static RAMs and other devices in which high resistivity is needed, the magnitude of the resistivity and its control are important. At low dopant concentrations, the resistivity changes only slowly with increasing dopant concentration. The measured resistivity remains in the mid-10^5 Ω-cm range and can be as much as six orders of magnitude greater than those of similarly doped, single-crystal silicon, as shown in Fig. 5.1. At intermediate dopant concentrations, the resistivity decreases rapidly and approaches that of single-crystal silicon. Control of the resistance of a high-value resistor is best in lightly doped material; achieving accurate values of resistance in the intermediate dopant-concentration range is difficult because of the rapid variation of resistance with dopant concentration. The large difference between polysilicon and single-crystal silicon is observed in material deposited over a wide range of deposition conditions, with deposition temperatures varying from less than 600°C to more than 1100°C. Similar behavior is seen when the dopant is added during deposition as when it is added after deposition by ion implantation or from a gas-phase dopant source.

In this section, we will examine the causes of the large difference between the resistivity of polysilicon and that of single-crystal silicon, and in a subsequent section we will discuss techniques to achieve better control with intermediate dopant concentrations. Two mechanisms — both involving grain boundaries — can cause the electrical behavior of polysilicon to depart from that of single-crystal silicon at low and moderate dopant concentrations. In Sec. 3.6 we saw that n-type dopant atoms physically segregate at grain boundaries because of their lower energy in the disordered, grain-boundary region [3.23, 5.6] and, therefore, do not contribute to the conduction process. In this section, we

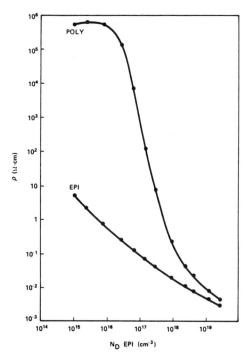

Figure 5.1: The resistivity of polysilicon is much greater than that of similarly doped, single-crystal, epitaxial silicon. It changes slowly at low dopant concentrations, but decreases rapidly at intermediate dopant concentrations. At high dopant concentrations it approaches the resistivity of single-crystal silicon. (Large-grain, ~15 μm-thick films deposited in a cold-wall reactor at ~1000°C.)

will see that carriers contributed by substitutional dopant atoms can be trapped in deep energy levels at the grain boundaries [5.7–5.11], again reducing the number of carriers available for conduction.

5.4.1 Carrier trapping at grain boundaries

To determine the number of free carriers in polysilicon films, the substitutional dopant concentration must first be found by considering the loss of dopant atoms that segregate to the grain boundaries and are electrically inactive. Even after this loss is considered, however, the conductivity of polysilicon is still much less than that of single-crystal silicon containing the same concentration of substitutional dopant atoms. Free

5.4. MODERATELY DOPED POLYSILICON

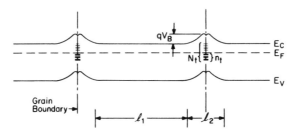

Figure 5.2: Many of the free carriers contributed by substitutional dopant atoms in the grains are quickly immobilized at the grain boundaries at traps of density N_T per unit area with energies within the bandgap.

carriers (either electrons or holes) are contributed to the conduction or valence band by substitutional dopant atoms located within the grains, just as in single-crystal silicon. However, many of these free carriers are quickly trapped at low-energy positions at the grain boundaries, as schematically shown in Fig. 5.2 [5.7] and, consequently, cannot contribute to the conduction [5.7, 5.8].

Dangling bonds at the grain boundaries and defects within the grains both lead to allowed states within the bandgap of the polysilicon (Fig. 5.3). The *deep states* associated with the dangling bonds have energies near the middle of the forbidden energy gap [5.9], where they are most effective in allowing generation and recombination of carriers. (The maximum density is slightly below mid-gap, as discussed below.) In addition to states near mid-gap resulting from broken bonds, strained bonds cause a high density of shallow *tail* states near the band edges.

In our discussion of carrier trapping, barrier formation, and conduction, we will consider n-type silicon. However, unlike dopant segregation, carrier trapping is similar in n-type and p-type polysilicon; barriers form in polysilicon of both conductivity types. (On the other hand, in germanium the grain boundaries are strongly p-type so that barriers form only in n-type material.) Some minor differences in the electrical behavior of n-type and p-type polysilicon will be considered later.

Potential barriers

The charge trapped at the grain boundaries is compensated by oppositely charged depletion regions surrounding the grain boundaries

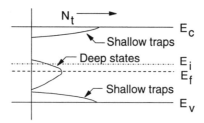

Figure 5.3: Traps are primarily located at grain boundaries, although defects within the grains also create some states. The shallow *tail* states are associated with strained bonds, and the deep states near mid-gap are caused by broken bonds.

(Fig. 5.4b) [5.10]. From Poisson's equation, the charge in the depletion regions causes curvature in the energy bands, leading to potential barriers that impede the movement of any remaining free majority carriers from one grain to another [5.7, 5.8]. The barrier height V_B (Fig. 5.4c) can be expressed in terms of the dopant concentration N and the width of the depletion region x_d using Poisson's equation:

$$\frac{d^2V}{dx^2} = \frac{qN}{\epsilon} \qquad (5.3)$$

where q is the magnitude of the charge on an electron, ϵ is the permittivity of silicon, and a one-dimensional case is considered. Solving for V_B, we find

$$V_B = \frac{qN}{2\epsilon} x_d^2 \qquad (5.4)$$

The barrier height depends strongly on the substitutional dopant concentration and the trap density and energy. Barrier formation in both n-type and p-type polysilicon indicates that both electron and hole traps exist at the grain boundaries. At low dopant concentrations the total number of carriers NL per unit area in a grain of length L is much less than the number of traps N_T per unit area at the grain boundary. If the energy levels of the traps are deep enough, virtually all the free carriers contributed to the conduction band by the substitutional dopant atoms are trapped at grain boundaries, and very few are free to contribute to conduction. Because the dopant concentration is low, the depletion regions surrounding the grain boundaries extend through the entire grain, and no neutral region remains. The low dopant concentration also means that the energy bands within the grain have little

5.4. MODERATELY DOPED POLYSILICON

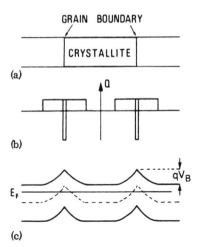

Figure 5.4: Depletion regions form in the grains surrounding the grain boundaries to compensate the charge trapped at the grain boundaries (b). The charged depletion regions cause curvature in the energy bands (c) and the resulting potential barriers. From [5.10]. Reprinted with permission.

curvature (Fig. 5.5a); the barriers to conduction are small, and any free carriers that are present can readily move from one grain to another. The small barrier and low carrier density are consistent with an observed resistivity of lightly doped polysilicon close to that of intrinsic silicon.

When the grains are completely depleted, the depletion region associated with a grain boundary extends a distance $L/2$ on each side of the boundary, and the barrier height can be expressed as

$$V_B = \frac{qN}{2\epsilon}\left(\frac{L}{2}\right)^2 = \frac{qNL^2}{8\epsilon} \tag{5.5}$$

As the dopant concentration increases, more carriers are trapped at the grain boundary; the curvature of the energy bands and the height of the potential barrier both increase (Fig. 5.5b), making carrier transport from one grain to another more difficult.

In the simplest version of the carrier-trapping model, the energy of the grain-boundary traps is assumed to be deep enough that they are completely filled when the dopant concentration exceeds a critical value $N^* = N_T/L$ [5.8, 5.10]. As the concentration of dopant atoms

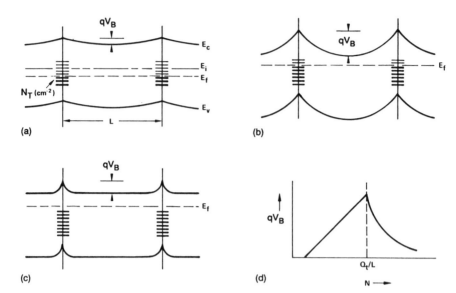

Figure 5.5: The barrier height V_B increases with increasing dopant concentration until all the traps are filled. Above the critical dopant concentration $N^* = N_T/L$, it decreases as neutral regions form within the grains.

increases above N^*, the number of trapped carriers per unit area of grain boundary remains constant at the value N_T. The added carriers, which are not trapped, form neutral regions within the grains, as seen in Fig. 5.5c; and, from charge neutrality, the width of the depletion regions decreases according to the relation

$$x_d = \frac{N_T}{2N} \tag{5.6}$$

The barrier height is then found from Eq. 5.4 to be

$$V_B = \frac{qN}{2\epsilon}\left(\frac{N_T}{2N}\right)^2 = \frac{qN_T^2}{8\epsilon N} \tag{5.7}$$

and the barrier to carrier transport between grains decreases as the dopant concentration increases above its critical value N^*. Thus, as shown in Fig. 5.5d, when dopant atoms are added to polysilicon, the potential barrier first increases, reaches a maximum, and then decreases. This nonmonotonic behavior has important consequences for the electrical conduction and can explain the markedly different electrical properties of polysilicon and single-crystal silicon seen in Fig. 5.1.

5.4.2 Carrier transport

At low dopant concentrations, the width of the depletion region on each side of the grain boundary is half the grain size. At moderate dopant concentrations, it decreases with increasing dopant concentration, but even at a dopant concentration of 10^{17} cm^{-3}, it is several tens of nanometers wide (for a typical grain-boundary trap density $\sim 10^{12}$ cm^{-2}). Tunneling is not significant for barriers of this width, and carriers usually travel from one grain to another by thermionic emission over the barrier. At high dopant concentrations the width of the depletion regions can be less than about 10 nm, and tunneling through the barrier can also contribute to the current flow. However, the barrier height decreases with increasing dopant concentration, so conduction across the barrier may no longer limit the current flow.

When thermionic emission dominates, the conduction process can be understood by considering the effect of an applied field on the barrier height. The thermionic-emission current density J can be written as

$$J = qnv_c \exp\left[-\frac{q}{kT}(V_B - V)\right] \qquad (5.8)$$

where n is the free-carrier density, v_c is the collection velocity ($v_c = \sqrt{kT/2\pi m^*}$), V_B is the barrier height with no applied bias, and V is the applied bias across the depletion region. With no applied bias, the barriers to carrier transport in the forward and reverse directions are equal. The forward current J_F of electrons emitted over the barrier by thermionic emission equals the reverse current J_R in the opposite direction, and no net current flows.

When a voltage is applied, the barrier to carrier transport in one direction decreases, while it increases in the opposite direction. In general, the applied voltage divides nonuniformly between the two sides of a grain boundary [5.12]; however, for small applied biases, approximately half the applied voltage appears across each depletion region. The barrier in the forward direction decreases by an amount $V \approx \frac{1}{2}V_G$ where V_G is the bias across one grain boundary (ie, the applied bias across the sample divided by the number of grains, assuming all grain boundaries are equivalent). The barrier increases by the same amount in the reverse direction [5.12].

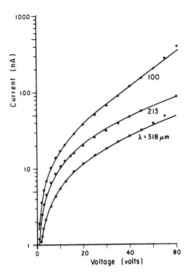

Figure 5.6: Current-voltage characteristics for 3 μm thick, polysilicon films doped with about 2×10^{16} cm^{-3} phosphorus atoms; the samples have lengths λ of 100, 213, and 318 μm, as indicated. From [5.13]. ©1978, Pergamon Journals, Ltd. Reprinted with permission.

By considering thermionic emission in the two directions, the net current density $J = J_F - J_R$ can be found [5.13]. With $V \approx \frac{1}{2}V_G$

$$J_F = qnv_c \exp\left[-\frac{q}{kT}\left(V_B - \frac{1}{2}V_G\right)\right] \qquad (5.9)$$

and

$$J_R = qnv_c \exp\left[-\frac{q}{kT}\left(V_B + \frac{1}{2}V_G\right)\right] \qquad (5.10)$$

The net current density is then given by

$$J = qnv_c \exp\left(-\frac{qV_B}{kT}\right)\left[\exp\left(\frac{qV_G}{2kT}\right) - \exp\left(-\frac{qV_G}{2kT}\right)\right] \qquad (5.11)$$

or

$$J = 2qnv_c \exp\left(-\frac{qV_B}{kT}\right) \sinh\left(\frac{qV_G}{2kT}\right) \qquad (5.12)$$

Current flow in polysilicon films fits this expression well, as shown in Fig. 5.6 [5.13].

5.4. MODERATELY DOPED POLYSILICON

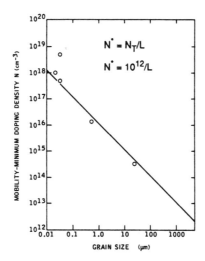

Figure 5.7: The critical dopant concentration N^* varies inversely with the grain size. From [5.14]. Reprinted with permission.

At low applied voltages the voltage drop across each grain boundary is small compared to the thermal voltage kT/q, and the general expression of Eq. 5.12 can be simplified by using the relation

$$\sinh\left(\frac{qV_G}{2kT}\right) \approx \frac{qV_G}{2kT} \tag{5.13}$$

to obtain a linear relation between current and applied voltage:

$$J = \frac{q^2 n v_c}{kT}\left[\exp\left(-\frac{qV_B}{kT}\right)\right] V_G \tag{5.14}$$

From Eq. 5.14 an expression for the average conductivity $\sigma = J/\mathcal{E} = JL/V_G$ can be obtained:

$$\sigma = \frac{q^2 n v_c L}{kT} \exp\left(-\frac{qV_B}{kT}\right) \tag{5.15}$$

Thus, conduction in polysilicon is an activated process with an activation energy of approximately qV_B, which depends on the dopant concentration and the grain size.

From this simple model we see the importance of the critical dopant concentration N^*, at which all the grain-boundary traps are filled with

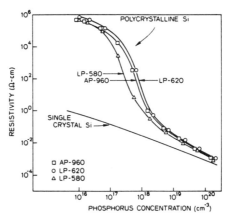

Figure 5.8: The resistivity of phosphorus-implanted polysilicon films is lower for films with larger grains, especially at intermediate dopant concentrations.

carriers from the substitutional dopant atoms. Below N^*, the free-carrier concentration is low, and the resistivity is high. As dopant is added, the increasing free-carrier concentration is compensated by the increasing barrier height, and the resistivity remains high. At dopant concentrations greater than N^*, the free-carrier concentration increases and the barrier height decreases, allowing the resistivity to decrease rapidly with increasing dopant concentration. Thus N^* divides the high resistivity region from the region with rapidly decreasing resistivity. From the definition of N^* ($N^* = N_T/L$), we expect it to vary inversely with grain size L. The inverse proportionality between critical dopant concentration and grain size has been observed for grain sizes varying from 20 nm to over 20 μm (Fig. 5.7) [5.14].

For the typical submicrometer-thick polysilicon films used in most integrated-circuit applications, the grain size is usually somewhat less than 100 nm. Using this value with a trap density somewhat greater than 10^{12} cm^{-2} of grain-boundary area, we find the critical dopant concentration to be of the order of 10^{17}-10^{18} cm^{-3}, as observed for thin polysilicon films deposited over a wide range of conditions (Fig. 5.8) [5.15]. For films deposited at a high temperature in a cold-wall reactor, the critical dopant concentration is slightly lower than for polysilicon films deposited at a much lower temperature in the low-pressure reactor. The films deposited at the higher temperature have a slightly larger grain size and, consequently, a lower critical dopant concentration. The

5.4. MODERATELY DOPED POLYSILICON

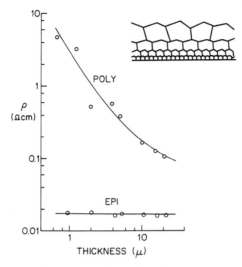

Figure 5.9: The *resistivity*, as well as the sheet resistance, of polysilicon films decreases as the thickness increases.

critical dopant concentration for low-pressure films deposited in an initially amorphous form and subsequently crystallized is appreciably lower than for either of the other two types of polysilicon, consistent with its larger grain size, as discussed in Sec. 2.11.4. The resistivity is most sensitive to the grain size in the intermediate-concentration range. It does not vary significantly with grain size for lightly or heavily doped films.

The effect of grain size is also seen by considering the variation of resistivity with increasing film thickness. The grain size increases as the film thickness increases [2.2], and the *resistivity* decreases, as seen in Fig. 5.9 [5.7]. The resistivity of single-crystal films deposited at the same time is independent of film thickness, as expected. Because the resistivity of the polysilicon films decreases with increasing thickness t, the sheet resistance decreases more rapidly than $1/t$.

Effective mobility

To express the conductivity in Eq. 5.15 in the usual linear form of Ohm's law, $\sigma = qn\mu$, an *effective mobility* μ_{eff} can be defined as

$$\mu_{\text{eff}} = \frac{qv_c L}{kT} \exp\left(-\frac{qV_B}{kT}\right) \quad (5.16)$$

The quantity μ_{eff} is a mobility in the sense that it describes the ease of carrier movement from one grain to another in the polycrystalline material. It is not the familiar microscopic mobility related to carrier scattering in a homogeneous piece of semiconductor. Instead, it describes the restriction of current flow by the grain-boundary barrier.

Because of its strong dependence on barrier height, the mobility depends sensitively on dopant concentration. As dopant atoms are added to undoped polysilicon, the barrier height increases (Fig. 5.5d), and the mobility decreases until the dopant concentration reaches its critical value N^*. Above this value the barrier height decreases, and the mobility increases. The predicted minimum in the mobility with increasing dopant concentration is observed, as shown in Fig. 5.10 [5.8]. (Note that Fig. 5.10 only shows the dependence of the mobility on dopant concentration over a limited range. At even higher dopant concentrations, grain-boundary barriers no longer limit the mobility, and ionized impurity scattering dominates, as in single-crystal silicon. In this region the mobility decreases with increasing dopant concentration, paralleling the behavior in single-crystal silicon; however, even at high dopant concentrations, the mobility is about a factor of two less than that in single-crystal silicon, as will be discussed in Sec. 5.6.)

Hall mobility

In the discussion above, we developed an expression for the effective mobility, which describes the ease with which a carrier moves through a material composed of grains and high-resistance grain boundaries. This macroscopic quantity differs fundamentally from the microscopic mobility characterizing carrier-scattering mechanisms. Although the *effective mobility* has been compared with the measured *Hall mobility*, the theoretical relation between these two quantities is not obvious.

In an inhomogeneous film both the grains and the space charge regions surrounding the grain boundaries contribute to the Hall voltage [5.16–5.20]. Detailed analysis of such an inhomogeneous structure shows that the carrier concentration corresponding to the measured Hall constant is always less than the carrier concentration in the grains [5.19]. The discrepancy is greatest when the barrier is high or the depletion regions occupy a large portion of the grain. At very high carrier concentrations, the barriers are small, and the measured Hall concentration approaches the actual carrier concentration. Similarly, in very lightly

5.4. MODERATELY DOPED POLYSILICON

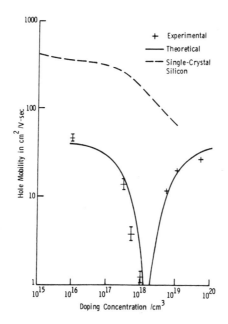

Figure 5.10: The room-temperature, Hall hole mobility as a function of boron concentration measured in approximately 1 μm thick polysilicon films. From [5.8]. Reprinted with permission.

doped material, the entire grain is depleted, and the barriers are again small so that the Hall concentration approaches the average carrier concentration. With certain approximations, the Hall mobility is related to the barrier height by the expression [5.20]

$$\mu_H \sim \exp\left(-\frac{qV_B}{kT}\right) \tag{5.17}$$

which has the same dependence on the barrier height as the effective mobility (Eq. 5.16) used in the expression for the conductivity.

Temperature dependence

From the expression for the low-field conductivity in Eq. 5.15, we can find the temperature dependence of the resistivity

$$\rho = \frac{kT}{q^2 n v_c L} \exp\left(\frac{qV_B}{kT}\right) \tag{5.18}$$

for dopant concentrations both below and above the critical dopant concentration N^*. At low dopant concentrations $NL < N_T$, and the grains are completely depleted. The maximum carrier concentration is then

$$n = n_i \exp\left(\frac{qV_B + E_F}{kT}\right) = N_c \exp\left(\frac{qV_B + E_F - \frac{1}{2}E_G}{kT}\right) \quad (5.19)$$

where n_i is the intrinsic carrier concentration, E_F is the Fermi level measured from the intrinsic Fermi level at the grain boundary, and N_c is the effective density of states at the conduction band edge. Using this expression in Eq. 5.18, we find the resistivity to be

$$\rho = \frac{kT}{q^2 N_c v_c L} \exp\left(\frac{\frac{1}{2}E_G - E_F}{kT}\right) \quad (5.20)$$

Assuming that the traps pin the Fermi level at the grain boundary near mid-gap, $E_F \approx E_i \equiv 0$, and the dominant temperature dependence of the resistivity has the form

$$\rho \propto \exp\left(\frac{E_G}{2kT}\right) \quad (5.21)$$

Thus, the resistivity exhibits an activated behavior, with an apparent activation energy of approximately half the bandgap.

For dopant concentrations above the critical dopant concentration, $NL > N_T$, and neutral regions exist within the grains. The carrier concentration n is approximately equal to the dopant concentration N, and the resistivity can be written

$$\rho = \frac{kT}{q^2 N v_c L} \exp\left(\frac{qV_B}{kT}\right) \quad (5.22)$$

In this region

$$V_B = \frac{qN_T^2}{8\epsilon N} \quad (5.23)$$

(Eq. 5.7), and the resistivity is proportional to

$$\rho \propto \frac{1}{N} \exp\left(\frac{q^2 N_T^2}{8\epsilon kT N}\right) \quad (5.24)$$

The resistivity is again an activated process; however, in this region the apparent activation energy E_a is no longer constant, but varies inversely with the dopant concentration N:

$$E_a = \frac{q^2 N_T^2}{8\epsilon N} \quad (5.25)$$

5.4. MODERATELY DOPED POLYSILICON

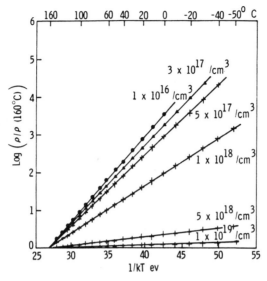

Figure 5.11: Logarithm of the normalized resistivity as a function of reciprocal temperature, showing a well-defined activation energy. From [5.8]. Reprinted with permission.

From Eqs. 5.21 and 5.25, we expect that the activation energy is about half the bandgap energy for undoped polysilicon, and remains at approximately this value for dopant concentrations below the critical dopant concentration. Above the critical dopant concentration, the resistivity and activation energy decrease until other conduction mechanisms limit the conduction process at very high dopant concentrations. Consistent with the model, the resistivity does, indeed, exhibit a well-defined activation energy which decreases with increasing dopant concentration, as shown in Fig. 5.11 [5.8]. (More detailed analysis [5.21] shows that the activation energy obtained from an Arrhenius plot differs from the barrier height because of the shift of the Fermi level with temperature. Calculations show that $E_a \approx qV_{B0} = qV_B(1 + \gamma T)$ with $\gamma \approx 1.5 \times 10^{-3}/\text{K}$ [6.114].) The rapid variation of resistivity with temperature in lightly doped films makes using such films for precise resistors difficult.

5.4.3 Trap concentration and energy distribution

The similar behavior of electrons and holes implies that either type of majority carrier can be trapped at the grain boundaries and that

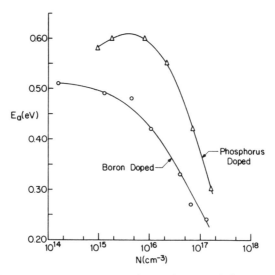

Figure 5.12: Dopant-concentration dependence of the activation energy of the conductivity in thick polysilicon films deposited in a cold-wall reactor.

the grain-boundary traps are located near mid-gap or are distributed symmetrically around mid-gap. More detailed observation reveals some subtle differences. When the p-type dopant boron is added, both the resistivity and the activation energy of the resistivity decrease monotonically with increasing dopant concentration, as shown in Fig. 5.12 [2.4]. However, for the common n-type dopants, the resistivity and its activation energy initially increase as dopant is added, before reaching a maximum and then decreasing [2.4, 5.22]. This asymmetry suggests that the dominant traps are located somewhat below mid-gap, pinning the Fermi level there in lightly doped polysilicon and is consistent with the conductivity-type measurements discussed in Sec. 5.2, which show that lightly doped polysilicon is slightly p-type even if small amounts of n-type dopant are added.

Other studies also indicate that the trap density is highest near mid-gap [5.23]; the Fermi level at the grain boundaries appears to be located slightly below mid-gap at about 0.62 eV below the conduction band edge [5.24]. This energy level is consistent with an electronic state related to dangling silicon bonds about 0.65 eV below the conduction band edge [5.25]. (Dangling bonds are related to trivalent silicon, which has been found at grain boundaries in polysilicon [5.26] and at individual

5.4. MODERATELY DOPED POLYSILICON

grain boundaries in silicon bicrystals.) In contrast to the behavior in polycrystalline silicon, the dominant traps in lightly doped, plasma-enhanced CVD amorphous silicon formed by decomposition of silane in a glow discharge appear to be located slightly above mid-gap [5.27]; undoped films are n-type, rather than p-type, and the resistivity and its activation energy initially increase as p-type dopant is added, before reaching a maximum and then decreasing.

From the measured electrical behavior and the carrier-trapping model, a typical grain-boundary density of traps is found to be about mid-10^{12} cm^{-2} of grain-boundary area. As the grain size increases, the number of traps per unit volume decreases. We saw in Sec. 2.11.4 that silicon layers deposited in amorphous form and subsequently crystallized can have larger grains than layers deposited in polycrystalline form. The electrical properties reflect the larger grains formed by crystallization and also the increase in grain size with increasing film thickness [5.28]. The resistivity is lower, the mobility is higher, and the fraction of active dopant is also higher.

An independent measure of the the density of grain-boundary states can be obtained by electron spin resonance measurements. The density is lower in films with a greater degree of preferred crystalline orientation [2.34], possibly because of the improved structure of grain boundaries between grains with similar orientation perpendicular to the film plane. As the grain-boundary tilt angle increases, the structure of the grain boundary becomes more complex, and the conduction becomes increasingly nonlinear [5.29].

The simplest carrier-trapping model suggests that a discrete level exists at the grain boundary and that the Fermi level is pinned close to this level until all the carrier traps there are filled as more dopant is added. These assumptions lead to a well-defined maximum in the barrier height as the dopant concentration increases. These approximations are appropriate for a moderate density of traps located near mid-gap. However, if the trap density is high or the traps are located far from mid-gap, the traps may be only partially filled when a neutral region begins to form. The bending of the energy bands may raise the trap level at the grain boundary above the Fermi level so that carriers from additional dopant atoms create a neutral region, with some of the trap levels remaining unfilled.

A further complication arises because of the varying properties of different grain boundaries: The traps can be distributed in energy within the bandgap, allowing the number of trapped carriers to continue increasing as the Fermi level moves through the trap distribution. In this case, the dopant concentration at which the traps are completely filled need not coincide with the formation of a neutral region within the interior of the grain [5.10]. A neutral region can form within a grain even though the traps are not completely filled, and there may be a range of dopant concentrations over which the barrier height varies only weakly with dopant concentration.

The basic model can be extended to consider both high trap concentrations, trap energies away from mid-gap, and traps distributed in energy. Of these more complex cases, we first consider the effect of increasing the density N_T of traps per unit area located at a discrete energy E_T within the band gap at the grain boundary [5.10]. In this case, the probability of a trap being occupied arises from the requirement that the combination of the grain boundary and compensating depletion regions must be neutral:

$$2Nx_d = N_T \left[1 + \frac{1}{2}\exp\left(\frac{E_T + qV_B - E_F}{kT}\right)\right]^{-1} \qquad (5.26)$$

The critical dopant concentration N^* is then found by letting $x_d = \frac{1}{2}L$:

$$N^* = \frac{8\epsilon}{q^2 L^2}\left[E_F - E_T + kT\ln\left(\frac{2N_T}{N^*L} - 2\right)\right] \qquad (5.27)$$

which can be solved iteratively. At low trap concentrations, all the traps are filled before any neutral region forms, and $N^* \propto L^{-1}$, as we found before. However, for higher trap concentrations, a neutral region forms before all the traps are filled, and $N^* \propto L^{-2}$.

For $N < N^*$, the corresponding barrier height is

$$qV_B = \frac{q^2 N L^2}{8\epsilon} \qquad (5.28)$$

as before. For $N > N^*$, however,

$$qV_B = E_F - E_T + kT\ln\left[\frac{qN_T}{(2q\epsilon NV_B)^{1/2}} - 2\right] \qquad (5.29)$$

Figure 5.13 [5.10] shows the barrier height for a grain size of 100 nm, typical of polysilicon films used in integrated circuits. For low trap densities, the barrier height reaches a maximum at the critical dopant

5.4. MODERATELY DOPED POLYSILICON

concentration and then decreases, as we found earlier. However, for higher trap concentrations, the neutral region begins to form before all the traps are filled. As additional dopant atoms are added, the corresponding carriers fill the remaining unoccupied traps. Because a neutral region exists for dopant concentrations greater than N^*, the intrinsic Fermi level in the center of the grain remains fixed in relation to the Fermi level. At the grain boundary, the Fermi level is pinned near the discrete mid-gap trap level, and the barrier height remains nearly constant. Not until all the traps are filled can the Fermi level move above the trap level to decrease the barrier height. In the intermediate region, the barrier height is approximately [5.10]

$$qV_B \approx \frac{1}{2}E_G - E_T + kT \ln\left[\frac{qN^{1/2}N_T}{N_c(2q\epsilon V_B)^{1/2}}\right] \quad (5.30)$$

which varies only slowly with dopant concentration.

If the grain-boundary states are distributed in energy, rather than being located at a discrete energy, the behavior is more complex. For a trap density D_T per unit area per unit energy (cm^{-2} eV^{-1}) distributed uniformly in energy, the charge-neutrality condition can be written [5.10]

$$2Nx_d = (E_F - qV_B)D_T \quad (5.31)$$

The critical dopant concentration for this case again varies as L^{-1} for small grains and low trap densities, while it changes as L^{-2} for large grains and high trap densities. Another analysis considered a Gaussian energy distribution of trap states at the grain boundary [5.30]. Detailed comparison of experimental data with models which assume either discrete or uniformly distributed trap levels indicates that the trap distribution appears to be strongly peaked near mid-gap, rather than being broadly distributed within the band gap [5.10].

Key points of the basic carrier-trapping models

The models predict the resistivity to be high below the critical dopant concentration and to decrease rapidly above this value, as observed (Fig. 5.1). Below the critical dopant concentration the free-carrier concentration is low, and the resistivity is high. As the critical dopant concentration is approached, the barrier height increases; the ease of carrier motion from one grain to another (*ie, the effective mobility*)

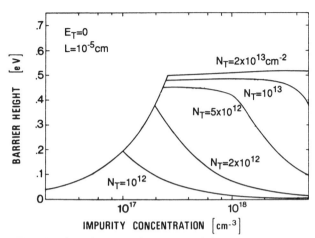

Figure 5.13: Barrier height as a function of dopant concentration when a neutral region forms before all the grain-boundary traps are filled. From [5.10]. Reprinted with permission.

decreases; and the resistivity remains high. The barrier height is maximum near the critical dopant concentration, and the effective mobility is minimum. Above the critical dopant concentration the free-carrier concentration increases and the barrier height decreases; in this range the resistivity decreases rapidly with increasing dopant concentration until it approaches that of single-crystal silicon.

5.4.4 Thermionic-field emission

The basic model presented above considers conduction only by thermionic emission of energetic carriers over the top of a parabolic potential barrier. However, near its top the barrier is narrow, and tunneling through the top portion of the barrier is likely for those carriers with energy slightly below the top of the barrier [5.31, 5.32].

The combination of thermionic emission and tunneling, called *thermionic-field emission*, can be modeled by calculating the number of carriers approaching the barrier with energy E and the probability $T(E)$ of each carrier tunneling through the portion of the potential barrier

$$V(x) = V_B \left(1 - \frac{|x|}{x_d}\right)^2 \qquad (5.32)$$

for which $V(x) > E$.

5.4. MODERATELY DOPED POLYSILICON

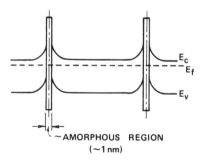

Figure 5.14: The impedance of the grain boundary is modeled by considering a high, narrow barrier at the grain boundary, in addition to the parabolic, depletion-region barriers surrounding the grain boundary.

For some ranges of dopant concentration, grain size, and temperature, thermionic-field emission can contribute a significant portion of the total current. For example, at room temperature with a dopant concentration of 5×10^{17} cm^{-3} and a grain size of 120 nm, the thermionic-field-emission current is estimated to be about 70% as large as the thermionic-emission current [5.32]. Thermionic-field emission becomes more important at lower temperatures, where fewer carriers have energies greater than the barrier height. This conduction mechanism will be discussed in more detail in the next section.

5.4.5 Grain-boundary barriers

In the basic carrier-trapping model the grain-boundaries themselves are assumed to be very narrow compared to the grains. Their only effect is to change the number of active dopant atoms and free carriers in the grains by acting as sites for dopant segregation and carrier trapping. However, the disordered nature of the grain boundaries and the discontinuities that they introduce into the periodic structure of the crystallites can also directly reduce the current flow. This additional impedance can be modeled by adding a high, narrow, potential barrier at the grain boundary, as shown in Fig. 5.14. This grain-boundary barrier is assumed to be present in addition to the parabolic, depletion-region potential barriers created by the uncompensated dopant atoms that neutralize carriers trapped at the grain boundary. The grain-boundary barrier can be modeled as a region with a width similar to that of the physical grain boundary (a few atom spacings) and a band gap different from that of crystalline silicon. To move from one crystallite to another, the carriers must either tunnel through the grain-boundary barrier or be sufficiently energetic to be thermally excited over the barrier.

To treat the problem over a fairly wide range of dopant concentrations, we consider the composite barrier shown in Fig. 5.14, consisting of both the parabolic potential barrier and the high, grain-boundary barrier. Because the composite barrier can be wide and high, we consider thermionic-field emission, in which a carrier is thermally excited before tunneling through a portion of the barrier. This intermediate case can be treated by finding the number of carriers with an energy E incident on the composite barrier from each side and the probability that each carrier can tunnel through the barrier [5.33]. The number of carriers having energy within the range dE_x incident on the barrier per second per unit area (the *supply function*) is given by [5.34].

$$N(T, \psi, E_x)dE_x = \frac{4\pi m^* kT}{h^3} \ln\left[1 + \exp\left(-\frac{E_x + \psi}{kT}\right)\right] dE_x \quad (5.33)$$

where $\psi = E_c - E_F$, E_x is the energy component in the x-direction, and h is Planck's constant. The probability that a carrier with energy E_x can tunnel through the barrier is given in the WKB approximation by

$$T(E_x) = \exp\left\{-\frac{4\pi}{h} \int_{x_1}^{x_2} [2m^*(qV(x) - E_x)]^{1/2} dx\right\} \quad (5.34)$$

where zero energy is taken as the bottom of the conduction band in the neutral regions of the grains, x_1 and x_2 are the classical turning points at which $E_x = V$, and m^*, the effective mass of the carriers, is assumed to be independent of the carrier energy.

The high, grain-boundary potential barrier is most important at intermediate and high dopant concentrations. At low dopant concentrations, the parabolic, depletion-region barrier limits the conduction. In the dopant-concentration region where both barriers must be included in the analysis, three different ranges of carrier energies are considered:

1. The carriers have energy less than that required to surmount the parabolic potential barrier in the depleted regions of the grains: Carrier transport occurs by tunneling through a portion of the parabolic potential barrier plus the grain-boundary barrier.

2. The carriers have sufficient energy to surmount the parabolic barrier in the depletion regions but not sufficient energy to travel

5.4. MODERATELY DOPED POLYSILICON

over the grain-boundary barrier: Transport occurs by tunneling through the grain-boundary barrier.

3. The carriers are sufficiently energetic to surmount the grain-boundary barrier: Carriers freely pass over the barrier.

The total current can be found by considering the energy distribution of the carriers and using the appropriate forms of the potential $V(x)$ in the expression for the net current — the difference between the currents traversing the barrier from the left and from the right. By comparing the model to experimental data, reasonable values of about 1 nm have been obtained for the width of the grain-boundary barrier [5.35].

The detailed behavior of the grain-boundary barrier can depend on the dopant being added [5.36]. For boron, the width of the barrier is independent of the dopant concentration, with a value in the range 0.7-0.9 nm. For phosphorus, however, the width appears to decrease with increasing dopant concentration above about 10^{18} cm^{-3}, changing from 1.1 nm at a dopant concentration of about 10^{18} cm^{-3} to about 0.5 nm for dopant concentrations in the mid-10^{19} cm^{-3} range. This different behavior of boron and phosphorus is attributed to their different tendencies to segregate to the grain boundaries. The incorporation of phosphorus atoms into the disordered sites at a grain boundary is postulated to decrease its effective width. Since boron does not tend to segregate to grain boundaries, the grain-boundary width is independent of dopant concentration for boron-doped films. As we will see in Sec. 5.6, the added barriers at the grain boundaries can model the higher resistivity of polycrystalline silicon compared to single-crystal silicon observed for very high dopant concentrations, at which the parabolic, depletion-region barriers are small.

5.4.6 Limitations of models

The carrier-trapping model can be further refined, but these more detailed treatments often require that physical parameters be the same for all grains and grain boundaries in the layer. For example, the density of carrier traps N_T at a grain boundary depends on the structure of the grain boundary, which can vary from one grain boundary to another [5.37]. Photoconductivity measurements suggest that the potential barriers are larger for high-angle grain boundaries than for low-angle grain boundaries. The properties can even vary from point to point along

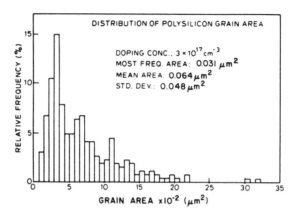

Figure 5.15: Distribution of grain sizes in a 1 μm thick polysilicon film. From [5.39]. ©1984, Pergamon Journals, Ltd. Reprinted with permission.

the same grain boundary. Boundaries with low densities of trap states, and therefore low barriers, are preferential current paths [5.38]. These variations limit the quantitative use of refined models.

Grain-size variations also limit detailed quantitative modeling. Although the "grain size" used to find the voltage across one grain boundary is consistent with the average measured grain size in some investigations, it differs in other studies and is often used as an adjustable parameter to match the experimental data; the models then only provide the functional form of the equation describing the conduction.

The number of grains N_G in the current path should be obtainable from the expression for the current flowing at high electric fields:

$$J = J_0 \sinh\left(\frac{qV}{2kTN_G}\right) \qquad (5.35)$$

where V is the voltage applied across the sample. If a uniform grain size is assumed, N_G should be just the sample length divided by the grain size L. However, values of N_G found from Eq. 5.35 appear to vary with dopant concentration [5.39] while transmission electron microscopy shows that the grain size does not change appreciably with dopant concentration for moderate doping.

Most models assume that all grains are the same size both along the current-flow direction and perpendicular to the film plane. However, experimental data indicate that the sizes of grains in a given sample are

5.4. MODERATELY DOPED POLYSILICON

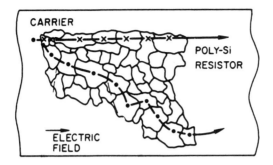

Figure 5.16: The nonuniform grain size in a polysilicon film leads to nonuniform current flow, making quantitative modeling difficult. From [5.39]. ©1984, Pergamon Journals, Ltd. Reprinted with permission.

fairly widely distributed, as shown in Fig. 5.15 [5.39], and the barrier height between grains can also vary markedly from one grain boundary to another. Consequently, the current flow is nonuniform, with the current traveling preferentially along lower resistance, but possibly longer paths, as indicated in Fig. 5.16 [5.39]. We also saw in Sec. 2.4 that the grain size varies through the thickness of the film, being greater near the top than near the bottom, so that the grain-boundary density is lowest near the top of the film. Thus, the resistance should be lower for paths closer to the film surface, and the current is expected to flow nonuniformly over the film thickness. If the current is a nonlinear function of the voltage, the current path can even vary as the applied voltage changes. The nonuniform current flow changes the effective number of grains in the path; the apparent variations in grain size inferred from electrical measurements are consistent with the effect of a distribution of grain sizes, both in the film plane and through the thickness of the film [5.39]. The effect of these nonuniformities is most noticeable near the critical dopant concentration, where some grains can be completely depleted while others contain a neutral region.

The nonuniform grain size in polysilicon films limits the usefulness of refined models which assume a uniform grain size. While differentiating between discrete and uniformly distributed energy levels for the traps seems to be possible, extracting more detailed information, such as the energy distribution of these levels may not be possible. Similarly, using experimental conduction data to differentiate between slightly different models is difficult. These refined models are only useful when an adequate statistical description of the grain-size distribution is available. However, this distribution depends sensitively on deposition conditions. Thus, the carrier-trapping model is a useful, semi-quantitative tool for describing conduction in polysilicon, but overly detailed refinement of this model requires more accurate values of physical parameters than can realistically be obtained.

5.4.7 Segregation and trapping

Although we have considered dopant segregation and carrier trapping independently, possible interactions can occur. We saw in Sec. 3.6 that dopant atoms become electrically inactive when they segregate at grain boundaries. However, more subtle modification of the electrical properties of the grain boundaries can be caused by the segregated dopant [3.24]. Dopant atoms segregating to the grain boundaries may interact with the dangling bonds or other defects which cause the grain-boundary traps, satisfying some of the unsaturated bonds and reducing the trap density and, consequently, the resistivity. For example, phosphorus segregation has been postulated to decrease the trap density and lower the barrier height [5.38]. The lower vibrational entropy implied by the decrease of A in Eq. 3.7 as dopant segregates suggests that segregation [3.24], as well as the heat treatment itself [5.11], causes the grain boundaries to become more ordered, possibly decreasing the number of carrier traps there [3.24]. Thus, as the dopant segregates, dopant loss from substitutional sites within the grains increases the resistivity, but reduced carrier trapping at grain boundaries tends to decrease it [5.40]. (These effects are most readily observed in moderately doped films, in which carrier trapping affects the electrical behavior most strongly.)

On the other hand, if dopant atoms at the grain boundary are readily ionized, they may cause additional Coulomb scattering and reduce the majority-carrier mobility. They may possibly change the minority-carrier transport as well. (These effects are most visible in highly doped samples.) However, loss of dopant atoms by segregation usually dominates, and modification of the carrier traps or scattering is secondary. Dopant segregation may also promote grain-boundary reconstruction, again changing the number of traps at the grain boundary. Unlike the other effects discussed, however, reconstruction is not reversible.

Although the disorder at grain boundaries makes complete bonding of dopant atoms there possible, magnetic resonance spectroscopy of phosphorus segregated at the grain boundaries in polysilicon suggests that the phosphorus is four-fold coordinated [5.41]. Four-fold coordination implies that the dopant is electrically inactivated by carrier trapping, rather than by changes in the number of bonds.

5.4.8 Summary: Moderately doped polysilicon

In this section, we have seen that three interrelated effects lead to the marked differences between the electrical properties of polysilicon and single-crystal silicon: First, atoms of some dopant species physically segregate to the grain boundaries and do not contribute to the conduction. Second, some of the carriers provided by substitutional dopant atoms within the grains are immobilized by carrier traps at the grain boundaries and are lost to the conduction process. Third, potential barriers at the grain boundaries impede the movement of the remaining free carriers from one grain to another.

As the dopant concentration increases beyond its critical value, the barrier height decreases until transport across the barrier no longer limits the conduction. Other conduction mechanisms then must be considered to understand the resistivity of polycrystalline silicon. For example, when the grain-boundary barriers are low and ionized impurity scattering within the grains limits the conduction, the conductivity decreases with increasing temperature. This temperature dependence contrasts with the increase in conductivity with increasing temperature found when thermionic emission over the grain-boundary barriers limits conduction. In Sec. 6.4 we will see that the change in the temperature coefficient of resistivity (TCR) from negative to positive with increasing dopant concentration is important for the low-value resistors often fabricated in polysilicon. At even higher dopant concentrations, other mechanisms limit the conduction. Thus, the carrier-trapping models explain the electrical behavior over an important dopant-concentration range, but do not apply to very lightly or very heavily doped polysilicon.

5.5 Grain-boundary modification

Because the traps located at grain boundaries strongly degrade the electrical properties of polysilicon and devices fabricated in it, attempts are made to reduce the density of active traps associated with the grain

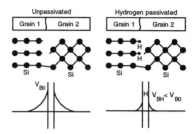

Figure 5.17: As the dangling bonds at the grain boundary are "passivated" with hydrogen, the depletion barrier surrounding the boundary decreases. From [6.122]. ©1985 IEEE. Reprinted with permission.

boundaries. Either the effectiveness of the trapping sites can be reduced or the grain boundaries themselves can be removed.

5.5.1 Grain-boundary passivation

In the first approach, the defects are *passivated* so that the states within the forbidden gap are not active and no longer trap carriers. As we have seen, traps are associated with the dangling bonds at the grain boundaries arising from the lattice discontinuities which form when differently oriented grains join. Dangling bonds at a Si/SiO_2 interface are often passivated by terminating them with hydrogen atoms. Similarly, hydrogen can passivate dangling bonds and other defects at grain boundaries [5.42] and reduce the number of active traps there. The passivation of grain boundaries has been directly observed by electron spin resonance [5.43]. The residual spin density, corresponding to unpassivated grain-boundary states, varies with the passivation temperature and is lowest after *hydrogenation* at 350°C. As the number of trapped carriers decreases, the potential barrier associated with the grain boundary also decreases (Fig. 5.17) [6.122]. The resistivity of moderately doped polycrystalline silicon is markedly reduced by incorporating hydrogen [5.44], and it becomes less sensitive to small variations in the dopant concentration. The properties of devices fabricated in the hydrogenated polysilicon are also improved [6.132], as we will discuss in Sec. 6.13.

5.5. GRAIN-BOUNDARY MODIFICATION

Passivation of grain-boundary traps has received increasing attention because of the use of polysilicon thin-film transistors in active-matrix, liquid-crystal displays (Sec. 6.13).

The atomic hydrogen needed to passivate the dangling bonds at grain boundaries cannot be readily obtained by annealing in molecular hydrogen or in forming gas (a N_2/H_2 mixture), but several other techniques can introduce atomic hydrogen into polysilicon.

A plasma dissociates molecular hydrogen; the resulting atomic hydrogen can then diffuses into the grain boundaries, interact with the defects, and effectively passivate the grain-boundary traps, lowering the grain-boundary-limited resistance [5.44, 5.45] and improving device characteristics [6.132]. Only a fraction of the added hydrogen passivates grain-boundary states [5.43] so that the amount of hydrogen added must be considerably greater than the number of grain-boundary states to be passivated. The efficiency of the hydrogenation process increases with increasing dissociation of H_2 or H_2^+ into atomic H and H^+ [5.46]. In addition to the concentration of hydrogen ions, the effectiveness of plasma hydrogenation depends on the temperature of the substrate, and the duration of the treatment [5.47]. The effectiveness of hydrogenation may possibly also be increased by coupling acoustic energy into the system [5.48]. Plasma hydrogenation is the most widely used hydrogenation technique, but it is time consuming.

In addition to passivating grain-boundary defects, plasma hydrogenation can also neutralize dopant atoms, especially boron, reducing the electrically active dopant concentration [5.49, 5.50]. For moderately doped films, the decreased dopant concentration may increase the grain-boundary barrier, adversely affecting the resistivity.

Hydrogen can also be added effectively by diffusing it from a hydrogen-containing, plasma-deposited, nitride layer (SiN_x:H) of the type often used to passivate integrated circuits [5.51]. The hydrogen content is these nitride layers is often 10-20%, and the hydrogen is weakly bound; it diffuses readily and can move from the nitride into an adjacent polysilicon film during a mild heat treatment — often at $\sim$400°C. The effectiveness of adding hydrogen by this technique depends on the annealing temperature and time. The technique is simple, but control and subsequent device stability are questionable.

Hydrogen passivation can also be obtained by implanting hydrogen into the polysilicon film. Subsequent annealing at a low temperature moves the hydrogen to the grain boundaries, where it can interact with the defects to reduce the effectiveness of the grain-boundary traps [5.45]. The hydrogen also appears to activate some of the dopant atoms segregated to grain boundaries [5.52], causing somewhat different interactions with dopants that segregate and those that do not. The effectiveness of hydrogenation by ion implantation depends on the hydrogen dose and the subsequent annealing temperature. The technique is very controllable, and throughput can be high, but implant damage can possibly degrade the devices. To reduce implant damage, the hydrogen can be implanted into the surrounding dielectrics and diffused into the polysilicon, but much higher hydrogen doses are then needed.

Although the hydrogen added to polysilicon can passivate the grain boundaries, it is not firmly bound. The hydrogen, and consequently the passivation, are rapidly lost from the grain boundaries by annealing at temperatures in the 400-500°C range [5.52, 6.132]. Therefore, it must be added after all high-temperature heat treatments in the device-fabrication process are complete. The hydrogenation process is repeatable, and hydrogen can be re-incorporated by another exposure to active atomic hydrogen. Positively charged hydrogen species may possibly move in electric fields within devices [5.53], causing concern about long-term reliability of hydrogen-passivated devices. The limited stability of hydrogen has led to the investigation of deuterium for similar passivation of grain boundaries, but with greater stability [5.54]. In addition to hydrogen and deuterium, lithium can also passivate grain boundaries [5.55] and neutralize recombination centers [5.56]. Although lithium is less compatible with integrated-circuit processing than hydrogen, lithium has been suggested for use in solar cells.

5.5.2 Recrystallization

Even when the grain boundaries are passivated with hydrogen to the maximum extent practical, the number of active grain-boundary states is still appreciable. To obtain devices of markedly higher quality, the grain boundaries themselves must be removed or at least reduced in number. As we saw in Sec. 2.11.2, solid-phase grain growth of moderately doped polysilicon requires extremely high temperatures for long times and is incompatible with integrated-circuit processing. However, large grains can be obtained by rapidly melting and solidifying the fine-grain polysilicon film with a scanned heat source such as a cw laser, electron beam, graphite strip-heater, or high-intensity incoher-

5.5. GRAIN-BOUNDARY MODIFICATION

ent lamp [2.119]. When the silicon solidifies, only a relatively small number of grains nucleate; these grains grow laterally into the rapidly cooling molten silicon to produce large-grain material containing few grain boundaries with their associated traps. The entire thickness of the fine-grain silicon must be melted so that the solidifying silicon does not regrow on the underlying small grains to again produce a fine-grain structure. The moving heat source provides the temperature gradient needed for lateral growth from a few nucleating grains [5.57]. Alternatively, the lateral thermal gradient can be obtained by locally varying the amount of power absorbed by partially covering the surface with layers of varying reflectivity [5.57]. The heat loss into the underlying substrate may also be varied locally by placing oxides of different thicknesses under different regions of the polysilicon [5.57].

Controlled lateral temperature gradients are not readily obtained with a pulsed laser. However, the availability of high-power lasers, especially excimer lasers, with relatively uniform light intensity over areas of ~ 1 cm^2 makes these lasers useful for significantly improving the properties of polysilicon films [5.58]. Although melting the layer with a pulsed laser may not increase the grain size optimally, it can reduce the defect density within the grains [5.59]. The grain size can be increased further by locally varying the power absorption by placing patterned layers on the surface to change the optical absorption [5.57, 5.60].

Crystallizing a film deposited in the amorphous form with a pulsed laser can be especially attractive when the film is deposited on a substrate with a low melting or softening temperature, such as inexpensive glass. Because the laser pulse is short, the silicon film stays at the melting temperature for ≤ 100 ns, limiting the heating of the underlying glass substrate.

In moderately doped, fine-grain polysilicon, the resistivity varies rapidly with dopant concentration as the grain-boundary traps become saturated with carriers. Because most of the grain boundaries are removed in recrystallized polysilicon, the resistivity is much less sensitive to variations in the dopant concentration. For example, for a resistivity in the intermediate range, melting with a laser can reduce the resistivity variation from a factor of ten to a factor of two for a 30% change in the implanted dose [5.61] if the resistor dimensions are large enough to sample many grains. If the dimensions are small, however, only a few grains are included, and significant statistical resistivity variations can occur.

The remaining grain boundaries in the recrystallized silicon film arise from random nucleation. In critical applications even these few grain boundaries must be removed. This can be accomplished by opening windows periodically in the oxide above the single-crystal substrate before the fine-grain polysilicon is deposited. As the molten silicon solidifies during recrystallization, the crystal structure of the substrate propagates into the solidifying material, first vertically, and then laterally over the oxide, to produce grain-boundary-free, single-crystal silicon over the insulating oxide. However, the distance that the grain-boundary-free material can propagate laterally is limited by competing random nucleation, restricting the use of this *seeding* technique.

As an alternate approach, transparent material of varying thickness can be placed above the silicon before recrystallization so that the amount of power absorbed from a narrow-band optical source varies from one region to another. The regions that absorb the least power are coolest and solidify first so that any grain boundaries that subsequently form are outside of these regions. Because the location of the grain boundaries can be controlled photolithographically using this technique, they can be placed outside of the active device regions (eg, in the isolation regions between devices). This technique can be coupled with the seeding technique discussed above to control the orientation of the grain-boundary-free material.

5.6 Heavily doped polysilicon

The major use of polysilicon films in ULSI circuits is for the gate electrode and one layer of interconnections in MOS integrated circuits. The material used for interconnections must have the lowest possible resistivity to increase circuit speed and avoid voltage drops along long lines. However, as increasing concentrations of dopant atoms are added to polycrystalline silicon (for example, by ion implantation), the resistivity decreases to a limiting value and then remains constant, as shown in Fig. 5.18 [5.62]. Because boron does not segregate to grain boundaries, it is the most effective and controllable dopant in moderately doped films; however, at high dopant concentrations the resistivity of phosphorus-doped polysilicon is lower. The limiting resistivities are about 400 $\mu\Omega$-cm in phosphorus-doped polysilicon and about 2000 $\mu\Omega$-cm in arsenic- or boron-doped polysilicon after annealing at 1000°C.

5.6. HEAVILY DOPED POLYSILICON

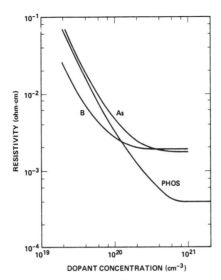

Figure 5.18: The resistivity of polysilicon after annealing at 1000°C is limited to approximately 2000 $\mu\Omega$-cm for boron and arsenic doping and about 400 $\mu\Omega$-cm when the polysilicon is doped with phosphorus. From [5.62]. Reprinted with permission.

Hall measurements [5.62–5.64] suggest that the electrically active carrier concentration saturates as the dopant concentration increases and remains constant as more dopant atoms are added. The dopant concentrations at which the number of electrically active carriers saturates are somewhat less than 1×10^{21} cm^{-3} for phosphorus, $\sim 2 \times 10^{20}$ cm^{-3} for boron and $\sim 6 \times 10^{20}$ cm^{-3} for arsenic [5.62, 5.63]. However, the number of electrically active carriers is considerably less than the dopant concentration needed to reach saturation in polysilicon, as in single-crystal silicon [5.65]. The carrier mobilities in heavily doped polysilicon are about half those in single-crystal silicon of the same dopant concentration. Similar behavior has been found for films deposited under widely varying deposition conditions, as discussed below.

5.6.1 Solid solubility

While the resistivity in moderately doped polysilicon is much greater than that in similarly doped single-crystal silicon, the resistivity in heavily doped polysilicon is only about twice that in correspondingly doped single-crystal silicon. Values of resistivity approaching those in single-crystal silicon suggest a common resistivity-limiting mechanism. In single-crystal silicon at high dopant concentrations, the amount of substitutional dopant that can be incorporated into the lattice is limited by the solid solubility of the dopant species in silicon [5.65, 5.66]. Solid solubility also appears to limit the amount of dopant that can be actively incorporated into polysilicon [5.67], with the same value of solid solubility in the crystalline grains of polysilicon as in single-crystal silicon [5.68]. At high phosphorus concentrations the carrier concentration saturates at that corresponding to solid solubility [5.69]; for arsenic, the saturation value appears to be less than the reported solid-solubility carrier concentration [5.70], possibly because of uncertainly in the value of arsenic solubility in silicon [5.63], as well as significant arsenic precipitation and segregation. The limiting resistivity is greater in arsenic-doped films than in phosphorus-doped ones for polysilicon, as for single-crystal silicon [5.67]. The solid solubility depends more strongly on temperature for boron than for arsenic over the 800–1000°C temperature range [5.71].

Solid solubility appears to limit the conductivity in films deposited over a wide range of conditions, including those deposited in cold-wall reactors over the temperature range from 700 to 1000°C [1.33], as well as films deposited in the more common, hot-wall, low-pressure reactors.

In somewhat less heavily doped polysilicon, the observed carrier concentration is less than the dopant concentration, probably because of dopant segregation at the grain boundaries. For phosphorus, about 80% of the dopant atoms are electrically active, while for arsenic, which segregates more strongly, only about 60% are active, contributing to the lower conductivity in arsenic-doped films even below solid solubility [5.63].

5.6.2 Method of doping

Gas-phase doping: Dopant atoms can be added to polysilicon during deposition by introducing a phosphorus-containing gas such as phosphine PH_3 into the deposition chamber or after deposition either by diffusion from a gaseous source such as PH_3 or $POCl_3$ or by ion implantation. When phosphorus is added to the films from a gaseous source at a relatively high temperature (so that thermal equilibrium is attained in a relatively short time), the amount of phosphorus incorporated appears to equal its solid-solubility value at the doping temperature. Auger

5.6. HEAVILY DOPED POLYSILICON

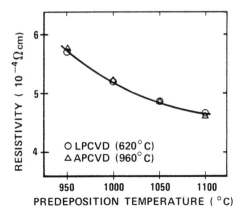

Figure 5.19: After doping from a gaseous source, the resistivity of polysilicon is slightly lower for higher doping temperatures.

analysis of films doped from a POCl$_3$ source suggests that the total phosphorus concentration increases monotonically with increasing doping temperature, corresponding to its increased solid solubility at higher temperatures. Relatively little additional dopant is incorporated within the grain boundaries or as precipitates *during the deposition process.* The amount of dopant needed to saturate the dopant segregation sites at grain boundaries appears to be only a small fraction of the total dopant concentration in very heavily doped films.

The electrical properties also reflect the increasing solid solubility with increasing temperature in the normal temperature range [5.65]. The resistivity decreases slightly with increasing doping temperature, as shown in Fig. 5.19 [5.72]. Figure 5.20 [5.72] shows that, even in single-crystal silicon, the limiting carrier concentration n_{SS} is lower than the solid-solubility dopant concentration C_{SS} [5.65, 5.66]. If the electrically active carrier concentration corresponding to solid solubility at each doping temperature is combined with the resistivity of the corresponding film (Fig. 5.19), a nearly constant mobility of about 32 cm^2/V-s is calculated over the entire doping-temperature range considered. This value is similar to that obtained from Hall measurements and is about half the single-crystal mobility at the same dopant concentration.

Ion implantion: Although little excess phosphorus is incorporated when polysilicon films are doped from a gaseous source, average concentrations of dopant atoms well in excess of solid solubility can be added by

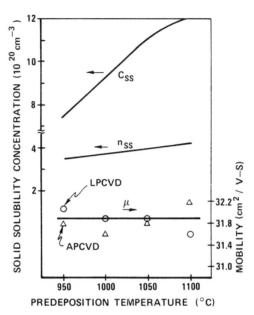

Figure 5.20: The resistivity is consistent with a free-carrier concentration n_{SS} corresponding to the solid solubility C_{SS} of phosphorus at the doping temperature and a mobility μ equal to about half that in similarly doped, single-crystal silicon.

ion implantation. However, dopant atoms at concentrations above solid solubility are not usually electrically active [5.71]. For at least phosphorus, the excess dopant atoms can precipitate at grain boundaries and may seriously degrade the integrated circuit. For example, as we saw in Sec. 4.2.3, phosphorus-containing precipitates can oxidize more rapidly than polysilicon to form a low-quality oxide [4.7]. Subsequent etching of the oxide can leave voids extending through much or all of the thickness of the polysilicon in these regions and attack the underlying gate oxide. For high dopant concentrations, conventional ion implantation from mass-separated sources can be slow. Alternative techniques are being investigated to add large amounts of dopant rapidly, especially for large-area polysilicon devices [5.73, 5.74].

Diffusion from doped oxides: Polysilicon films can also be doped by diffusion from a doped oxide formed by chemical vapor deposition or from a spin-on glass containing a high concentration of the dopant species [5.75]. Using a spin-on glass requires careful evaporation of the

5.6. HEAVILY DOPED POLYSILICON

solvent before heating to the dopant diffusion temperature to preserve the film integrity. The portion of the doped oxide or glass adjacent to the polysilicon must be heavily doped; a thin undoped region can block the diffusion of boron. For phosphorus, the phosphorus-containing region can "melt through" a thin undoped region.

In situ **doping:** The kinetics of *in situ* doping were discussed in Sec. 1.8. Excess dopant can be incorporated when the dopant is added during deposition. At the low temperatures at which polysilicon films are usually deposited, the time needed for the dopant to reach equilibrium can be much longer than the time to deposit a few atom layers of silicon, and nonequilibrium dopant concentrations in excess of solid solubility can be incorporated. The excess dopant may be electrically active immediately after deposition, contributing carriers to the conduction process [5.68, 5.76].

Because of the nonunity segregation coefficient of the dopant from the gas phase into the depositing silicon film, the dopant-gas concentration can decrease at a different rate than the silicon-gas concentration along the length of the reaction chamber. Therefore, the concentration of dopant incorporated into the film can vary along the length of the chamber. When the dopant concentration is above solid solubility, the amount of excess dopant in different films can vary even though the carrier concentration of all the films corresponds to solid solubility. Consequently, the amount of precipitation and the related film degradation occurring later in the device-fabrication process can vary with the position of the wafer in the reactor.

5.6.3 Stability

Although carrier concentrations corresponding to high solid solubilities can be introduced at high doping temperatures, the dopant may not remain electrically active after further, lower-temperature, heat treatments. Because the solid solubility is usually lower at the lower temperature, the dopant is rejected from active substitutional sites. It may precipitate within the grains, or more likely at the grain boundaries, where the disordered structure provides sites for easy precipitation. If the time of the lower-temperature heat treatment is long enough for thermal equilibrium to be reached, the dopant in excess of solid solubility at the lower temperature precipitates and no longer contributes to the conductivity of the film. However, during very short heat treatments, the dopant may not have time to reach equilibrium at the new temperature, and active dopant concentrations in excess of solid solubility can be obtained.

The effect of annealing on dopant deactivation can be seen for polysilicon films doped with phosphorus from a gas-phase source at 1000°C and subsequently annealed at lower temperatures for a limited time [5.63]. High active dopant concentrations can be obtained by rapidly heating to a high temperature with a laser and cooling quickly. Subsequent furnace annealing can reduce the active carrier concentration [5.63]. At very low annealing temperatures, the carrier concentration changes little, although the dopant concentration is substantially above solid solubility at these temperatures; the annealing time is not long enough for excess phosphorus to leave solution. At intermediate annealing temperatures, less time is needed, and the carrier concentration decreases as excess phosphorus is able to leave solution during the heat treatment. Some of the excess dopant is likely to leave solution even during the final integrated-circuit annealing process in the 400°C temperature range. At high temperatures, the carrier concentration again increases because of the increased solid solubility. Assuming that the amount of dopant added during gas-phase doping equals the solid solubility at the doping temperature, the phosphorus concentration should be below solid solubility at temperatures higher than the doping temperature, and the carrier concentration should not depend strongly on the annealing temperature.

In addition to the nominal annealing temperature and time, the rate of temperature transitions can affect the final resistivity. The cooling rate is especially important after annealing heavily doped films at high-temperatures. At high temperatures the dopant solid solubility is high. As the films are cooled the equilibrium solid solubility decreases. If the cooling is slow, much of the dopant has time to leave solution; the active carrier concentration decreases; and the resistivity increases, as discussed in Sec. 3.6.2 [3.92]. The slow cooling used to avoid wafer deformation increases the loss of dopant. The fast cooling used after rapid thermal processing minimizes the dopant loss.

Competing changes during annealing can lead to complex changes in the resistivity. *In situ* doped films deposited at low temperatures can contain dopant concentrations above solid solubility at the deposition temperature. When they are annealed at higher temperatures, the grains may grow, with a corresponding increase in mobility and de-

5.6. HEAVILY DOPED POLYSILICON

crease in resistivity. At the same time, excess dopant can leave solution, increasing the resistivity.

5.6.4 Mobility

Although the barrier to conduction created by the depletion regions surrounding the grain boundaries decreases at high dopant concentrations (as expected from Poisson's equation), the mobility at the highest dopant concentrations is only about half that in single-crystal silicon. Similar values are found from Hall-mobility measurements [5.15, 5.62] and inferred from the limiting resistivity and the carrier concentration corresponding to solid solubility [5.72] (Fig. 5.20). The high, narrow, grain-boundary barrier, through which the carriers must tunnel [5.32, 5.33], can explain this difference. A barrier thickness of about 1 nm and a barrier height corresponding to amorphous silicon are consistent with the experimental data [5.35, 5.36].

The mobility is also different for different n-type dopant species [5.67, 5.68], and is lower in arsenic-doped films than in phosphorus-doped films [5.63, 5.64]. A value close to 30 cm^2/V-s is obtained for the highest phosphorus concentrations that can be achieved by furnace annealing, but the mobility is only about 20 cm^2/V-s for arsenic-doped polysilicon [5.15, 5.63, 5.64]. The lower mobility in arsenic-doped films is attributed both to stronger ionized-impurity scattering by arsenic, as in single-crystal silicon [5.63], and to more effective grain-boundary scattering in arsenic-doped polysilicon than in phosphorus-doped films. Melting the film with a laser reduces the strong grain-boundary scattering in arsenic-doped polysilicon [5.63]. The resistivity of furnace-annealed films containing high concentrations of both arsenic and phosphorus is higher than that for films doped only with phosphorus. This behavior, attributed to the increased grain-boundary scattering caused by arsenic, degrades the performance of some MOS integrated circuits in which the arsenic implantation used to dope the source and drain regions also enters the polysilicon gate that has previously been doped with phosphorus.

5.6.5 Trends

The trend toward lower processing temperatures for ULSI fabrication decreases the conductivity that can be achieved in polysilicon.

Using short, high-temperature, heat cycles near the end of the device-processing sequence can improve the conductivity [5.77]. Because grain boundaries and other defects limit the mobility in even heavily doped polysilicon, the mobility can be improved somewhat by reducing the number of grain boundaries. Larger grains can potentially be obtained by solid-state grain growth or by liquid-phase recrystallization, but the improvement in mobility is limited to approximately a factor of two for very heavily doped polysilicon.

Although the limited conductivity of polysilicon is becoming an increasingly severe constraint on the performance of small-geometry, high-performance integrated circuits, polysilicon is often augmented by an overlayer of a more highly conductive material, such as a refractory metal silicide, as we saw in Sec. 3.5.3. Retaining the superior electrical properties and excellent stability of MOS transistors with polysilicon directly over the gate insulator is strong motivation for augmenting, rather than replacing, polysilicon in the near future.

However, as will be discussed in Sec. 6.2.6, the modulation of the carrier concentration by the gate voltage within the polysilicon near the underlying gate dielectric can create a depletion region that decreases the effective coupling of the gate voltage to the transistor channel and degrades the maximum transistor current. This limitation may eventually lead to the replacement of polysilicon with a metal [6.11]. Even in this case, however, the gate may be initially formed from polysilicon — to utilize its compatibility with high-temperature processing — and replaced with metal later in the process [6.12].

5.7 Minority-carrier properties

5.7.1 Lifetime

In our discussion of conduction in polycrystalline silicon, we saw that the current was limited by potential barriers that impede the flow of majority carriers. Minority carriers, on the other hand, see an attractive, rather than a repulsive, potential, and can readily move to the grain-boundary defect states and recombine with trapped majority carriers. Because of the efficient recombination at grain boundaries, the minority-carrier lifetime in polysilicon is expected to be markedly less than in single-crystal silicon. In addition, the defect concentration within the grains of polysilicon is higher than in high-quality, single-crystal sili-

5.7. MINORITY-CARRIER PROPERTIES

con, and recombination at defects within the grains again reduces the minority-carrier lifetime.

The effective lifetime τ_{eff} can be written in terms of the volume lifetime τ_b in the grains and the grain-boundary recombination velocity s [5.24, 5.78, 5.79]. Alternatively, it can be expressed as [5.80]

$$\tau_{\text{eff}} = \frac{2d \, \exp(-qV_B/kT)}{3\sigma v D_T(E_{fn} - E_{fp})} \tag{5.36}$$

where D_T is the interface trap density (cm^{-2} eV^{-1}), σ is the capture cross section, d is the grain size, v is the thermal velocity of the carriers, V_B is the height of the potential barrier at the grain boundary, and E_{fn} and E_{fp} are the quasi-Fermi levels for electrons and holes, respectively.

Another approach estimates the lifetime by treating the surface recombination sites as if they were distributed uniformly throughout the bulk of the material [5.14]. In this approximation

$$\tau_{\text{eff}} = \frac{1}{\sigma v N_{sr}} \tag{5.37}$$

where N_{sr} is the effective density of recombination centers (per unit volume). For cubic grains, the number of recombination centers per unit volume is given by $N_{sr} = 3N_{it}/d$, where N_{it} is the concentration of recombination centers per unit area at the grain boundary.

Using the latter approach, the lifetime can be found from diffusion length measurements, from photoconductive decay, or from the reverse characteristics of p-n junction diodes. The reverse characteristics of diodes formed within material with submicrometer grain size are dominated by generation within the space-charge region of the p-n junction, and the corresponding lifetime is of the order of 20–100 ps [6.106]. In thicker films with larger grains, the lifetime inferred from forward diode characteristics is about 100–300 ps in polysilicon, much less than the microsecond lifetime found in similarly doped single-crystal films.

Because the total grain-boundary area per unit volume depends on the grain size, the lifetime also varies as the grain size changes. Figure 5.21 [5.14] shows the predicted linear dependence of lifetime on grain size. From the empirical relation $\tau_{\text{eff}} = 5 \times 10^{-6} \, d$ found from Fig. 5.21, and $v = 10^7$ cm/s, the product $\sigma N_{it} = 6.7 \times 10^{-3}$. With the value of $\sigma = 2 \times 10^{-16}$ cm^2 measured for the capture cross section of surface states, N_{it} is found to be 3×10^{13} cm^{-2} — about an order of magnitude

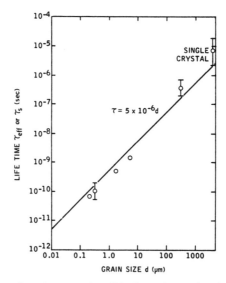

Figure 5.21: The minority-carrier lifetime in polysilicon increases approximately linearly with increasing grain size. From [5.14]. Reprinted with permission.

greater than the density of carrier traps found from the majority-carrier conductivity. However, the different effect of lifetime in neutral regions and in depletion regions should also be considered in interpreting the experimental data [5.81].

5.7.2 Switching characteristics

The properties of the parabolic, depletion-region barriers between grains not only depend on the physical nature of the grain boundary, but they can also be modulated by the electrical bias applied and the current flowing. Applying a high voltage can cause the current to increase abruptly. The change may persist only while the voltage is applied (*threshold switching*), or it can be permanent (*memory switching*).

Temporary threshold switching is usually caused by changes in the net charge in the grain-boundary traps and the surrounding depletion regions by excess carriers traveling across the sample [5.82–5.85]. As the charge that causes the depletion-region barriers is neutralized by injected charge, the depletion-region barriers decrease, and the majority-carrier current flowing in the sample increases. The resulting abrupt change in the conductivity with applied voltage leads to the observed

5.7. MINORITY-CARRIER PROPERTIES

threshold switching. The dynamics of charge trapping and emission can also cause anomalous capacitive current [5.86].

In the simplest case, *minority carriers* injected into the grains cause threshold switching by neutralizing some of the majority carriers trapped at the grain boundaries, with the consequent barrier-height decrease and current increase. Consider a p^+-n junction within a polysilicon layer [5.87]. At low voltages, only a few holes are injected from the p-type polysilicon into the n-type polysilicon. The trapped charge at the grain boundary is not appreciably changed, and the resistivity of the n-type region remains high. However, when a large number of holes is injected, the holes can interact strongly with the grain-boundary traps. The holes can recombine with trapped electrons or themselves be trapped, depending on the trap energies. In either case, the net negative charge in the grain boundary decreases, and the amount of compensating charge in the surrounding space-charge region also decreases, lowering the barrier height. This reduction of the barrier height increases the conductivity of the polysilicon, so that more voltage is available to forward bias the p^+-n junction. More holes are then injected into the polysilicon, leading to regenerative feedback above a critical *switching voltage*. When sufficient charge has been injected to neutralize the grain boundaries in the lightly doped region, only a small *holding voltage* is required to provide the current needed to maintain the grain boundaries in their low-resistance, neutral state. Similar behavior can occur if charge is injected from the electrodes into the polysilicon. In either case, the resistance of the polysilicon appears to "switch" from a high-impedance state to a low-impedance state. When the voltage is removed, the excess minority carriers recombine; majority carriers are again trapped; and the resistance of the polysilicon returns to its initial high value determined by the zero-bias barrier height.

Injection of excess *majority carriers* can also cause threshold switching. Initially, some of the injected carriers are trapped at the grain boundaries (assuming that not all of the traps were previously filled), and the barrier height increases. However, once the grain-boundary traps are filled, additional majority carriers are not trapped at the grain boundaries, but remain in the grains to neutralize the uncompensated dopant atoms near the grain boundaries, decreasing the barrier height.

In addition to charge injection, the applied voltage itself can cause threshold switching. When a voltage is applied, a portion of the voltage appears across the depletion region on each side of each grain boundary. The resulting high electric field across the reverse-biased side of the barrier can narrow and lower the barrier. Some of the trapped carriers can then escape from the grain-boundary traps either by tunneling through the narrowed barrier, or by thermionic emission over the barrier, which has been lowered by the applied voltage (or by thermionic-field emission). As the traps at one grain boundary empty, the barrier height decreases, allowing more of the applied voltage to appear across other grain boundaries, and similar *field-assisted trap emptying* becomes likely at the other grain boundaries. This regenerative effect can lead to a rapid increase in the conductivity of the polysilicon as the voltage increases.

The switching behavior depends strongly on the temperature, with the probability of switching increasing rapidly with increasing temperature. The switching voltage and the holding voltage both increase significantly when the number of grains across which the voltage is applied increases, but the switching and holding currents are relatively insensitive to the number of grains. In fine-grain polysilicon, the applied voltage is divided across many grain boundaries so that any effect of the statistical inhomogeneities from one grain to another is small. However, in large-grain material, formed by laser recrystallization for example, only a few grain boundaries appear in the conduction path, and multiple, discrete, negative-resistance switching voltages are observed [5.84].

At higher currents, a large, permanent reduction in the resistivity of the polysilicon can occur. This *memory switching* is probably related to the formation of a permanent, relatively conductive filament by resistive heating of the polysilicon when a high current flows [5.82]. Dopant atoms from the heavily doped contact regions can diffuse rapidly along a very hot solid path or a molten region to form a heavily doped, conductive channel that remains when the applied voltage is removed [5.88, 5.89]. Memory switching can decrease the resistance by a factor of 10^6 or more because of the highly nonlinear dependence of the resistivity of polysilicon on the dopant concentration. More complex memory switching behavior is observed in polysilicon doped with both boron and phosphorus. It has been suggested that n-type regions within the grains are converted into p-type regions as the phosphorus content within the grains is reduced by the tendency of phosphorus to segregate to grain boundaries [5.90].

5.8 Summary

Grain-boundary effects cause the electrical properties of polycrystalline silicon to differ from those of single-crystal silicon. At low and moderate dopant concentrations, dopant segregation to and carrier trapping at grain boundaries reduces the conductivity of polysilicon markedly compared to that of similarly doped, single-crystal silicon. Not only is the number of carriers reduced, but movement of the carriers from one grain to another is impeded by potential barriers surrounding the grain boundaries. Because the properties of moderately doped polysilicon are limited by grain boundaries, modifying the carrier traps at the grain boundaries by introducing hydrogen to saturate dangling bonds improves the conductivity.

When polysilicon is used as a gate electrode or an interconnection in integrated circuits, its limited conductivity can degrade circuit performance. At high dopant concentrations, the active carrier concentration is limited by solid solubility of the dopant species in crystalline silicon. The amount of active dopant is influenced by the time and temperature of the last fabrication step at which the dopant is mobile enough to interact with grain boundaries or approach its equilibrium solid-solubility value.

The minority-carrier lifetime in polysilicon is markedly lower than that in single-crystal silicon, but it increases with increasing grain size. High applied voltages can lead to temporary or permanent reduction in the resistance of a polysilicon structure.

Chapter 6

Applications

6.1 Introduction

In previous chapters we considered the properties of polysilicon that make it useful in integrated circuits. In this chapter, we want to apply that information to specific applications. Because of the wide range of integrated-circuit devices in which polysilicon is now used, this discussion cannot be all inclusive. A few of the more important applications are considered in detail, and others are briefly mentioned.

The dominant application of polysilicon is for gate electrodes in CMOS (*complementary metal-oxide-semiconductor*) integrated circuits, and that technology will be discussed in some detail. Related devices, such as nonvolatile, floating-gate, memory elements and high-value load resistors will also be examined. In addition to CMOS applications, polysilicon is used in most bipolar integrated circuits, where it both provides compact, low-capacitance structures and modifies the basic transistor physics. As use of polysilicon for bipolar integrated circuits became more widespread, the technologies used to fabricate MOS and bipolar integrated circuits have tended to converge, and many of the processes developed for one technology are applied to the other.

Polysilicon is also widely used in dynamic, random-access memories (DRAMs). In some DRAMs multiple layers of polysilicon are placed above the surface of the silicon substrate to form complex *stacked* capacitor cells; in other cases, a *trench* is formed in the substrate, coated with a dielectric, and then filled with polysilicon to form the storage capacitor. Polysilicon-filled trenches are also being increasingly used for

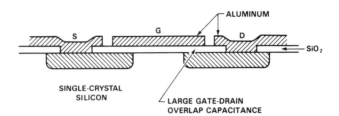

Figure 6.1: In the aluminum-gate MOS transistor, the gate must overlap the source and drain by one alignment tolerance to ensure that a continuous channel is formed when the transistor is turned on.

isolation between adjacent devices of high-performance circuits.

We will also examine structures in which the active portion of the device is placed *within* the polysilicon layer itself. These devices include diodes with both the *p*-type and *n*-type regions within the polysilicon and MOS transistors with their *channels* in polysilicon films. Thin-film transistors are being increasingly used in active-matrix displays. Novel use of polysilicon to form structural elements of *microelectromechanical systems* (MEMS) is seeing increasing attention. This application requires careful control of the mechanical properties of polysilicon.

Use of polysilicon in optical devices, such as solar cells, is not considered extensively here, primarily because of space limitations, but also because the polysilicon used for solar cells often differs significantly from that used in integrated circuits.

6.2 Silicon-gate MOS transistor

Development of polysilicon technology was motivated by the potential use of polysilicon as a gate electrode for MOS integrated circuits.[1] In the mid-1960s most gate electrodes were made from aluminum (Fig. 6.1), which was deposited after the source and drain regions were doped. The aluminum gate must overlap the source and drain regions by at least one alignment tolerance to insure that a continuous conducting channel is formed from the source to the drain when the gate is biased to turn on the transistor. The required alignment tolerance causes a significant

---

[1]The term *Metal*-oxide-semiconductor (MOS) transistor is used to describe the device even when the gate is made from heavily doped polysilicon.

6.2. SILICON-GATE MOS TRANSISTOR

Miller feedback capacitance between the gate and the drain, reducing the circuit speed.[2]

By employing a gate-electrode material that allows the source and drain regions to be formed using the gate electrode as a mask (*ie*, to be *self-aligned*), this capacitance can be greatly reduced. Self-aligned gates not only reduce the capacitance, but they also make it more uniform from one device to another; this device matching is often critical for improved circuit performance. The material selected for a self-aligned electrode must be compatible with the high temperatures required to activate the source and drain dopant and with other high-temperature processes that follow gate-electrode formation.

Although other high-temperature materials, such as tungsten and molybdenum were investigated, the compatibility of polycrystalline silicon with integrated-circuit fabrication led to its rapid adoption. In the initial attempts to use thin films of silicon for gate electrodes, the silicon layers were deposited by evaporation. However, evaporated films could not adequately cover the steps between the gate oxide and the field oxide on the integrated-circuit surface. At the time silicon-gate technology was being developed, chemical vapor deposition (CVD) was routinely used to form the epitaxial silicon needed for bipolar integrated circuits. The epitaxial CVD technique was readily extended to polysilicon deposition by depositing on an oxide-coated substrate, instead of exposed single-crystal silicon, and by using a silicon source gas that did not contain chlorine. Because of the readily available technology and the conformal step coverage of films produced by chemical vapor deposition, this method of forming the silicon layers was quickly adopted [6.1] for the silicon-gate MOS transistor shown in Fig. 6.2. The ability of polysilicon to be oxidized and its compatibility with further high-temperature processing has allowed development of complex device structures employing several layers of polysilicon. In addition, polysilicon can be used to interconnect devices, allowing more flexible layout than obtained with only aluminum interconnections. Polysilicon can also be used for high-value resistors, further reducing chip area.

Initial silicon-gate integrated circuits used *p*-channel transistors because the turn-on voltage or *threshold voltage* was easier to control both in the active device and in the isolation regions. As more experience was

---

[2]The *Miller effect* increases the effective capacitance by multiplying the geometrical capacitance by the gain of the transistor.

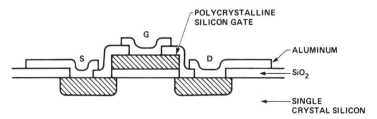

Figure 6.2: In a silicon-gate transistor, the gate electrode can be used to define the location of the source and drain dopant atoms, thereby eliminating the need for a large overlap of the gate electrode. This self-aligned structure reduces the overlap capacitance.

gained, n-channel circuits became more popular because of the higher mobility of electrons than holes. Having both n-channel and p-channel transistors available on the same chip allows building logic elements with very low dc current flowing when the circuit is not switching. Today, most integrated circuits are made using this complementary MOS (CMOS) technology.

6.2.1 Complementary MOS

In the basic CMOS circuit, an n-channel and a p-channel transistor are in series, with the same gate voltage applied to both. The source of the n-channel transistor is grounded; the source of the p-channel transistor is connected to the supply voltage (V_{DD}), as shown in Fig. 6.3. Under dc conditions either the n-channel transistor or the p-channel transistor should be normally off regardless of the gate voltage applied to the pair of transistors to limit the dc current flowing and, therefore, power dissipation.

To satisfy this requirement, the threshold voltage of the n-channel transistor should be positive, and that of the p-channel transistor should be negative. When the gate voltage is lower than the threshold voltage of the n-channel transistor, that transistor is nonconducting (OFF), and no dc current flows in the circuit even though the p-channel transistor is conducting (ON). When the gate voltage is higher than the threshold voltage of the n-channel transistor, that transistor conducts. However, the gate voltage is now close to the supply voltage so that the effective magnitude of the gate voltage on the p-channel transistor ($V_{DD} - V_G$) is small (less than the magnitude of its threshold voltage). The p-channel

6.2. SILICON-GATE MOS TRANSISTOR

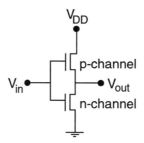

Figure 6.3: The n-channel and p-channel transistors of a CMOS circuit are in series so that one of the transistors is nonconducting when the circuit is not switching to limit the dc current.

transistor is nonconducting, and again negligible current flows in the circuit.

Appreciable current only flows when the transistor is switching. The reduced power dissipation of *complementary* MOS (CMOS) circuits is critical when integrated circuits contain millions or tens of millions of transistors. For maximum current flow when the transistor is switching the magnitude of the threshold voltage should be fairly low. The threshold voltage magnitude is made as low as practical, consistent with the requirement for low *leakage current* when the gate voltage is zero. It is typically ~0.5 V and decreases with more advanced processing. Unfortunately, as the threshold voltage decreases, the leakage current at zero gate voltage increases, so the leakage current increases as CMOS technology advances.

6.2.2 Threshold voltage

In addition to reducing the parasitic capacitance of an MOS transistor, using polysilicon provides a threshold voltage more compatible with the voltages used to bias integrated circuits. The threshold voltage V_T of an MOS transistor can be written [6.2]

$$V_T = V_{FB} \pm 2\mid \phi_B \mid \pm \mid Q_d \mid /C_{ox} \qquad (6.1)$$

where the upper signs refer to n-channel transistors and the lower ones refer to p-channel transistors. ϕ_B is the bulk potential [$\phi_B = (kT/q)(E_f - E_i)$, where E_f is the Fermi level and E_i is the intrinsic Fermi level]; Q_d is the charge in the depletion region separating the

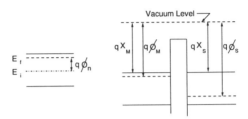

Figure 6.4: Definitions of terms used in transistor threshold voltage.

channel and the neutral bulk material; and C_{ox} is the capacitance per unit area of the gate dielectric.

The *flatband voltage* V_{FB} is the gate voltage required to remove the curvature from the energy bands in the silicon [*ie*, the gate voltage needed to compensate for the metal-semiconductor work function difference ϕ_{MS} between the gate material and the substrate material ($\phi_{MS} = \phi_M - \phi_S$) and the interface charge Q_f]:

$$V_{FB} = \phi_{MS} - Q_f/C_{ox} \qquad (6.2)$$

The work function ϕ_S of a semiconductor depends on its doping concentration: $\phi_S = \chi_S + (E_c - E_f)/q$, where χ_S is the electron affinity of the semiconductor (*ie*, the energy between the vacuum level and the bottom of the conduction band), and E_c is the energy of the bottom of the conduction band.

For aluminum the work function is about 4.1 V. For silicon the electron affinity is about 4.05 V. To obtain maximum conductivity in the gate electrode, the polysilicon is usually doped as heavily as practical, so that the Fermi level is near the edge of the conduction or valence band for *n*-type or *p*-type polysilicon, respectively. The gate work function is then about 4.1 V for *n*-type polysilicon and about 5.2 V for *p*-type polysilicon.

The ability to have different work functions for the gate electrode of the *n*-channel and *p*-channel transistors can be useful in providing the desired threshold voltages for each type of transistor. For the *n*-channel transistor, a small positive threshold voltage (typically ~ 0.5 V) is desired, while the threshold voltage of the *p*-channel transistor should be similarly small in magnitude, but negative.

6.2. SILICON-GATE MOS TRANSISTOR

For an n-channel transistor with a p-type silicon substrate containing a dopant concentration of about 10^{16} cm^{-3}, the work function of the substrate is 4.9 V. With either aluminum or n-type polysilicon as the gate electrode, the gate work function is about 4.1 V, so ϕ_{MS} contributes about -0.8 V to the threshold voltage V_{Tn} of the n-channel transistor, partially compensating the other terms in Eq. 6.1 and reducing the threshold voltage to approximately the desired value.

For the p-channel transistor, the work function of the n-type substrate is about 4.3 V, only moderately different from the work function of either an aluminum or an n-type polysilicon gate. The magnitude of ϕ_{MS} is small and does not significantly compensate the other terms in Eq. 6.1. The threshold voltage V_{Tp} is more negative than desired, making the transistor difficult to turn on. To reduce the magnitude to the desired value, the surface must be *counterdoped* with p-type dopant, degrading the mobility and process control and causing other difficulties. If the gate of the p-channel transistor is made from heavily doped, p-type polysilicon, the gate work function is about 5.2 V. ϕ_{MS} is then significantly positive, compensating the other terms in the expression for the threshold voltage, and reducing its magnitude to the slightly negative value desired.

6.2.3 Silicon-gate process

Basic process: A basic silicon-gate MOS transistor can be built with four masking steps. However, building both n- and p-channel transistors on the same chip complicates the process and increases the number of masking steps needed. Further refinements needed for transistors with very small dimensions and for multiple layers of metal to interconnect them can increase the number of masking levels to well over twenty and add significant complexity to the fabrication process.

The four basic mask levels shown in Fig. 6.5 are

1. *Isolation* to separate the individual transistors. Usually LOCOS (LOCal Oxidation of Silicon) or trench isolation (Sec. 6.10).

2. *Polysilicon definition* to form the gate of each transistor and polysilicon interconnections.

3. *Contact openings* to the source, drain, and gate of each transistor.

4. *Metal definition* to form the interconnections between transistors.

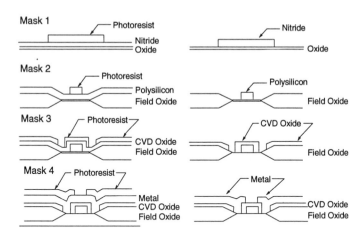

Figure 6.5: The four basic masks needed to form an MOS transistor: (1) Isolation. (2) Polysilicon. (3) Contact openings. (4) Metal.

For complementary transistors, selected regions of the substrate must be doped to form a *well* of the the opposite conductivity type that acts as an effective substrate for the second type of transistor. Defining this well adds a masking step. To dope the source and drain regions of each type of transistor with the appropriate dopant element, one or two further masking steps must be added to block (or overcompensate) the other dopant species. Adjusting the threshold voltage of each type of transistor by selectively adding dopant to the channel adds one or two more masks. Another mask may be needed to form lightly doped regions in the source and drain adjacent to the channel to reduce the electric fields and improve the reliability (the *lightly doped drain* or LDD structure). Other masks include one to add extra dopant near the surface of the isolation (field) region surrounding the n-channel transistor to prevent inversion of this lightly doped, p-type region from shorting adjacent transistors together.

Finally, additional metal layers are needed to effectively interconnect transistors. Two masks are needed for each additional metal level: One to open *vias* through the the dielectric separating two metal layers, and one to pattern the metal itself. With five or even more metal interconnection layers, the total number of masks increases significantly. For a complex CMOS process, the number of masks can exceed 20.

6.2. SILICON-GATE MOS TRANSISTOR

Conductivity type of gate electrode

We saw in Sec. 5.6 that the resistivity of n-type polysilicon doped with phosphorus can be lower than that of p-type polysilicon doped with boron. For maximum conductivity, the polysilicon for both n-channel and p-channel transistors can be doped with phosphorus, with the phosphorus added before the polysilicon is patterned. Phosphorus can be added to the polysilicon either during deposition or after deposition by diffusion from a gas-phase dopant source (usually $POCl_3$) or by ion implantation. Adding dopant during deposition is especially attractive if the polysilicon is to be covered with a layer of chemically vapor deposited silicide to lower the resistance of interconnections. The silicide can be deposited in another chamber of a cluster tool without exposing the surface of the polysilicon to air, eliminating native-oxide formation, as discussed in Sec. 1.5.2.

With n-type polysilicon and a p-type substrate, the threshold voltage of the n-channel transistor is somewhat positive, as desired. However, with n-type polysilicon and an n-type substrate, the threshold voltage of the p-channel transistor is more negative than desired for high conductivity when it is turned on. To obtain a threshold voltage closer to zero, the effective dopant concentration near the surface must be reduced by *counterdoping* with p-type boron. The heavy counterdoping needed lowers the carrier mobility and increases the effective distance of the controlled charge from the gate electrode.

On the other hand, as described above, the desired, slightly negative, threshold voltage can be readily obtained for the p-channel transistor by using p-type polysilicon as the gate electrode. The work function of the heavily doped, p-type polysilicon is higher than that of heavily doped n-type polysilicon by approximately the 1.1 eV bandgap of silicon, making the threshold voltage of p-channel transistors less negative. Thus, the threshold-voltage magnitude can be reduced without the otherwise-required, counterdoping of the channel region. Therefore, high-performance CMOS processes use n-type polysilicon for the gate of the n-channel transistor and p-type polysilicon for the gate of the p-channel transistor.

Complementary doped gates can be conveniently fabricated. The initially undoped polysilicon is patterned before doping, and the n-channel and p-channel transistors are doped separately. The same ion-implantation used to dope the source and drain regions also dopes the gate electrode. This is especially beneficial for the p-channel transistor: When an n-type gate is used on a p-channel transistor, the p-type source-drain dopant counterdopes the n-type gate electrode, reducing its conductivity. Using p-type doping in the gate electrode avoids this conductivity reduction. However, the arsenic implantation that forms the source and drain of the n-channel transistor is also used to dope the gate, and the dopant concentration is usually lower than when a separate doping cycle is used. With the lower doping, the Fermi level may not be at the polysilicon conduction-band edge, changing the effective work function of the gate electrode. For reasonably heavy source-drain implant doses, however, the flat-band voltages obtained are comparable to those of devices in which the polysilicon is doped from the gas phase in a separate cycle [6.3].

6.2.4 Polysilicon interconnections

In metal-gate, MOS integrated circuits the signals can only propagate in diffused single-crystal silicon regions and in the metal. In silicon-gate circuits the polysilicon can also be used as an additional partial level of short interconnections. Placing the polysilicon interconnections over thick field oxide also reduces the interconnection capacitance. As integrated circuits become more complex, interconnecting the device elements becomes more critical, and the additional flexibility provided by polysilicon interconnections becomes more important. However, this additional level of interconnections is not totally flexible because the processing sequence of a typical silicon-gate integrated circuit prevents a heavily doped region of single-crystal silicon from crossing under the polysilicon line.

In addition, the resistance of polysilicon interconnections can be high. In Sec. 5.6.1 we discussed the limited conductivity of polysilicon resulting from the solid solubility of the dopant in silicon. Even with the maximum electrically active dopant concentration, the limiting resistivity of polysilicon is about 400 $\mu\Omega$-cm, much higher than the 3 $\mu\Omega$-cm resistivity of aluminum.

When the polysilicon is doped by the source and drain implant, the amount of dopant introduced is limited by the requirement for a shallow source and drain in the single-crystal silicon. The limited amount of dopant restricts the conductance of the polysilicon interconnections. To increase the conductance, a silicide or other low-resistance conducting

6.2. SILICON-GATE MOS TRANSISTOR

material is usually placed over the polysilicon gates, both to lower the interconnection resistance and to short the p^+-n^+ diode in the polysilicon between the gates of the p-channel and the n-channel transistors. As discussed in Sec. 3.5.3, a *salicide* (self-aligned silicide process) can be used to form silicide simultaneously over the source, drain, and gate electrodes of an MOS transistor. An oxide or nitride spacer is first formed on the sidewall of the gate electrode to prevent shorting of the gate to the other electrodes as the silicide is formed. A thin layer of the metal to be used to form the silicide (usually titanium or cobalt) is deposited. The metal is then annealed at a moderate temperature so that it reacts with the exposed regions of single-crystal or polycrystalline silicon to form silicide. The unreacted metal over oxide or nitride regions is then selectively removed by wet chemical etching without removing the silicide. The silicide is then annealed at a higher temperature to lower its resistance. This method of forming silicide lowers the resistance of the source, drain, and gate electrodes considerably and is stable during further processing at moderate temperatures [6.4]. However, heavy n-type doping in the polysilicon retards silicide formation. Forming $TiSi_2$ also becomes more difficult as the width of the polysilicon lines decreases. Titanium silicide is most frequently used. However, because of the limitations of $TiSi_2$, $CoSi_2$ has been investigated as an alternative [6.5].

Even when the conductivity of the polysilicon is augmented by a silicide, the conductance is much less than that of typical aluminum metallization. Because the resistance of a long polysilicon interconnection can limit high-performance circuits, polysilicon is primarily used for short interconnections.

6.2.5 Gate-oxide reliability

In addition to the advantages of polysilicon for design and fabrication, using a polysilicon gate electrode can improve long-term device reliability. During operation of an MOS integrated circuit, high fields are applied across the gate dielectric, leading to possible charge injection and trapping in the dielectric, which increases the internal electric field. When the internal field exceeds the dielectric breakdown field, catastrophic device failure occurs. The *time-to-breakdown* of the gate oxide is higher for structures which use polysilicon as the gate electrode than for those with aluminum gates [6.6, 6.7]. The increased time-to-breakdown

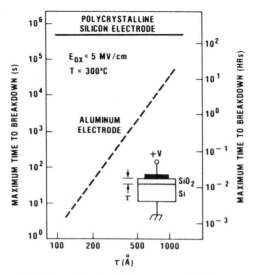

Figure 6.6: The reliability of polysilicon-gate MOS transistors is greater than that of aluminum-gate devices, and the difference increases as the gate-oxide thickness decreases. From [6.8]. ©1979 IEEE. Reprinted with permission.

is primarily observed with positive bias on the gate electrode; only a small improvement is seen with negative bias. For positive gate bias the *wear-out* time at 300°C can 3 to 4 orders of magnitude greater with a polysilicon gate electrode. The difference is larger for thinner gate oxides, as shown in Fig. 6.6 [6.8]. While the wear-out time is independent of oxide thickness with polysilicon electrodes, it decreases as the oxide thickness decreases with aluminum electrodes, making polysilicon more attractive in scaled devices. In addition, the activation energy for wearout is greater with polysilicon (2.4 eV) than with aluminum (1.4 eV), so that the reliability is improved even more near room temperature.

The improved reliability is postulated to be related to the different interactions of polysilicon and aluminum with silicon dioxide [6.6]. Aluminum is thought to reduce SiO_2 by the reaction

$$4\,\text{Al} + 3\,\text{SiO}_2 \rightarrow 2\,\text{Al}_2\text{O}_3 + 3\,\text{Si} \tag{6.3}$$

to form stable Al_2O_3. Alternate explanations involve the hydrogen introduced during polysilicon deposition or the formation of a thin phosphosilicate-glass layer during doping of the polysilicon.

6.2. SILICON-GATE MOS TRANSISTOR

6.2.6 Limitations

The silicon-gate MOS transistor has allowed unprecedented transistor density to be achieved during the three decades that it has been used as the gate electrode of MOS transistors. However, the limitations of polysilicon gate electrodes are becoming more serious as transistor dimensions decrease. Both n-channel and p-channel transistors become more difficult to build as the gate-oxide thickness decreases along with the lateral device dimensions, and the limited conductivity of polysilicon increasingly restricts routing signals long distances in polysilicon.

n-channel transistors: The n-channel transistor is limited by the difficulty of heavily doping the entire thickness of the polysilicon gate electrode. When a metal is used as the gate electrode of an MOS transistor, the high density of states near the Fermi level ensures that adequate carriers are always present adjacent to the gate dielectric to prevent carrier depletion and the associated voltage drop in the gate electrode. When n-type polysilicon is used, the carrier density in the polysilicon near the gate dielectric may be low enough that electrons can be pulled away from the bottom of the polysilicon adjacent to the gate dielectric at the positive biases used to turn on the transistor, forming a depletion region in the polysilicon. For a given gate voltage, the voltage drop across this depletion region reduces the voltage across the gate dielectric and the number of electrons induced in the channel of the transistor, degrading the maximum current that can flow in the channel of the transistor.

The depleted region of polysilicon effectively acts as an additional dielectric in series with the gate dielectric, increasing the gate voltage needed to turn on the transistor (the threshold voltage) and decreasing the drain current by decreasing the coupling of the gate voltage to the channel. The drain current I_D in the saturation region can be expressed by the simplified equation

$$I_D = k\, C_G \,(V_G - V_T)^2 \qquad (6.4)$$

where k depends on the transistor layout and carrier mobility, C_G is the effective gate dielectric capacitance and V_T is the transistor threshold voltage. Both the increase of V_T and decrease of C_G degrade the saturation current. The depletion also increases the switching time of the circuit [6.9, 6.10] and increases the energy per cycle for the same gate delay. Short-channel effects also increase because the bulk charge under the channel is reduced.

This carrier depletion is especially serious when the gate electrode of the n-channel transistor is doped with arsenic at the same time as the source and drain junctions are formed. The amount of arsenic that can be added to the polysilicon is limited by the need for a shallow source and drain, so the amount is less than optimum for doping the polysilicon. In addition, to form shallow source and drain junctions, the heat cycle after implantation is limited, restricting the diffusion of the dopant through the thickness of the polysilicon. Even though dopant diffusion in polysilicon is much greater than in single-crystal silicon (Sec. 3.3), high concentrations of arsenic may not have adequate time to diffuse through the entire thickness of the polysilicon to dope the bottom of the polysilicon heavily enough that it cannot be depleted by an applied gate voltage. Avoiding gate depletion requires higher implanted arsenic concentrations (consistent with the amount that can be added to the source and drain regions), maximizing the diffusion heat cycle (consistent with forming shallow source and drain junctions), and optimizing the structure of the polysilicon to allow rapid diffusion. This *polysilicon depletion* can be important for arsenic concentrations less than about mid-10^{19} cm^{-3}, with the effective dopant concentration somewhat less than the total dopant concentration, as discussed in Sec. 5.6.1.

Carrier depletion is more serious for the thinner gate dielectrics used in advanced CMOS circuits. As the gate dielectric becomes thinner, its effective coupling capacitance to the channel increases, as desired. A given series capacitance associated with the depleted region of the gate electrode then causes a greater fractional decrease in the effective gate capacitance. With gate dielectrics only a few nanometers thick, even slight depletion at the bottom of the polysilicon seriously degrades the driving current capability of the transistor.

Because carrier depletion is an increasingly important limitation in n-channel transistors, alternate materials are being investigated for the gate electrode [6.11]. However, most metals are less compatible with silicon processing than is polysilicon. To reduce process incompatibility, in one proposed technique [6.12] the polysilicon gate is formed in the normal manner and retained during most of the device processing. After the high-temperature steps are completed, the polysilicon is removed and replaced by a metal.

6.2. SILICON-GATE MOS TRANSISTOR

***p*-channel transistors:** While carrier depletion can limit *n*-channel transistors, a different limitation seriously degrades *p*-channel transistors. Boron diffuses rapidly through polysilicon, and can possibly diffuse through the underlying gate dielectric to change the dopant concentration in the channel region of the transistor. Boron diffusion through the gate oxide occurs readily if hydrogen or fluorine is present. Avoiding a hydrogen ambient during the diffusion reduces the penetration. However, boron is often added to the source, drain, and gate regions of the *p*-channel transistor by implanting BF_2 (to obtain a shallow junction in the source and drain regions). The fluorine incorporated into the polysilicon near the gate oxide enhances boron diffusion through the gate oxide and into the channel of the transistor, changing its threshold voltage [6.13]. Using B^+ as the implant source eliminates the fluorine, but at the expense of a lower implant current and an implant voltage that is harder to control. Alternatively, nitrogen can be added to all or a portion of the thickness of the gate dielectric to reduce the boron diffusion even when fluorine is present [6.14–6.16]. Fluorine can also degrade reliability by promoting interface-state generation during operation [6.15].

Adding nitrogen to the polysilicon near the gate dielectric has also been suggested [6.17–6.19]. Although the nitrogen in the polysilicon can effectively block the diffusion-enhancing effect of fluorine, it also reduces the electrical activity of boron. The carrier concentration may then not be adequate to avoid carrier-depletion near the bottom of the polysilicon, leading to reduced coupling of the gate voltage to the channel of the transistor.

The structure of the polysilicon can also be modified to reduce boron penetration of the gate dielectric. The columnar structure, which allows rapid vertical diffusion, can be avoided (although at the expense of less effective doping of the gate electrode of the *n*-channel transistor). For example, depositing amorphous silicon and crystallizing it creates larger, non-columnar grains. A two-step polysilicon deposition process can also decrease the diffusion if the grain boundaries are not continuous through the thickness of the polysilicon.

6.2.7 Process compatibility

The compatibility of polysilicon with high-temperature processing allows complex fabrication sequences involving multiple layers of polysilicon deposition, along with implantation and annealing processes. The surface can also be smoothed to allow better coverage of steps by aluminum and improve fabrication yield. When density and shallow junctions do not limit the heat treatment, a silicon-dioxide layer containing phosphorus and possibly boron can be deposited over the polysilicon gate electrodes. It can be heated until its viscosity decreases and it "flows," reducing the angle of the steps that the aluminum must cover. This phosphosilicate glass (PSG) or borophosphosilicate glass (BPSG) can be *flowed* before the contact windows are opened, or it can be *reflowed* after the windows are opened to provide a more easily covered step at the edges of the contact windows.

The flexibility of polysilicon processing can be used in an additional way. As device dimensions decrease, the gate dielectric thickness must decrease also. Fragile gate dielectrics of 5–10 nm are often used in advanced processes. If additional processing is needed between the formation of the gate dielectric and the deposition of the polysilicon gate electrode (*eg*, masking and implantation to adjust the threshold voltage of one type of transistor), a thin, protective layer of polysilicon can be placed over the gate dielectric after its formation. After the intermediate processing, the remainder of the thickness of the polysilicon gate electrode is deposited.

6.2.8 New structures

The high-temperature capability of polysilicon also permits the development of more complex device structures that provide denser integrated circuits with the same minimum feature size. For example, the size of the one-transistor, dynamic, random-access-memory (RAM) cell to be discussed in Sec. 6.11 cell can be reduced significantly by using two layers of polysilicon. The first level is patterned and oxidized, and then the second is deposited so that the effective spacing between these electrodes is only the thickness of the oxide on the sidewall of the first level of polysilicon.

Independent, closely spaced polysilicon electrodes above single-crystal silicon are also used in charge-coupled devices. As in the dynamic RAM cell, different voltages are applied to adjacent electrodes, and no potential barrier can be tolerated in the substrate between these electrodes when they are biased to allow charge transfer. Two or more layers of polysilicon are typically used to form these electrodes. Although the

6.3. NONVOLATILE MEMORIES

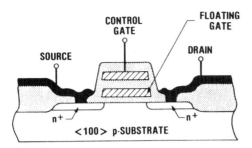

Figure 6.7: A nonvolatile memory element can be constructed by using a *floating* polysilicon gate between the silicon and a second *control* gate. Courtesy of J. Andrews

electrodes are laterally separated by the thickness of the oxide grown on the sidewall of the first layer of polysilicon, fringing fields reduce the barrier between the two electrodes when both are biased to induce inversion layers in the substrate.

6.3 Nonvolatile memories

Using two layers of polysilicon can be advantageous in many other devices also. Nonvolatile, electrically erasable, programmable, read-only memories (EEPROMs or E^{2}PROMs) can be fabricated by using one level of polysilicon above another in the gate region of the device, as shown in Fig. 6.7. The first level of polysilicon is not electrically connected and "floats." The conductance of the underlying channel between the source and drain regions depends on the charge stored on this *floating gate*. If it contains excess electrons, it repels electrons from the channel and turns off an n-channel transistor. A positive charge on the gate induces a channel and allows current to flow between source and drain. Substantial amounts of this type of memory are frequently included in specialized microprocessor chips to allow storage of code specific to a particular application and other information that must be infrequently modified or must remain when power is removed from the system. Similar devices are used in *flash memory,* although the devices are organized in the circuit differently.

Applying a voltage of the appropriate potential to the top *control gate* allows charge to be transferred through the gate oxide to charge the *floating gate* and write the cell. The charge can be generated by avalanching the drain-substrate junction to create free carriers, which are then attracted to or repelled from the floating gate, depending on the potential of the control gate. The charge can be removed from the floating gate by tunneling through the oxide separating the two layers of polysilicon. When this method of erasing the cell is used, the surface texture of the lower layer of polysilicon is critical. We have already seen in Sec. 4.3 that conduction through oxide grown on polysilicon depends strongly on the surface roughness. Consequently, controlling the deposition conditions and the resulting structure of the polysilicon is probably more critical for this application than for any other. To reduce stringent control of the structure formed during deposition, the floating silicon layer can be deposited in the amorphous form and subsequently crystallized to provide a smooth surface. The surface can then be controllably roughed by oxidizing and removing the oxide before the tunnel oxide is grown [6.20].

Charge may also be removed from the floating gate by forcing it back through the underlying gate oxide into the substrate. The discharging characteristics depend on the structure of the polysilicon [6.21]. Erasing may be more effective along the phosphorus-doped oxide under the grain boundaries [6.21], making the grain size and grain-boundary density critical. Erasing a number of devices uniformly may depend on minimizing statistical variations between devices, favoring a fairly large number of grains within each device to minimize statistical variations. As the device size decreases, the grain size should also decrease to retain a certain number of grains (perhaps ~ 20 in each device); this need may limit scaling.

Using hemispherical-grain polysilicon (HSG) with its rough surface as the floating electrode increases the capacitive coupling to the control gate and improves device performance [6.22]. Separating the floating gate into small, individual nuclei, rather having than a continuous layer of polysilicon, may allow operation at lower voltages and increase endurance [6.23]. Each small island of silicon contains one or a very few electrons.

6.4 Polysilicon resistors

After silicon-gate technology was developed for MOS devices, polysilicon found other uses in integrated circuits. Because it is readily available,

6.4. POLYSILICON RESISTORS

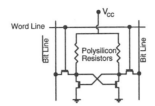

Figure 6.8: Two, high-value, polysilicon resistors can be used as the load devices in a static RAM cell to reduce the cell size.

it is frequently used to form resistors over a wide range of resistances [6.24].

Polysilicon resistors are often used as the high-value load resistors in static random-access memory cells. A typical static RAM cell uses a cross-coupled structure that allows current to flow in one or the other of two parallel current paths; the leg that the current flows through and the associated voltages at the intermediate nodes of the two legs define the state of the cell. This static RAM cell can be formed by using two transistors connected in series in each leg of the circuit and two access transistors, one connected to the intermediate note of each leg, so that the resulting memory cell has six transistors. To conserve area the two load transistors can be replaced by high-value resistors, forming the 4-transistor, 2-resistor cell shown in Fig. 6.8 [6.25]. The nominal electrical requirements on the resistors are not severe. Their resistance must be low enough that adequate current can flow to retain a voltage on the intermediate node and prevent stray charge or electrical noise from causing the cross-coupled cell to change state. It must be high enough to limit the current flowing through the driver transistors so that power dissipation in the chip remains acceptable. Resistors in the range of 10^9 Ω are generally satisfactory, and control of the resistance is not critical.

Lightly doped polysilicon can readily meet these requirements. Polysilicon is especially attractive because of its compatibility with integrated-circuit processing and the high resistance possible [6.26]. To fabricate the polysilicon resistors, lightly doped polysilicon needs only to be shielded from dopant introduction during doping of the source, drain, and polysilicon gate of nearby transistors.

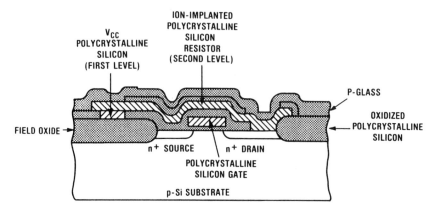

Figure 6.9: Placing the polysilicon load resistors of a static RAM cell above the transistors reduces the area required for the cell. Courtesy of J. Andrews.

As we saw in Sec. 5.2, undoped polysilicon has a resistivity in the mid-10^5 Ω-cm range. With a typical thickness of a few hundred nanometers, a sheet resistance of about 10^{10} $\Omega/\square$ is easily obtained. In some cases, this value is satisfactory for the load resistors. More frequently, however, a somewhat lower sheet resistance is needed to insure that adequate current flows to avoid stray charge at the the intermediate nodes from switching the state of the cell. A small amount of dopant can readily be added by ion implantation. Although control of the resistivity in the lightly doped region is difficult (Fig. 5.1), the wide resistance range tolerated by this circuit makes the use of lightly doped polysilicon practical. However, the rapid temperature variation of the resistivity of lightly doped polysilicon is a disadvantage, causing the power dissipation to increase at higher temperatures. The temperature variation can possibly be reduced by doping with both phosphorus and boron [6.27].

The 4-transistor, 2-resistor cell can reduce area, especially if the resistor is placed above a transistor in a second layer of polysilicon, as shown in Fig. 6.9 [6.28]. The size of the memory cell is determined by the area of the four transistors. The resistors do not require additional area, except possibly for connection to the underlying transistors.

The rapid diffusion of dopant atoms in polysilicon limits the minimum physical length of the resistors. Heavily doped contact regions are usually formed in the polysilicon on either side of the lightly doped resistor for good ohmic contact. During subsequent heat cycles, this

6.4. POLYSILICON RESISTORS

dopant diffuses rapidly along grain boundaries from the contacts into the resistor. If the resistor is not long enough, a highly doped, conducting filament can form along a grain boundary, creating a low-resistance path between the two contact regions at opposite ends of the resistor. As processing temperatures and times decrease, however, lateral grain-boundary diffusion becomes less limiting.

As we saw in Sec. 5.5, hydrogen can saturate dangling bonds at grain boundaries, changing the grain-boundary barrier height and, therefore, the resistivity of a polysilicon film. Because lightly doped films are sensitive to the barrier height, these films are affected strongly by adding hydrogen. Therefore, the resistance of polysilicon load elements can be inadvertently changed by hydrogen introduced during the last stages of the fabrication cycle or during circuit operation. Hydrogen from a plasma-nitride passivation layer is of special concern because it can move during device operation and possibly change the value of the resistor. If circuit margins are not adequate, these changes can degrade long-term circuit reliability.

Although polysilicon load resistors are adequate for many static RAM cells, improved circuit performance is obtained if the load element connected to the transistor being turned off can conduct enough current to charge the intermediate node rapidly during a transient, but draws little dc current when in series with the driving transistor that is on. A voltage-variable load element is, therefore, advantageous. If a gate electrode can modulate the conductivity of a polysilicon resistor to even a moderate degree, the circuit performance of the static RAM can be improved significantly. We have already seen that circuit density is improved if the polysilicon resistor is placed above the gate electrode of the bulk MOS transistor. This location also potentially allows one gate electrode to control both the bulk transistor and the voltage-variable polysilicon resistor. Polysilicon thin-film MOS transistors with their *channels* in a layer of polysilicon are widely used in static random-access memories and other applications. These polysilicon transistors will be discussed in Sec. 6.13.

In addition to high-value resistors for static random-access memories, polysilicon is also widely used to form moderate- and low-value resistors. Moderate-value resistors can be formed by implanting controlled amounts of dopant into polysilicon. However, operating on the steep portion of the resistivity-concentration curves shown in Figs. 5.1 and 5.8 limits control of the resistance. In the moderate-concentration range, the n-type dopants tend to segregate to the grain boundaries. The amount of segregation depends on the temperature and time of the last high-temperature processing steps, again making the final value of resistance harder to control. Boron does not segregate to the grain boundaries, so the resistance is easier to control when boron is used. For moderately doped resistors, fixed charges associated with the surrounding oxides can enhance or deplete the carrier concentration near the edges of the resistor [6.29, 6.30]. The resistance no longer scales directly with width; its value also depends on the quality of the surrounding oxide, making it harder to control.

For very low values of resistance, the dopant concentration may approach solid solubility. The effective carrier concentration corresponding to solid solubility is greater for phosphorus than for boron [5.65], so doping with phosphorus may allow lower values of resistance. The variation of the resistivity with varying processing temperature may also be less for phosphorus. High concentrations of phosphorus promote grain growth during subsequent high-temperature processing, changing the resistance somewhat.

For high- and moderate-value resistors, the grain boundaries limit the conductivity by trapping carriers and by forming a depletion-region potential barrier in the grains surrounding the grain boundary. When the conductivity is limited by the grain boundaries, the resistance decreases with increasing operating temperature; *ie,* the *temperature coefficient of resistance* (TCR) is negative. At high dopant concentrations, the grain-boundary barriers are comparable to or smaller than the thermal energy near room temperature and no longer limit the conduction. Other scattering mechanisms dominate, and the resistance increases with increasing operating temperature (positive TCR). Near the transition between the regions with different conduction-limiting mechanisms, the temperature coefficient of resistance passes through zero. Judicious choice of dopant concentration and processing conditions can lead to a very small temperature coefficient of resistance, but it is difficult to adjust the TCR and conductivity independently.

Because the value of polysilicon resistors can be difficult to control precisely, the value of critical resistors is sometimes adjusted after fabrication. Either electrical current [5.88, 6.31-6.34] or a laser [6.35] can be used to trim the resistor. Trimming may involve selectively melting regions of the resistor, probably near the grain boundaries, and changes in the segregation of the dopant between the grains and grain boundaries,

6.5. FUSIBLE LINKS

either in the molten or solid state. Because the temperature coefficient of resistance is different in the grains and at the grain boundaries, changing the relative importance of grains and grain boundaries also changes the TCR of the resistor [6.36]. For some applications the noise in polysilicon resistors is also important, and $1/f$ noise is related to fluctuations [6.37].

6.5 Fusible links

As the number of components on an integrated circuit increases, the ability to select current paths after device fabrication becomes more important. A desired logic function can be obtained by programming an uncommitted *field-programmable gate array* (FPGA), allowing rapid prototyping of integrated circuits. If the volume of circuits needed is high, mask-programmed versions of the same circuit can be produced after the prototype is developed using FPGAs. The programming capability can also be used to select functioning portions of a large integrated circuit, significantly increasing the effective yield of complex chips. In array structures, such as RAMs, a few extra rows or columns of memory cells can be placed on the chip, and these can be used to replace malfunctioning elements in the main portion of the array after testing. This *redundancy* can greatly increase the effective yield of complex circuits.

Polysilicon links can be used to insert or remove elements of the integrated circuit. A low-resistance polysilicon link can be vaporized by a high-power laser pulse to remove a conducting path. A high current forced through a highly conductive polysilicon link can also break the connection. Alternatively, polysilicon can be used as an *anti-fuse*. As has already been discussed in Sec. 5.7.2, a high current forced through a high-value polysilicon resistor can lower the resistance by many orders of magnitude, effectively connecting circuit elements [5.90, 6.38, 6.39].

The motion of dopant atoms in the solid phase under long-term operation at high currents can also be observed and, under extreme conditions, can lead to failure of a conductor by *electromigration* [6.40–6.42]. Momentum transfer from the moving carriers in a temperature gradient causes net movement of the dopant atoms away from one end of the film, locally depleting the dopant. Because the resistivity of polysilicon depends strongly on the dopant concentration, the resistivity in this region increases significantly, leading to runaway Joule heating and finally to catastrophic failure. The direction of dopant motion depends on the sign of the majority carriers.

6.6 Gettering

Gettering of heavy metals away from critical device regions is necessary to produce highly reliable integrated circuits in high-volume manufacturing facilities. For example, removing metals contaminants from the charge-storage regions of a dynamic RAM reduces the number of generation-recombination centers there and increases the time allowed between refresh cycles. The high-temperature capability of polysilicon allows efficient gettering. The single-crystal silicon on the back of the wafer is often exposed by removing any oxide or deposited layers before the polysilicon on the front of the wafer is heavily doped with phosphorus from the gas phase. During this doping cycle, phosphorus diffuses into the back of the single-crystal wafer, as well as into the polysilicon on the front. The phosphorus-containing region on the back of the wafer can attract heavy metals during all subsequent high-temperature device processing. However, *backside gettering* is becoming less important as integrated-circuit fabrication temperatures are reduced and thicker wafers are used; alternate gettering techniques, such as *intrinsic gettering* of impurities to oxide precipitates in the bulk of the silicon wafer, do not rely on diffusion of the heavy metals through the entire thickness of the silicon wafer.

6.7 Polysilicon contacts

6.7.1 Reduction of junction spiking

During the final annealing cycle used to improve the Si/SiO$_2$ interface properties and the contact between aluminum interconnections and silicon, some reaction occurs between the silicon and the aluminum in the contact regions. Because silicon has a finite solubility in aluminum (about 1.5% at the melting temperature of aluminum), it tends to be drawn out of the silicon contact regions and diffuse into pure aluminum interconnections [3.56], as already discussed in Sec. 3.5.1. The resulting voids in the silicon are then filled with nearby aluminum, leading to aluminum *spiking* through shallow p-n junctions. Such spiking can short the interconnections to the substrate. Spiking is accentuated near the

6.7. POLYSILICON CONTACTS

edges of contacts because more silicon must be supplied from these regions to diffuse laterally into the aluminum interconnections over silicon dioxide [6.43]. Although spiking is reduced by adding a small amount of silicon to the aluminum during deposition, the potential for spiking remains significant for the very shallow junctions used in scaled devices.

Placing a polysilicon layer between the single-crystal silicon contact region and the aluminum interconnection can reduce spiking by providing the silicon needed to saturate the aluminum line [6.44]. The low specific contact resistance between aluminum and heavily doped polysilicon [6.45] also makes polysilicon contacts attractive. In addition, when phosphorus is used, the gettering action provided can improve the electrical properties of nearby junctions.

6.7.2 Diffusion from polysilicon

As lateral device features are scaled to smaller dimensions, the junction depth should decrease also. However, with shallower junctions, control of the dopant introduction and subsequent diffusion becomes more difficult. Dopant is typically added by ion implantation into single-crystal silicon, followed by a heat treatment to reduce the lattice damage caused by the implantation and to move the depletion regions away from the remaining lattice damage. As junctions become shallower, however, the dopant can diffuse excessively during the heat treatment. For very shallow junctions, the heat cycle must be limited, and the proximity of the depletion region to the implantation damage remaining after annealing can degrade the junction. The crystal damage caused by the implantation creates point defects, which greatly enhance the dopant diffusion during moderate thermal cycles (*transient-enhanced diffusion*). To obtain very shallow, high-quality junctions, this implantation damage in the single-crystal silicon must be avoided. In addition, even at low implantation energies, boron penetrates a significant distance into silicon, and achieving shallow p-type junctions is difficult. The alternative doping method using gas-phase diffusion sources does not allow very shallow phosphorus or boron junctions, and gas-phase doping is generally impractical for arsenic.

Very shallow, damage-free junctions can be obtained using polysilicon. The dopant is implanted into a layer of polysilicon deposited directly on single-crystal silicon. The implant damage is contained in the polysilicon, preserving high crystal quality in the underlying single-crystal silicon. A short heat cycle then moves the dopant from the polysilicon into the single-crystal silicon. Because the dopant diffusivity is much lower in single-crystal silicon than in polysilicon, the dopant diffuses rapidly through the polysilicon. Its subsequent movement into the single-crystal silicon is limited, forming a very shallow junction. Contact is generally made by depositing metal interconnections on the polysilicon, reducing the possibility of spiking through shallow junctions, as discussed above. In addition, the polysilicon may extend over thick oxide layers adjacent to the silicon region being doped so that contact to the metal can be made over oxide with its low relative permittivity, reducing the area of the diffused region and the associated capacitance. Extending the polysilicon over oxide also provides some additional local interconnection capability.

6.8 Vertical *npn* bipolar transistors

Although polycrystalline silicon is most frequently used in silicon-gate, MOS integrated circuits, its use in bipolar transistors has been demonstrated for many years [6.46], and it is used in virtually all modern, high-performance, bipolar integrated circuits [6.47, 6.48]. It is used to make contact to the base region, reducing resistance and capacitance to improve both device performance and packing density. More importantly, using polysilicon to contact the emitter region of the transistor improves the basic transistor operation by reducing the reverse hole injection from the base into the emitter which often degrades the gain of high-performance, bipolar integrated-circuit transistors. In addition, implanting into polysilicon and diffusing the dopant into the underlying single-crystal silicon reduces the crystal damage in the single-crystal silicon that leads to the transient-enhanced diffusion that limits forming shallow junctions.

6.8.1 Fabrication: Polysilicon contacts

In modern bipolar transistors, the minority-carrier transit time through the thin base region is very short, and the speed of the transistor is often limited by parasitic elements. The resistance of the *extrinsic base* (*ie*, the *p*-type region between the external base contact and the *intrinsic base* under the emitter) can degrade device performance significantly, as can the capacitance associated with the depletion region between the

6.8. VERTICAL NPN BIPOLAR TRANSISTORS

extrinsic base and the collector. In conventional bipolar transistors, the extrinsic base region is heavily doped to reduce its resistance and is large so that metal contact can be made to it. The heavy doping and large area both increase the capacitance. Moreover, in the conventional structure the extrinsic base must be separated from the emitter region by more than one alignment tolerance to avoid low emitter-base breakdown voltages. This separation and its variability limit the effectiveness of the heavily doped, extrinsic base region.

To minimize the conflicting requirements of low base resistance, low capacitance, and ease of mask alignment, the major part of the extrinsic base region can be taken out of the single-crystal silicon and placed in a layer of heavily doped polysilicon above an oxide, as shown in Fig. 6.10a. Contact between metal and the base region is made through the polysilicon, with the actual contact between the metal and the polysilicon placed over the oxide (Fig. 6.10d) to reduce the p-n junction area in the silicon and the associated parasitic capacitance. The polysilicon over the thick, low-permittivity oxide can be heavily doped without adding significantly to the capacitance. In addition to providing a convenient contact to the base, polysilicon can be used to self-align the emitter region to the extrinsic base so that the two are separated only by the thickness of an oxide layer formed on the sidewalls of the base polysilicon. This self-alignment reduces the parasitic base resistance and eliminates the variability resulting from inaccurate mask alignment. A second level of polysilicon is usually used to make contact to the emitter region, with the significant advantages to be discussed below.

To fabricate the self-aligned, *polysilicon-emitter,* bipolar transistor, the base polysilicon is first deposited, doped p-type with boron, and defined to remove it from the future emitter region (Fig. 6.10a). The single-crystal portion of the extrinsic base is then formed by diffusing dopant from the polysilicon. This first layer of polysilicon is oxidized, and the oxide is removed from the single-crystal silicon (Fig. 6.10b). The intrinsic base dopant, and possibly the emitter dopant, can be added by implantation to the exposed single-crystal silicon at this point, or they can be added later by diffusion from the second layer of polysilicon.

Next, the second polysilicon film is deposited and defined so that it remains over the emitter region (Fig. 6.10c). If not already added, the base dopant is implanted into this second layer of polysilicon and diffused into the underlying single-crystal silicon. The n-type emitter

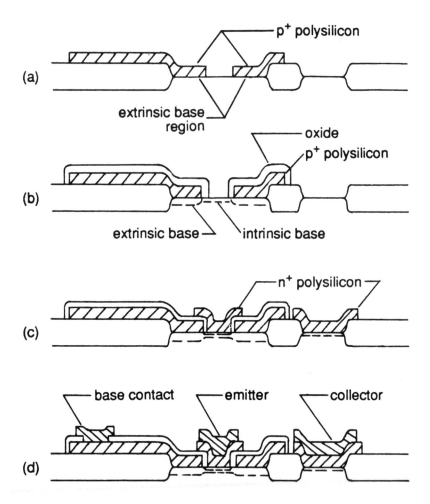

Figure 6.10: Two layers of polysilicon can be used in the polysilicon-emitter transistor, with an oxide on the sidewall of the first layer separating the two. (a) The extrinsic-base region is formed by the first polysilicon layer, which extends over the field oxide. (b) Boron is diffused from the polysilicon into the single-crystal silicon to connect the polysilicon to the intrinsic base, which can be added by implantation. The polysilicon is oxidized, and the oxide is removed from the single-crystal region that is to become the emitter. (c) A second layer of polysilicon is deposited and defined either before or after adding the emitter doping to the single-crystal silicon. (d) Contact to the base region is formed over the oxide, reducing the area of the single-crystal extrinsic-base region.

6.8. VERTICAL NPN BIPOLAR TRANSISTORS

dopant is then implanted into the polysilicon and briefly heat treated to form a very shallow n-type region in the single-crystal silicon. Implanting the dopant into the polysilicon minimizes the *transient-enhanced diffusion* caused by point defects created when dopant is implanted directly into single-crystal silicon, as discussed in Sec. 6.7.2. Significant transient-enhanced diffusion can prevent shallow junction formation and limit transistor performance. The point defects created by implanting into polysilicon are likely to recombine before they reach the single-crystal silicon, so they cause little acceleration of the diffusion there. Alternatively, the second layer of polysilicon can be doped n-type during deposition and the single-crystal emitter formed by a short heat treatment. The polysilicon can be deposited in polycrystalline form or deposited as amorphous silicon and subsequently crystallized [6.49, 6.50]. Arsenic is usually used as the emitter dopant when it is added by implantation. Phosphorus is usually used when it is added to the polysilicon during deposition, although arsenic is sometimes used [6.51].

Adding dopant during the polysilicon deposition is especially attractive for very small emitters. As the emitter size decreases, the edges of the emitter opening can shadow implants into the polysilicon, limiting device scaling [6.52]. When the emitter opening is less than twice the thickness of the second layer of polysilicon, polysilicon deposition on the sides of the opening can completely fill the opening, forming an *emitter plug* [6.53]. Diffusion of implanted dopant through this thick plug makes adding the emitter dopant by implantation and subsequent diffusion difficult. Adding n-type dopant during the polysilicon deposition provides a uniformly doped polysilicon region to serve as a dopant source for the subsequently formed shallow emitter. The emitter and collector contacts are usually located on extensions of the n-type polysilicon over the oxide, just as the base contact is made to p-type polysilicon over the oxide.

During fabrication of the structure shown in Fig. 6.10, the first layer of polysilicon must be etched away from the emitter region, stopping the etching process when the underlying single-crystal silicon is reached. Because polysilicon and single-crystal silicon etch similarly, removing all the polysilicon without damaging the critical region of the underlying single-crystal silicon is difficult. Refinements of the polysilicon-emitter structure avoid this crystal damage by using a single layer of polysilicon for both the base and emitter contact regions [6.54–6.56].

6.8.2 Physics of the polysilicon-emitter transistor

Not only does the polysilicon provide a convenient source of dopant atoms, but a polysilicon-emitter, bipolar transistor can have much a higher current gain than a similar transistor in which metal makes direct contact to the single-crystal emitter region. One of the factors degrading current gain is the injection of holes from the p-type base into the n-type emitter. This *reverse injection* adds to the base current a component which dilutes the useful base current composed of electron injection from the emitter into the base. In conventional bipolar transistors with metal contacts, the excess minority-carrier hole density in the emitter approaches zero at the contact. For the shallow emitters used in high-performance transistors, the hole gradient in the emitter is large, as is the resulting hole current $I_h = qD_p(dp/dx)$, degrading the *emitter injection efficiency* γ [$\gamma = I_e/(I_e + I_h)$] and the current gain. If the base doping is reduced to lower reverse hole injection into the emitter, the base resistance increases, again degrading the transistor performance. Placing a layer of polysilicon above the emitter reduces the reverse injection, increasing the injection efficiency and the current gain of the transistor. The base dopant concentration can then be increased to lower the series base resistance without unacceptably increasing the reverse injection. The emitter diffusion in the single-crystal silicon can also be shallower without a corresponding increase in the reverse injection and the base current.

Four different mechanisms have been proposed to explain the reduced reverse injection of holes from the base into the emitter in the polysilicon-emitter transistor. First, the interfacial oxide between the polysilicon and the single-crystal silicon appears to be critical [6.57]. Immersion in a chemical oxidizing agent, such as H_2O_2, forms a thin layer of oxide (of the order of 2 nm thick) during the last stages of the cleaning process before polysilicon deposition. This layer is non-stoichiometric SiO_x, with an oxygen content equivalent to an SiO_2 layer about half as thick as the chemical oxide. Even if no chemical oxide is purposely grown and a dilute HF-containing etch is used near the end of the cleaning process, a thin native oxide can grow as the wafers are transported to the polysilicon deposition system and loaded into a hot-wall reactor. Typical deposition temperatures of about 620°C are not adequate to desorb the thin oxide, and using HCl as a gaseous etchant is not effective at this temperature, although HCl may have more subtle

6.8. VERTICAL NPN BIPOLAR TRANSISTORS

effects on the polysilicon emitter [6.58]. Consequently, an unintentionally formed, thin oxide containing about 1–2×10^{15} oxygen atoms/cm^2 can be present between the single-crystal silicon and the polysilicon. Although this native oxide layer is usually continuous after deposition of the polysilicon, it breaks up during annealing at high temperatures and when high dopant concentrations are present in the polysilicon, as discussed in Sec. 2.13 [2.140]. The base current flows primarily through those areas of the interface that are free of oxide [6.48]. Fluorine also appears to accelerate break up of the interfacial oxide [6.59–6.61], possibly because the Si-O-F complexes formed are less stable than Si–O bonds.

As the oxide becomes discontinuous, a portion of the polysilicon realigns epitaxially on the silicon substrate. Because of the improved contact, breaking up the interfacial oxide reduces the emitter resistance. Low-frequency noise is predominately generated at the interfacial oxide at the polysilicon/single-crystal interface. Breaking up the interfacial oxide reduces the noise. Fluorine segregation at the interface can also reduce the low-frequency noise [6.62].

When the oxide is continuous, it impedes the reverse injection of holes into the polysilicon. The carriers are postulated to tunnel through the thin interfacial oxide layer, and some band bending in the single-crystal region of the emitter has also been suggested [6.57]. When a thin, continuous, oxide layer is present, the gain of the polysilicon-emitter transistor can be as much as a factor of 30 or 40 greater than that of a conventional metal-contact transistor. If the oxide becomes discontinuous during annealing, the gain decreases [6.63, 6.64].

As the oxide becomes thicker, the tunneling probability of electrons, as well as holes, is reduced [6.65]. However, the effective barrier height for electrons is much less than that for holes [6.66], and electrons can penetrate the oxide barrier more easily. Thus, there is a range of oxide thicknesses for which the base current can be suppressed without severely decreasing the emitter current [6.67]. However, if the oxide is thicker than optimum, electron tunneling through the oxide decreases, and the emitter resistance increases, degrading the transistor performance. In addition, an excessively thick interfacial oxide can retard arsenic diffusion from the polysilicon into the single-crystal silicon, producing an emitter junction that is too shallow to have ideal diode characteristics. Therefore, to obtain the highest gain, an extremely thin, continuous oxide layer must be formed and possibly prevented from

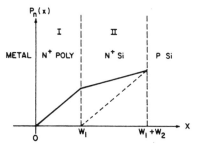

Figure 6.11: The polycrystalline silicon portion of the emitter can be considered to be an extension of the single-crystal emitter region, decreasing the hole gradient at the base edge of the emitter and, consequently, reducing the reverse hole injection. From [6.69]. ©1980 IEEE. Reprinted with permission.

breaking up during subsequent device fabrication. When the transistor characteristics are controlled by the interfacial oxide, the same thickness and other characteristics of the oxide must be obtained reproducibly [6.68].

A second mechanism proposed to explain the behavior of the polysilicon-emitter transistor involves hole transport within the polysilicon layer itself, rather than the presence of an interfacial oxide layer [6.69]. Holes injected from the base into the emitter diffuse toward the emitter contact, where they recombine. The polysilicon layer effectively lengthens the emitter region; the hole gradient is reduced; and less hole current flows. The low hole diffusion length in the polysilicon, however, makes it less effective in lengthening the emitter than a single-crystal region of equal length would be.

When this mechanism dominates, the hole current in the emitter, and hence the gain of the transistor, can be found from the emitter *Gummel number* G_e, the number of holes injected from the base into the emitter, through the equation

$$J_p = \frac{qn_i^2}{G_e} \exp\left(\frac{qV_{BE}}{kT}\right) \qquad (6.5)$$

The emitter Gummel number is related to the hole distribution in the single-crystal and polycrystalline regions of the emitter (Fig. 6.11) [6.69]. For a shallow emitter, the ratio of the Gummel number in a polysilicon-

6.8. VERTICAL NPN BIPOLAR TRANSISTORS

emitter transistor to that in a standard transistor is given by the relation

$$\frac{G_{ep}}{G_{ec}} = \frac{\beta_p}{\beta_c} = 1 + \frac{D_{p2}L_{p1}}{D_{p1}W_2} \tanh\left(\frac{W_1}{L_{p1}}\right) \qquad (6.6)$$

where G_{ep} and G_{ec} are the Gummel numbers in the polysilicon-emitter and conventional transistors, respectively, β_p and β_c are the corresponding current gains, the subscripts 1 and 2 refer to the polycrystalline and single-crystal regions of the emitter structure, D_p is the minority-carrier diffusion coefficient, L_p is the minority-carrier diffusion length, and W is the length of each region; for the conventional emitter $W_1 = 0$.

The dependence of the gain on the thickness of the polysilicon layer is cited as evidence for the importance of this mechanism [6.69]. The gain increases with increasing polysilicon thickness before saturating for films about 50 nm thick, indicating that the minority-carrier diffusion length in the polysilicon is of this magnitude. The minority-carrier lifetime found from the diffusion length is of the order of 10 ps, consistent with the value expected from Sec. 5.7.1. Increasing the effective length of the emitter increases the transistor gain by a factor of about 3 or 4, much less than the increase attributed to a continuous oxide layer at the polysilicon/single-crystal interface. When minority-carrier transport in the polysilicon affects the transistor performance, the inhomogeneous structure of the polysilicon layer should also be considered [6.70]. The opposite effect of the potential barriers on majority and minority carriers can degrade the gain at high currents and low temperatures. Charge storage in the polysilicon can also decrease the cut-off frequency.

The reduction of base current in the polysilicon-emitter transistor is also strongly correlated with the segregation of arsenic at the polysilicon/single-crystal interface. A high arsenic concentration appears to form a barrier to minority-carrier hole transport across the interface and recombination in the polysilicon [6.71]. Three possible mechanisms have been suggested by which segregation of arsenic to the interface and nearby grain boundaries can affect the minority-carrier behavior [6.71]: (1) Electrically active arsenic contributes to a potential barrier to minority-carrier transport into the polysilicon. (2) Saturation of the dangling bonds by arsenic reduces the trap density. (3) Grain-boundary reconstruction promoted by arsenic reduces the trap concentration at the grain boundaries. All three of these mechanisms reduce injection and recombination as the arsenic concentration increases.

Values of the effective interface recombination velocity as low as 10^4 cm/s have been reported [6.72].

The interface recombination velocity is important because a high interface recombination velocity reduces the hole concentration near the interface. The low hole concentration, in turn, increases the hole gradient in the single-crystal emitter, increasing the reverse injection of holes into the emitter and degrading the transistor gain. Passivating the dangling bonds at the polysilicon/single-crystal interface can reduce the interface recombination velocity and improve the electrical performance of the transistor [6.61, 6.73].

As arsenic segregation increases, the interface recombination velocity and the base current decrease, but only up to average arsenic concentrations in the low 10^{20} cm^{-3} range. At higher dopant concentrations and also after annealing at high temperatures, the base current increases. This increase is attributed to changes in the structure of the polysilicon/single-crystal interface [6.71]. During annealing, the native oxide at the interface can become discontinuous, and minority carriers can penetrate into the polysilicon more readily, increasing the base current and decreasing the gain. In addition, when the oxide is discontinuous, both higher dopant concentrations and more severe annealing promote epitaxial regrowth of the polysilicon, moving the polysilicon/single-crystal interface farther from the base-emitter junction. The polysilicon/single-crystal interface still influences the base current, but its location, as well as the emitter doping profile, changes during the heat treatment.

Finally, stress in the polysilicon layer may create an additional potential barrier for holes at the polysilicon/single-crystal interface and increase the gain of the transistor [6.50].

As expected, the transistor characteristics depend strongly on the condition of the polysilicon/single-crystal interface and, consequently, any factors that modify this interface. Maximum gain enhancement is obtained with an interfacial oxide layer of optimum thickness; however, reproducibly achieving such an oxide in a manufacturing environment can be difficult. Although less significant gain enhancement is obtained when no oxide is intentionally formed between the single-crystal silicon and the polysilicon, fabrication is more controllable, and only limited improvement in manufacturing practices and equipment are needed to make this process manufacturable.

When moderate gain enhancement is adequate but low emitter resistance is needed, the single-crystal surface is usually etched in a dilute HF solution as the last step in the cleaning process before polysilicon deposition, and the time between cleaning and loading the wafers into the reactor is minimized. When arsenic is added by implanting into the polysilicon after deposition, the arsenic is first annealed at a low temperature to distribute it through the thickness of the polysilicon and slightly into the single-crystal silicon, forming the emitter junction. A short, higher-temperature, rapid thermal anneal is then used to break up the interfacial oxide and lower the emitter resistance.

This discussion shows that several different mechanisms can influence the gain of the polysilicon-emitter transistor and their coupled effects must be considered [6.74]. The computer modeling techniques discussed in Sec 3.7 allow a number of these coupled effects to be analyzed numerically.

As an alternate method of reducing minority-carrier injection from the base into the emitter, a material with a wider bandgap than silicon can be used to form a barrier to minority-carrier injection from the base into the emitter contact. Oxygen-doped polycrystalline silicon (*semi-insulating polycrystalline silicon* or SIPOS) has been used for this purpose [6.75]. Because of the relative alignment between the bandgaps of single-crystal silicon and SIPOS, the barrier to electron injection across this heterojunction is low, and electron flow is not significantly impeded, but the unwanted reverse injection of holes into the emitter is greatly reduced, improving the injection efficiency. However, the higher resistivity of SIPOS degrades the emitter resistance.

6.9 Lateral *pnp* bipolar transistors

Circuit design can be greatly simplified by using pairs of *complementary* bipolar transistors (*ie*, a combination of *npn* and *pnp* transistors). In a typical bipolar integrated-circuit process, however, forming high-performance vertical *pnp* transistors is difficult because the layers available are usually chosen to optimize the more common *npn* transistors. Therefore, a *lateral pnp* transistor is generally used. Two *p*-type regions are added at the same time as the base of the *npn* transistor to serve as the emitter and collector of the *pnp* transistor, as shown in Fig. 6.12a; the *n*-type epitaxial layer serves as the base region, and the current

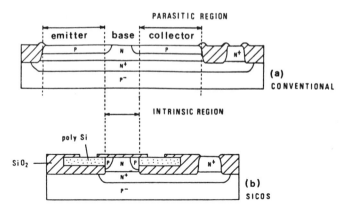

Figure 6.12: The large parasitic resistance and capacitance associated with the extrinsic base in the conventional, lateral *pnp* transistor (a) are greatly reduced in the advanced, lateral *pnp* transistor (b) in which the major portion of the emitter and collector are placed over an oxide layer. From [6.76]. ©1986 IEEE. Reprinted with permission.

flows laterally. The *p*-type regions must be large to allow contact to the metal interconnections, and the performance of this transistor is poor because of the large parasitic effects. Many carriers are injected vertically from the large emitter region; they contribute to base current but do not reach the collector region to provide a useful component of the collector current, thus decreasing the transistor gain. The parasitic capacitances associated with the large emitter and collector junctions also degrade the high-frequency performance.

The major drawbacks of the lateral structure can be reduced if most of the emitter and collector area can be isolated from the single-crystal substrate by an oxide layer. This isolation prevents vertical charge injection and reduces the capacitance associated with the horizontal portions of the emitter-base and base-collector junctions. The isolation can be achieved by using polysilicon over silicon dioxide to make *lateral* contact to the edges of narrow, *p*-type, single-crystal emitter and collector regions. Figure 6.12b shows the cross section of a transistor, called the *sidewall-contact structure* (SICOS), realized in this manner [6.76]. The intrinsic transistor region is a mesa-etched structure; the *p*-type emitter and collector regions are formed in the single-crystal silicon by diffusion from the adjacent, boron-doped, polysilicon regions. The layers in this structure are compatible with the standard, vertical *npn* transistor. The

6.10. DEVICE ISOLATION

reduced parasitic vertical injection decreases the diffusion capacitance, and the low permittivity of silicon dioxide reduces the junction capacitance; both effects increase the speed of the transistor. The reduced parasitic injection also increases the gain.

6.10 Device isolation

In addition to enhancing the performance of active devices, polysilicon is also employed in several isolation techniques used to reduce electrical interaction between adjacent devices. An older technique, which is still employed for radiation resistance and power applications, uses thick layers of polysilicon deposited under very different conditions than the thin layers we have primarily been discussing. Two other isolation methods use polysilicon deposited under more typical conditions.

6.10.1 Dielectric isolation

The oldest application of polysilicon still being manufactured is the use of polysilicon as the supporting layer for the single-crystal islands in dielectrically isolated integrated circuits [6.77]. In this technique the majority of a single-crystal silicon substrate is removed; the few remaining regions of single-crystal silicon are embedded in a thick layer of polysilicon and completely separated from it by an oxide layer. Devices are then fabricated in the isolated single-crystal-silicon pockets. Excellent isolation is obtained between adjacent pockets of silicon, but the process is complex. It is difficult to apply to high-volume or dense integrated circuits and is used only in special devices, such as radiation-tolerant circuits.

To fabricate a circuit using this technology, a single-crystal substrate of the appropriate conductivity type (usually n-type silicon to serve as the collector of an *npn* bipolar transistor) is first oxidized, and the oxide is removed from the isolation regions. Isolation moats are etched into the substrate to a depth greater than the thickness of the final silicon pockets. For (100)-oriented silicon an anisotropic etch can be used so that the etched regions are bounded by {111} planes inclined from the surface by about 55°. If the isolation regions being etched are narrow, etching terminates when the sloped {111} planes bounding the etched grooves intersect. The depth of the grooves is, therefore, determined by the width of the oxide windows. After silicon etching, the remaining

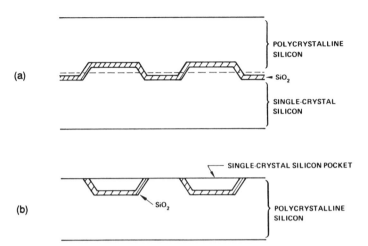

Figure 6.13: (a) In the dielectric-isolation process a thick layer of polysilicon is deposited after device regions are defined in the silicon and an oxide layer is formed. (b) The majority of the single-crystal region is removed, leaving isolated single-crystal pockets embedded in a polysilicon support.

oxide is removed. Heavily doped regions, which serve as the buried layers or subcollectors of the bipolar transistors, are next added to the exposed silicon surface, and the surface is oxidized. A thick layer of polysilicon is then deposited, as shown in Fig. 6.13a.

Because the polysilicon serves as a mechanical support, it must be approximately as thick as a silicon wafer, and high deposition rates are needed. To obtain these high rates, the polysilicon is usually deposited at a high temperature using a chlorinated silicon source gas, such as dichlorosilane or silicon tetrachloride, after a thin nucleating layer is formed with silane. Using a high temperature also reduces wafer deformation (*warpage*) during deposition by increasing the desorption of gaseous impurities from the silicon surface so that fewer are included in the deposited film, as discussed in Sec. 2.2 [2.14]. After polysilicon deposition, most of the original single-crystal wafer is removed by mechanically lapping and polishing the wafer until the embedded, oxide-coated grooves appear. The structure is then inverted, and transistors are fabricated in the exposed, single-crystal silicon pockets (Fig. 6.13b).

Because of wafer distortion during the deposition process, the pockets in the final structure are often displaced substantially from their

6.10. DEVICE ISOLATION

original locations. Registration of devices to the pockets is difficult, and this technique can only be used for low-density circuits. However, it does provide the excellent isolation required for very-high-voltage or radiation-tolerant integrated circuits. Some of the limitations can be removed by using more advanced, self-terminating etching techniques [6.78–6.80], so that grooves do not have to be formed before polysilicon deposition. Registration of masks applied before and after polysilicon deposition and substrate removal is then avoided. However, control of the polysilicon deposition and the substrate removal remain difficult, restricting the technique to specialized circuits.

In another isolation technique that uses polysilicon formed at high temperatures, polysilicon and epitaxial silicon are simultaneously deposited on a wafer, with the polysilicon in the isolation regions between device areas [6.81–6.84].

6.10.2 Poly-buffered LOCOS

MOS integrated circuits and most bipolar circuits do not need dielectrics completely surrounding each transistor region. For example, in a bulk MOS transistor, the conductivity type of the induced channel is opposite to that of the substrate so that a depletion region forms between the channel and the substrate. This depletion region isolates each conducting channel from the substrate so that no separate vertical isolation is needed between transistors of the same conductivity type. Only the lateral isolation provided by conventional isolation techniques is needed. Thin layers of CVD polysilicon can be used to refine standard isolation techniques.

Conventional LOCOS (LOCal Oxidation of Silicon) isolation technology has been used successfully for many years. However, as device dimensions decrease, lateral oxidation under the nitride mask that protects the device areas can create a transition region almost as large as the the nominal device dimension. To reduce the lateral oxidation, the nitride can be made thicker and the underlying *stress-relief* oxide can be made thinner. However, the increasing stress on the silicon substrate during oxidation can create defects in the silicon.

To minimize the stress-induced defects a *buffer* layer of polysilicon can be inserted between the thin stress-relief oxide and the thick oxidation-blocking nitride [6.85, 6.86]. This *poly-buffered LOCOS* isolation technique produces narrower transition regions. However, in ex-

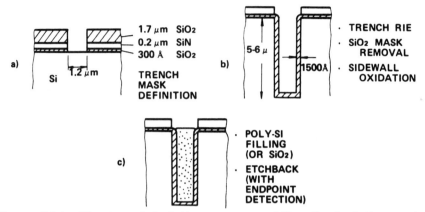

Figure 6.14: The trench-isolation process. After the isolation region is defined (a) and a groove is etched into the silicon, the walls of the groove are oxidized (b); it is then filled with polysilicon or SiO_2 (c). The isolation width scales well to dimensions much smaller than those shown in the figure. From [6.92]. ©1982 IEEE. Reprinted with permission.

treme cases, excessive stress on the polysilicon during field oxidation enhances silicon atom migration, creating voids in the polysilicon [6.87–6.89]. Subsequent processing replicates these voids as defects in the underlying single-crystal silicon. These defects occur especially near edges and corners, where the stresses are highest. Depositing the silicon layer in an amorphous form may reduce the void formation [6.89]. Doping the silicon layer with nitrogen may further extend the use of poly-buffered LOCOS [6.90].

6.10.3 Trench isolation

Another isolation technique suitable for small-dimension devices also uses thin layers of CVD polysilicon. In this *trench-isolation* technique (Fig. 6.14) [6.91, 6.92] a groove or *trench* is etched in the single-crystal silicon where the isolation is needed. This trench is formed by anisotropic dry etching to obtain a narrow, deep trench with nearly vertical walls. The walls of the trench are then oxidized, and polysilicon is deposited to fill the trench completely. The deposition conditions are chosen to be well within the surface-reaction-limited regime so that the silicon-containing gas (usually silane) can readily travel down the entire length of the narrow trench. Good control of the etching process is

6.11. DYNAMIC RANDOM-ACCESS MEMORIES

essential to form a trench that can be successfully filled with polysilicon; in particular, the angle of the trench walls must be a few degrees from the vertical, and a subsurface *bulge* must be avoided. Otherwise, filling may be incomplete, leaving a void that produces a mechanically unstable structure. If the etching and deposition processes are well controlled, the trench is completely filled once a layer half as thick as the width of the trench is deposited. A polysilicon layer of about the same thickness is simultaneously deposited on the oxide-covered top surface of the wafer, and this layer is subsequently removed without using a mask by anisotropic dry etching or chemical-mechanical polishing until the underlying oxide surface is exposed. After the oxide is removed from the single-crystal regions and a protective oxide is formed over the tops of the polysilicon-filled trenches, transistors are fabricated in the single-crystal regions.

Trench isolation can be used for either bipolar or MOS integrated circuits. It can reduce *latch-up* in dense, small-geometry, CMOS integrated circuits by increasing the effective path length between adjacent devices. Combining trench isolation with an epitaxial layer produces a CMOS structure especially resistant to latch-up.

As a modification of the trench-isolation technique, a selective deposition process can be employed, in which silicon is deposited in the trenches, but not on the top surface of the wafer [6.93]. In this case, the top surface and the walls of the trench are covered with oxide, but the oxide is removed from the bottom of the trench by anisotropic etching. Silicon is then deposited under carefully controlled conditions on the exposed silicon at the bottom of the trench, but not on the oxide-coated surfaces. A chlorine-containing ambient prevents nucleation of silicon on the oxide. Thus, the trench is filled from the bottom without having silicon nucleate on the oxide-coated walls or on the top surface of the wafer. If desired, an isolation diffusion can be added to the exposed single-crystal silicon at the bottom of the trench before refilling it. Other modifications of the basic trench-isolation process can decrease the time needed to fill the trench with polysilicon [6.94].

6.11 Dynamic random-access memories

Polysilicon can be used in a number of different ways to improve the packing density of dynamic random-access memories. Many of these improvements rely on the compatibility of polysilicon with sub-

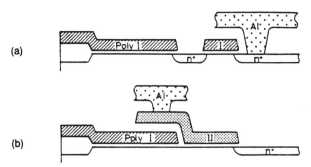

Figure 6.15: Dynamic RAM cells achieved with (a) a single layer of polysilicon and (b) two layers of polysilicon. Using two layers eliminates the need for the lateral separation between the gate and storage electrodes and allows a more compact cell. From [6.92]. ©1985 IEEE. Reprinted with permission.

sequent high-temperature processing. The size of the one-transistor, one-capacitor, dynamic, random-access-memory (DRAM) cell can be reduced significantly by using two layers of polysilicon—one as the gate of the access transistor and one as the counterelectrode of the storage capacitor. Figure 6.15a [6.92] shows that the single-layer polysilicon dynamic RAM cell requires space between the gate electrode and the capacitor counterelectrode for electrical isolation. This space is filled with a diffused region to prevent a barrier between the inversion layers formed under the electrodes from impeding charge flow to or from the storage region when the access transistor is turned on. When two levels of polysilicon are used (Fig. 6.15b), no mask separation is needed between the access transistor and storage region. They are separated laterally only by oxide grown on the edge of the first layer of polysilicon, reducing the cell size appreciably, although at the expense of more complex processing.

This use of two levels of polysilicon takes advantage of the different oxide thicknesses grown on lightly doped single-crystal silicon and on heavily doped polysilicon, as previously discussed in Sec. 4.2.2. The oxide under the second level of polysilicon serves as the capacitor dielectric for the storage cell and should be as thin as possible to increase the amount of charge stored per unit area. On the other hand, the oxide between the two levels of polysilicon should be as thick as possible to reduce the parasitic capacitance between these two electrodes

6.11. DYNAMIC RANDOM-ACCESS MEMORIES

and increase the breakdown voltage. Oxidation conditions are chosen to maximize the difference in oxide thicknesses grown in the two regions.

As the minimum feature size decreases, planar capacitors become impractical. While the size of the access transistor scales with decreasing minimum feature size, the storage capacitor does not scale. The sensitivity of the charge-detecting amplifiers and noise requirements dictate that the value of the capacitor remains relatively constant or decreases only slowly as the minimum feature size is scaled to smaller dimensions. The cell size is, therefore, not strictly limited by the minimum feature size; as minimum features become smaller, a planar capacitor between the substrate and an overlying polysilicon electrode occupies an increasing fraction of the cell area and limits the memory density. To obtain a greater storage area in a given cell size, a nonplanar structure is used. The nonplanar capacitor can extend either below the surface in a trench or be stacked above the surface.

6.11.1 Trench capacitor

We saw in Sec. 6.10.3 that polysilicon can be used to fill isolation regions between adjacent devices. The technique has been refined and applied to the storage capacitor of a one-transistor, dynamic RAM cell. The process used to form the trench capacitor is similar to the trench-isolation process: A recessed trench or well is etched in the surface of the wafer; dielectric is formed on the nearly vertical walls of the well; and the well is filled with heavily doped polysilicon. Charge is stored on the capacitor formed by the nearly vertical polysilicon/dielectric/silicon structure, markedly increasing the effective storage area without increasing the surface area of the cell [6.95]. Using a *trench capacitor* in the one-transistor, dynamic RAM cell allows the cell size to be reduced while retaining adequate capacitance for reliable operation. In addition, the processing used for the trench capacitor is similar to that used for isolation in logic circuits, allowing reasonably straightforward integration of dynamic RAM and logic on the same chip (*embedded* DRAM). A key difference between trench capacitors and trench isolation is that the polysilicon for the trench capacitor must be heavily doped. As we saw in Sec. 1.9, forming heavily doped polysilicon in deep recesses is difficult.

In the simplest realization of this structure (Fig. 6.16) [6.92], the conventional combination of inversion layer, oxide, and polysilicon counter-

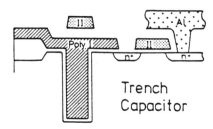

Figure 6.16: With the basic trench capacitor of a dynamic RAM cell, the inversion layer is formed outside the well, and a polysilicon counter-electrode inside the well is separated from the inversion layer by a thin dielectric. From [6.92]. ©1985 IEEE. Reprinted with permission.

electrode is placed vertically, rather than horizontally, so that the charge is stored in an inversion layer in the single-crystal silicon outside the well. This arrangement simplifies connection between the access transistor and the storage region. Controlling the shape of the bottom of the etched well is especially important for the trench capacitor. The maximum local electric field across the insulator separating the substrate from the polysilicon depends on the shape of the well near its bottom. Sharp, angular corners can cause high local electric fields and decrease the effective dielectric breakdown strength and reliability. The capacitor quality, therefore, depends strongly on the details of the etching process [6.96], and the bottom of the well is typically rounded to reduce the local electric field.

Although this trench capacitor offers the possibility of a large capacitance in a small area, the wells cannot be placed arbitrarily close together because the depletion regions surrounding the wells of adjacent cells can interact. The capacitor is also sensitive to alpha-particle-generated charge. These limitations led to the development of more complex structures in which the charge-storage region is placed *inside* the well [6.97]. The counterelectrode can be placed outside the well since its potential is common to all cells. Interaction between the storage regions of adjacent cells is significantly reduced in this structure, although at the expense of a more complex connection between the access transistor and the storage region.

To further reduce interactions between adjacent cells and increase the storage capacitance, two separate, isolated regions can be formed within the well [6.98, 6.99], as shown in Fig. 6.17. The charge-storage

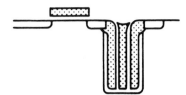

Figure 6.17: The *isolation-merged-vertical-capacitor* cell uses two layers of polysilicon inside the well, with charge usually stored in the first layer.

region is a thin layer of polysilicon on the walls of the well and is separated from the substrate by a dielectric layer. A second dielectric layer is formed on the exposed surface of the polysilicon inside the well, and the inner well is subsequently filled with conducting polysilicon to serve as the counterelectrode of this *isolation-merged-vertical-capacitor* (IVEC) cell.

To reduce the size of the memory cell to one minimum feature size plus associated isolation regions, both the access transistor and the storage capacitor can be placed on the nearly vertical walls of an etched well, as shown in Fig. 6.18 [6.100]. The access transistor encircles the well just below the surface, with the channel in the single-crystal silicon outside the well and the polysilicon in the well serving as the gate electrode. The storage capacitor is contained in the same well and is below the access transistor. To minimize interaction between adjacent cells, the charge is stored in a heavily doped region inside the well; the counterelectrode of the storage region is the single-crystal silicon outside the well. The charge flows vertically *outside* the well through the access transistor, then moves *inside* the well into a polysilicon storage region, which is isolated from the overlying gate electrode of the access transistor. This arrangement provides a very compact cell, which, in principle, can have an area as small as four minimum geometry areas (one for the active components and three for the surrounding isolation and the separation between the interconnections needed to read, write, and address the cell). This substantial area reduction is, of course, achieved at the expense of much more complex processing.

6.11.2 Stacked capacitor

An alternative to the trench capacitor is the *stacked capacitor*, which is built above the plane containing the access transistors. Unlike the

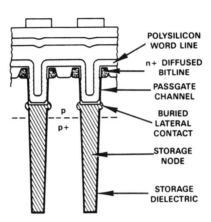

Figure 6.18: Cross section of the trench-transistor, cross-point dynamic RAM cell. From [6.100]. ©1985 IEEE. Reprinted with permission.

process used to fabricate the trench capacitor, the process used for stacked capacitors is very specialized to dynamic RAMs and is not readily compatible with logic processes, making its integration into embedded DRAMS difficult. However, the basic idea of maximizing the capacitor area while minimizing the projected area on the surface of the chip is the same. Instead of placing the capacitor below the surface by etching into the substrate, stacked capacitors are formed by building several layers of polysilicon and dielectrics above the surface. The effective capacitor area can be increased significantly above that of a planar capacitor, but at the expense of much more complex processing. Two examples of stacked capacitors are shown in Fig. 6.19. The finned capacitor (Fig. 6.19a) is formed by depositing several sequential layers of dielectric and polysilicon to form the lateral *fins* of the structure and then etching through these layers to the substrate contact and filling this region with a post of polysilicon, which makes electrical contact to the fins. After removing some sacrificial dielectric layers, the capacitor dielectric is formed; depositing the polysilicon counterelectrode completes the capacitor. The cylindrical capacitor (Fig. 6.19b) is built by forming the polysilicon connection to the substrate and the horizontal plate of the capacitor. After a sacrificial dielectric layer is formed and patterned over the post, polysilicon is deposited on the vertical and horizontal surfaces and then removed from horizontal surfaces by anisotropic dry etching to leave polysilicon only on the vertical sides of the sacrificial dielectric, which is then removed. The capacitor dielectric and polysil-

6.11. DYNAMIC RANDOM-ACCESS MEMORIES

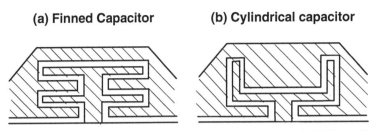

Figure 6.19: Stacked capacitors for dynamic RAMs: (a) Finned structure; (b) cylindrical structure. The polysilicon regions are shaded.

icon counterelectrode are then formed. The long, narrow, polysilicon post connecting the substrate and the capacitor is used in both of these stacked capacitor structures. Both increase the effective capacitor area significantly without increasing the projected area on the chip.

However, an even larger increase is needed as minimum feature sizes continue to decrease. The effective surface area can be further increased by roughening the surfaces of the capacitor plates. The surface of the polysilicon can be roughened by using the different oxidation rates of differently oriented grains (Sec. 4.2.2). First an oxide is grown under conditions that maximize the difference in oxidation rate between differently oriented grains and between grains and heavily doped grain boundaries. This oxide is removed, leaving a roughened surface on the polysilicon, and then the capacitor dielectric is formed on this rough surface. The effective capacitance can be increased by 30% or more using this technique [6.101].

Alternatively, the effective surface area can be increased significantly by forming the hemispherical-grain polysilicon (HSG) discussed in Sec. 2.12 on the polysilicon electrode of the stacked capacitor. More complex, non-planar electrodes can also increase the effective area of the capacitor. Using a cylindrical electrode, rather than a plane electrode, almost doubles the capacitance; roughening the surface with HSG polysilicon doubles the capacitance again [6.102, 6.103].

Using a dielectric with a higher permittivity than oxide or nitride can further increase the capacitance. For example, chemically vapor deposited Ta_2O_5 has a relative permittivity of about 22. Although only a fraction of this higher permittivity is realized in a practical structure, the capacitance can be increased by about 70% by using this higher-permittivity dielectric [6.104]. Very-high-permittivity dielectrics, such as barium strontium titanate (BST) are being studied, but the integration of such different materials into a silicon integrated-circuit process is difficult. Materials compatibility is a major concern; finding suitable barrier and electrode materials is a significant challenge. The dielectrics are often deposited in oxidizing atmospheres, so the electrode material either must not form an oxide or the oxide must be conductive. Materials such as ruthenium, platinum, and iridium [6.105] are being investigated.

6.12 Polysilicon diodes

In the applications discussed above, polysilicon is used to form conducting, resistive, or structural elements of an integrated circuit. Devices, such as diodes and transistors, can also be placed totally within the polysilicon. When active devices are fabricated in the polysilicon, its detailed electrical properties become critical.

The simplest of these devices is a lateral p-n junction diode [6.106, 6.107]. A structure in which both the p-type and the n-type polysilicon regions are heavily doped can readily be realized in a CMOS process that employs p^+ and n^+ polysilicon gates. As mentioned in Sec. 6.2, these diodes can form undesired parasitic elements between the gates of the two types of transistors unless a silicide is formed over the polysilicon or the two gate electrodes are connected together by other means. Parasitic polysilicon diodes can also appear in the polysilicon-resistor SRAM cells discussed in Sec. 6.4. However, for selected applications, the diode can be used advantageously. Polysilicon diodes are also key elements of the thin-film transistors to be discussed in Sec. 6.13.

For some applications polysilicon diodes can be used in place of high-value resistors to simplify device fabrication and save area. Using these diodes in high-performance, diode-transistor logic (DTL) has been suggested [6.108], as has their use to maintain logic levels. Because both sides of a polysilicon diode are isolated from the substrate, they can readily be connected in series, unlike diodes in single-crystal silicon where complicated isolation techniques must be used if series-connected diodes are needed. Series-connected diodes can be used for level shifting or in photosensor applications. Because polysilicon diodes can be placed over thick layers of oxide, parasitic capacitances can be small. However, polysilicon diodes have relatively high reverse leakage currents compared

6.12. POLYSILICON DIODES

to p-n junction diodes in single-crystal silicon [6.109], restricting their use when reverse leakage is critical. High series resistance can also limit their use. Polysilicon diodes have been suggested as the load elements of a static random access memory [6.110].

Diodes fabricated in polysilicon have also been used as gate-protection elements in vertical MOS power transistors [6.111] and could be similarly applied in other integrated circuits. In an MOS power transistor, the substrate serves as the drain. Typical input-protection techniques, which form a diode between the input and the substrate, interfere with operation of this device, and another protection technique is needed. To provide this protection a pair of back-to-back n^+-p diodes is fabricated in a polysilicon layer deposited on oxide. One n^+ region is connected to the gate of the MOS transistor, and the other is connected to the source, protecting the device from excess voltage in either direction. When a voltage greater than the rating of the device appears across its input terminals, one of the polysilicon junctions breaks down nondestructively and shunts the excess current. The breakdown voltage of a typical diode is about 8 V; for higher voltages a series of similar diodes is used, again taking advantage of the oxide isolation of both sides of the polysilicon diode from the substrate.

For diodes fabricated in a typical CMOS or bipolar process, both sides of the junction are heavily doped, and the reverse breakdown voltage is low. The breakdown voltage increases as the dopant concentration on one side of the junction decreases, varying from 6–8 V for dopant concentrations greater than about 10^{18} cm^{-3} to about 20–25 V for a dopant concentration of 2×10^{17} cm^{-3} [6.106]. However, because the resistivity of polysilicon increases rapidly with decreasing dopant concentration, series resistance can limit electrical performance in the forward direction when the dopant concentration is low.

In the forward direction the total current is the sum of the diffusion current and the recombination current in the space-charge region. Rather than resolving the two components separately, their relative importance is often considered by using the equation

$$I_D = I_{D0} \left[\exp\left(\frac{qV_D}{mkT}\right) - 1 \right] \tag{6.7}$$

where the diode *ideality* or *quality factor* m increases from 1 to 2 as space-charge-region recombination becomes increasingly important. Studies of diodes fabricated in polysilicon show similar behavior for p^+-n

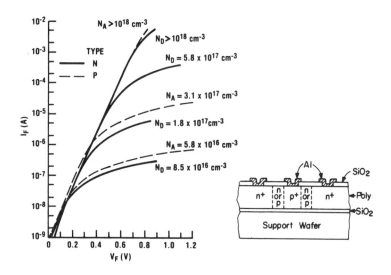

Figure 6.20: Forward I-V characteristics of lateral, polysilicon p-n diodes. The solid curves correspond to p^+-n diodes, and the dashed curves refer to n^+-p diodes.

and n^+-p diodes [6.106]. In the forward direction the current increases logarithmically with a diode quality factor of about 1.8–2.0 (110–120 mV/decade of current), as shown in Fig. 6.20 [6.106], indicating that recombination in the space-charge region dominates until series resistance limits the current. The series resistance tends to be more severe for p^+-n diodes than for n^+-p diodes. Most polysilicon diodes are lateral structures. Vertical n^+-p junctions have been fabricated in thicker polysilicon layers; however, the rapid diffusion of dopant along grain boundaries can create a very nonuniform diffusion front when a junction is formed parallel to the film surface. The low lifetime in polysilicon films (typically 10–300 ps) degrades some device properties, but it also reduces minority-carrier storage; therefore, polysilicon diodes are expected to switch rapidly. The capacitance of polysilicon diodes depends on the measurement frequency, with a stronger voltage dependence at higher frequencies. This behavior is consistent with the presence of a large number of deep traps [6.112].

The inhomogeneous structure of polysilicon complicates detailed understanding of the behavior of polysilicon diodes [6.107, 6.113]. The interaction of the junction with the neutral regions of the grain, the

6.12. POLYSILICON DIODES

space-charge regions surrounding the grain boundaries, and the grain boundaries themselves must all be considered. At low dopant concentrations, few of the traps are filled, and they serve as additional recombination-generation centers. Traps with energy levels near midgap have the greatest effect on both the forward and the reverse behavior of polysilicon diodes [6.114].

At higher dopant concentrations, the effect of the traps is more subtle. For example, majority carriers trapped at the grain boundaries within the junction space-charge region screen the electric field in the junction and reduce the built-in potential. At high electric fields, carriers can be pulled out of the grain-boundary traps by field-enhanced emission and possibly by impact ionization, leading to a rapid increase of the current. The strong dependence of current on voltage in the reverse direction often appears to be consistent with the Poole-Frenkel effect [6.114], in which thermionic emission from Coulombic traps is enhanced as the barrier is lowered by an electric field. Thermally assisted tunneling through the barrier (thermionic-field emission) and the dependence of the barrier thickness on the nature of the trap are used to refine the basic Poole-Frenkel model. Other results [6.115] suggest that the field dependence of the emission rate e_n is stronger than predicted by the Poole-Frenkel effect and can be better modeled by a power law of the form $e_n \approx e_{n0} \exp(E/E_0)^n$, with the exponent n generally between 0.5 and 1.5. Because the traps modify the electric field, the local electric field, rather than the average field, should be considered, usually by numerical methods [6.116], to understand the current flow in detail.

Because the traps are concentrated at the grain boundaries, they are relatively close together and may interact. (A density of 10^{12} traps/cm^2 implies a spacing of 10 nm if uniformly distributed, and less if irregularly spaced.) Overlap of the potential wells can reduce the potential barriers, increasing the emission rate; trap-to-trap tunneling can also play a role. Alternately, the disordered structure at the grain boundary can possibly increase the vibrational *attempt-to-escape* frequency [6.117].

Because the electrical properties of polysilicon are dominated by the grain-boundary traps, the diode characteristics depend sensitively on any factor which modifies the number or nature of the traps. For example, hydrogen incorporation decreases the reverse current of polysilicon diodes by decreasing the number of active grain-boundary states and the electric field in the junction space-charge region [6.118]. The device behavior is also sensitive to small amounts of gaseous contaminants inadvertently introduced into the reactor during deposition, as has been demonstrated by adding N_2O or HCl [6.109]. The reverse current also decreases as the diodes are annealed.

Schottky diodes can be formed by evaporating metal on polysilicon and forming silicide by annealing. The characteristics of the diodes depend on the structure of the polysilicon. For platinum-silicide/polysilicon diodes, the ideality factor m is about 2 for a columnar structure. Diodes formed after annealing to epitaxially realign the deposited silicon film on the underlying single-crystal silicon, are more ideal with $m \sim 1.2$ [6.119].

The potential application of polysilicon in solar cells has led to extensive study of diodes fabricated in large-grain polysilicon formed by various techniques. The larger grains make the diffusion component of current more readily observable in solar-cell material than in the fine-grain polysilicon generally used in integrated circuits [6.120, 6.121].

6.13 Polysilicon thin-film transistors

The majority of integrated circuits rely on fabricating complementary pairs of transistors in a single plane near the surface of a high-quality silicon wafer. Because of the cost of the silicon wafer and the processing to fabricate transistors, the transistor size is continually decreased and the density of transistors is increased. Removing the restriction of fabricating the transistor in the top surface of a single-crystal silicon wafer offers the possibility of radically different applications, as well as integration with conventional monolithic silicon integrated circuits. The transistor can be fabricated in a thin film of polycrystalline silicon, either in a separate layer above a conventional integrated circuit, or on a low-cost, usually transparent substrate.

When *thin-film transistors* are placed above other elements of an integrated circuit, the effective transistor density can be increased significantly. Because of the limited quality of thin-film transistors, they are best used where moderate electrical performance is adequate; for example, as the load elements in a static random-access memory cell [6.122]. As the performance of thin-film transistors improves, the possibility of stacking multiple layers of transistors above a silicon substrate offers potentially greater improvements in density and reduction in the length of interconnections.

6.13. POLYSILICON THIN-FILM TRANSISTORS

The ability to use the third dimension by employing several layers of transistors allows fabricating novel device structures and should also reduce the interconnection length between devices, improving overall circuit speed. Isolating the transistors from each other by insulating layers also eliminates the interaction of nearby n-channel and p-channel transistors that causes latch-up of CMOS circuits, reduces interactions by charge injected at one node of a circuit traveling to another node through the substrate, and improves radiation tolerance.

When thin-film transistors are placed above a silicon wafer, however, the physical size of the circuit is still limited by the cost and size of the silicon wafer. For applications that serve as an interface between a computer and people, the area must be large, so the cost per unit area must be low [6.123]. The prime example of *large-area electronics* is the transistor *active-matrix* used to drive a liquid-crystal display. In this application, one transistor is used to drive each pixel (*picture element*) of the display. The area of the the array is equal to the size of the display, and larger displays require larger substrates. Reducing the transistor dimensions is not critical for this application; the cost of the substrate and processing for a sparse array of moderate-quality devices is critical.

The transistors in the active matrix itself can be fabricated with their channels in amorphous silicon. The low effective mobility of amorphous-silicon transistors is adequate to provide the needed current. However, the number of interconnections between the active matrix and the driving electronics can be greatly reduced if the driving circuitry can be fabricated using thin-films transistors formed on the same glass substrate. For this driving circuitry, better transistor performance is needed than amorphous silicon transistors can provide. Polysilicon-channel transistors with their higher mobility can supply adequate current for this application.

The basic fabrication steps of a polysilicon thin-film transistor are analogous to those of the basic silicon-gate transistor. First, a layer of polycrystalline or amorphous silicon is deposited on an insulating substrate. If the silicon is deposited in amorphous form, it is crystallized to form polysilicon with moderately large grains. The gate dielectric is formed, often by deposition rather than thermal oxidation, especially if the thermal stability of the substrate is limited. The gate electrode is formed, and the source, drain, and gate are doped. A dielectric is deposited; contact regions are opened; and metal is deposited and patterned.

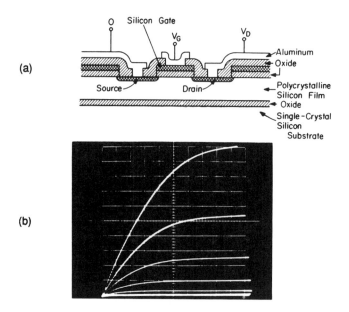

Figure 6.21: (a) Cross section of a thin-film transistor with its *channel* within a polysilicon film and a second polysilicon layer forming the gate electrode. (b) Corresponding drain characteristics of a *p*-channel transistor: I_d=−50 μA/div, V_d=−2 V/div; V_g=−2.5 V/step, 8 steps.

6.13.1 Device physics

The concept of placing the conducting channel of an MOS transistor within a thin layer of polysilicon has been explored for a number of years [6.124, 6.125]. Figure 6.21 [6.125] shows a transistor with its channel in a layer of CVD polysilicon, and the corresponding drain characteristics.

The basic expressions for current in a thin-film transistor are similar to those of MOS transistors in single-crystal silicon:

$$I_{D\text{lin}} = \mu\, C_{ox} \frac{W}{L} (V_G - V_T) V_{DS} \tag{6.8}$$

in the linear region (low drain voltage), and

$$I_{D\text{sat}} = \mu\, C_{ox} \frac{W}{2L} (V_G - V_T)^2 \tag{6.9}$$

in the saturation region (higher drain voltage); μ is the carrier mobility in the channel; C_{ox} is the capacitance of the gate dielectric, which couples the gate voltage to the channel; V_T is the threshold voltage; and W and L are the transistor width and length.

6.13. POLYSILICON THIN-FILM TRANSISTORS

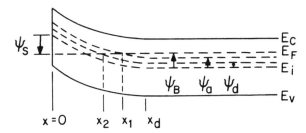

Figure 6.22: Schematic representation of energy-band diagram of a polysilicon transistor with both deep donor and deep acceptor states within the forbidden gap.

Although the shape of the characteristics in Fig. 6.21 is similar to that of a bulk MOS transistor and the drain current saturates reasonably well, a large gate voltage must be applied before significant drain current can flow; the threshold voltage is high; and the transistor has a very low transconductance. The inefficiency of the gate voltage in inducing a conducting channel is caused by the high concentration of traps within the forbidden gap of the polysilicon. As we saw in Chapter 5, these states arise from crystalline defects in the polysilicon, especially at the grain boundaries. Both *deep states* near the middle of the gap and shallow *tail states* near the band edges (Fig. 5.3) degrade the transistor characteristics. The deep states arise from dangling bonds and primarily affect the threshold voltage and subthreshold slope while the shallow, tail states are related to strained bonds and influence the off-state leakage current and the field-effect mobility [6.126].

Threshold voltage: As a gate voltage is applied, the energy bands must be bent to induce a conducting inversion layer near the silicon surface (Fig. 6.22) [6.125]. In material containing a high concentration of deep states within the forbidden gap, the energy levels corresponding to many of the traps must be moved through the Fermi level before a conducting inversion layer can be induced; that is, the charge state of many of the traps must be changed. Much of the applied gate voltage is used to charge or discharge these traps, rather than inducing free carriers in an inversion layer, and excessively high gate voltages must be applied before a conducting channel can form between source and drain. Thus, the characteristics of transistors built in polysilicon films are dominated by the traps. The magnitude of the *threshold voltage* is high because of the mid-gap states. With a higher density of mid-gap states, more of the applied gate voltage is used to charge or discharge these states, rather than to induce free carriers. Because the charge trapping causes a gradual increase in drain current, rather than a well behaved turn on, the onset of strong inversion is difficult to determine, and the threshold voltage is usually defined at a fixed drain current scaled by the device dimensions:

$$I_{DT} = I_{DN} \times \frac{W}{L} \qquad (6.10)$$

The normalized threshold current I_{DN} is the same for all devices and is often defined as 100 nA for $V_{DS} = 10$ V.

Because most of the traps are located near the grain boundaries, the transistor can be visualized as several transistors in series, as shown in Fig. 6.23. Comparatively high-quality grains are separated from each other by highly defective grain boundaries. Consequently, the transistors over the central portions of the grains have a threshold voltage approaching that in single-crystal silicon, but the transistors over the grain boundaries have a much higher threshold voltage. A moderate gate voltage induces conducting channels over the central regions of the grains (Fig. 6.23b), but barriers still exist at the surface near the grain boundaries, and no continuous conducting channel forms. A considerably higher gate voltage is needed to induce channels over the grain boundaries and form a continuous conducting path (Fig. 6.23c). Consequently, the observed threshold voltage is much higher than expected from the dopant concentration in the polysilicon grains. Even after a continuous conducting channel forms, the transistor characteristics are still limited by the traps near the surface. If the trap density is high, the charge state of many traps must still be changed to increase the surface potential even slightly, and the *field-effect mobility* is low.

Polysilicon films with the largest grains are expected to produce transistors with the best properties. As we saw in Sec. 2.11.4, initially amorphous films deposited at relatively low temperatures have the fewest nuclei for grain growth as the films crystallize and, therefore, produce polysilicon films with the largest grains. The characteristics of transistors fabricated in these larger-grain films are appreciably better than those produced in films deposited in a polycrystalline form, with their smaller grains [6.127]. Implanting with silicon to amorphize polysilicon films and then annealing to recrystallize them also improves the transistor properties [6.128].

6.13. POLYSILICON THIN-FILM TRANSISTORS

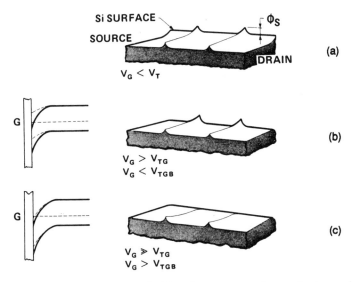

Figure 6.23: The effective threshold voltage near the grain boundaries is higher than that in the center of the grains. Because the transistors are in series, the observed threshold voltage is approximately the threshold voltage corresponding to the gain boundaries.

Accumulation-mode operation: The need to change the charge state of many traps can be partially overcome by using *accumulation-mode* transistors in place of the more common *inversion-mode* transistors used in bulk silicon integrated circuits. In a bulk silicon transistor (Fig. 6.24a), an inversion layer must be formed near the surface of the silicon to obtain a conducting channel isolated from the underlying wafer. In a thin-film transistor, however, an oxide layer lies immediately beneath the silicon film, and parallel conduction in the bulk of the film can be small if the film is thin and lightly doped; the majority of the source-drain current then flows in an induced accumulation layer. The energy bands need not be bent as much to form an accumulation layer (Fig.6.24c) as an inversion layer (Fig. 6.24b), so the applied gate voltage does not need to change the charge state of as many traps. The gate voltage is more effective in controlling the current, and the threshold voltage of the transistor is lower, making the device more compatible with conventional integrated-circuit voltages. Because of the parallel conducting path below the channel, the characteristics of an accumulation-mode transistor depend strongly on the thickness of

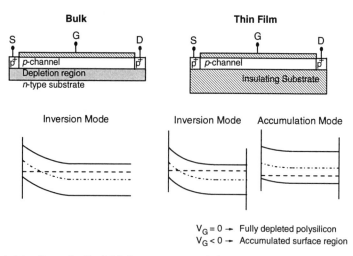

Figure 6.24: In a bulk MOS transistor (a) an inversion layer must be induced to form a channel isolated from the substrate. In a thin film either an inversion-mode transistor (b) or an accumulation-mode transistor (c) can be used.

the polysilicon film. A thin, lightly doped film can be completely depleted at zero gate voltage so that little drain current flows, while the neutral region in a thicker film can cause significant leakage current [6.122].

Reducing the thickness of the polysilicon film also reduces the defect state density per unit area, improving the carrier mobility, the leakage current, the subthreshold behavior, and the threshold voltage. The minimum polysilicon film thickness is determined by the uniformity of the film, the continuity of the film, and the selectivity of the contact hole etch.

Mobility: As we saw in Sec. 5.4.2, the mobility is degraded by grain boundaries. Many of the free carriers are trapped at allowed states at the grain boundary and cannot contribute to the conduction. The charge trapped at the grain boundaries is compensated by oppositely charged depletion regions surrounding the grain boundaries, creating potential barriers that impede carrier movement from one grain to the next. Because the mobility decreases exponentially with increasing barrier height, reducing the barrier height rapidly increases the mobility. The barrier height depends on the number of traps at the grain boundary, so the traps must be passivated or removed to increase the mobility.

6.13. POLYSILICON THIN-FILM TRANSISTORS

Subthreshold behavior: The subthreshold behavior determines the turn-on and turn-off characteristics of the transistor. For polysilicon TFTs the gate voltage must change the charge state of traps, as well as inducing free carriers, so the current increases only slowly as the gate voltage increases. Therefore, the subthreshold characteristics are strongly degraded by the traps in the polysilicon grain boundaries, especially by the dangling-bond states near mid-gap. The inverse subthreshold swing S (the gate voltage change needed to change the drain current by a factor of 10) is at least several times the 60-90 mV found for a single-crystal MOS transistor.

The subthreshold swing is given by

$$S = \frac{kT}{q} \ln(10) \left[1 + \frac{C_d + C_t}{C_{ox}} \right] \qquad (6.11)$$

where C_d is the capacitance below the channel, $C_t = qN_T$, is the capacitance associated with the trap density N_T, and C_{ox} is the gate-oxide capacitance.

Leakage current: The leakage current of polysilicon TFTs is much greater than that of single-crystal silicon transistors, primarily because of carrier generation at the grain-boundary defects. The leakage current arises primarily within the space-charge region between the channel and the drain. The dominant generation mechanism is thermionic-field emission assisted by the high density of tail states within the polysilicon energy bandgap.

At high drain voltages, the high electric field in the pinch-off region near the drain causes impact ionization. Polysilicon TFTs are especially susceptible to impact ionization because carriers trapped at the grain boundaries can be freed by the high fields resulting from the high voltages needed to operate polysilicon TFTs [6.129]. The combination of impact ionization and the floating substrate leads to the *kink* effect. The kink effect occurs in inversion-mode transistors and results in an increase of the drain current in the saturation region as the drain voltage increases. It is caused by the leakage current through the neutral region of the film beneath the channel modulating the potential of this region and lowering the threshold voltage. Linear circuits, such as amplifiers, are especially degraded by the change in current in the saturation region caused by the kink effect.

The leakage current of polysilicon TFTs can be reduced by decreasing the density of defect states and by reducing the electric field at the drain depletion region. Because the leakage current depends strongly on drain voltage, it can be greatly reduced by placing two gates in series, so that only a portion of the drain voltage appears across each of the two drain depletion regions [6.130]. However, the added complexity and area consumed limit use of this approach. Alternatively, the doping in the drain region adjacent to the channel can be reduced by forming an *offset-gate* structure analogous to the *lightly doped drain* (LDD) used in single-crystal silicon MOS transistors [6.131]. This offset-gate structure reduces the electric field in the drain depletion region. If the lightly doped region is too long, the ON current is reduced unacceptably, but there is a range of offset lengths for which the leakage current decreases markedly with only a small decrease of the ON current.

Polysilicon TFTs usually have leakage currents 10-100 times those of amorphous-silicon TFTs because of their smaller energy bandgap (1.1 eV for crystalline silicon, compared to ~1.7 eV for amorphous silicon).

In summary, the physics of operation of the polysilicon thin-film transistor is governed by the same basic MOS transistor equations as the single-crystal silicon MOS transistors, but with substantial modifications to incorporate the effects of grain boundaries.

6.13.2 Methods of improving polysilicon for TFTs

Even in accumulation-mode transistors fabricated in conventional, fine-grain polysilicon, the threshold voltage is too high to be readily compatible with the voltages commonly used in integrated circuits. The characteristics of the polysilicon must be modified to make polysilicon transistors more useful. Obtaining a polysilicon layer with large grains, few electrically active defects, and a smooth surface is critical to optimizing polysilicon TFTs. These characteristics must often be obtained within the very limited temperature regime set by the thermal stability of the substrate.

As discussed in Sec. 5.5, methods of obtaining suitable films include the following:

- Low-temperature solid-phase crystallization
- Rapid thermal crystallization
- Laser crystallization or recrystallization
- Hydrogenation or deuteration

6.13. POLYSILICON THIN-FILM TRANSISTORS

- Using polycrystalline alloys of germanium and silicon
- Chemical mechanical polishing of as-deposited polysilicon

Crystallization: Films deposited in an amorphous form have smooth surfaces. If the grain growth is properly controlled, they can have large grains after crystallization. The amorphous film can be deposited using SiH_4 at about 550°C or using Si_2H_6 at about 450°C; the lower temperature limit is determined primarily by the need to achieve an economically practical deposition rate. If few nuclei form during the deposition process, crystallization after deposition is a thermally activated process described by classical nucleation theory. As discussed in Sec. 2.11.4, the activation energy for nucleation is greater than that for grain growth. Therefore, to favor growth of existing nuclei into large grains, instead of additional nucleation, lower temperatures (*eg,* 600°C) are used to crystallize silicon films for thin-film transistors. When glass substrates are used, the low glass softening point also requires low temperatures. The slow crystallization rate at lower temperatures requires crystallization times of as much as 24 hours, making crystallization a bottle-neck in the transistor-fabrication process.

Hydrogen passivation: As we saw in Sec. 5.5, the active trap density in polysilicon can be reduced by *passivating* the traps with hydrogen or deuterium. As the dangling bonds at the grain boundaries become terminated with hydrogen, a smaller fraction of the applied gate voltage is used to change the charge state of the traps, and more is available to induce and modulate a conducting channel. Therefore, the threshold voltage decreases, and the mobility increases, improving the transistor characteristics [6.122, 6.132–6.134]. For thin-film transistors, the hydrogen can be added by creating atomic hydrogen in a plasma, from a hydrogen containing nitride layer deposited near the polysilicon, or by hydrogen implantation. During hydrogenation, the deep states related to dangling bonds are passivated first, followed by the shallow tail states [6.126]. Although hydrogen improves the electrical properties by passivating defects, movement of the hydrogen in the polysilicon and gate oxide during electrical stress can re-activate some of the traps [6.135].

Even when the grain boundaries are passivated with hydrogen to the maximum extent practical, however, the number of active grain-boundary states is still appreciable, and the polysilicon transistor characteristics remain markedly inferior to those of transistors in single-crystal silicon. Hydrogenated polysilicon transistors are, therefore, most useful in applications which do not demand the highest transistor performance.

Recrystallization: Transistor characteristics approaching those in bulk-silicon devices are obtained when the majority of the grain boundaries are removed by melting and recrystallizing the film, often with a laser [6.136–6.138]. In addition to increasing the grain size, recrystallization appears to lower the trap density per unit area of grain boundary [6.139] and also to reduce the active defect density within the grains. Transistors in recrystallized polysilicon can be employed in a number of applications, but any remaining grain boundaries limit their use in ULSI circuits. If a grain boundary is parallel to the current flow and intersects the source and drain regions, the source-drain dopant can readily diffuse along the grain boundary during fabrication [3.11] to reduce the effective channel length or even cause a short circuit in small-geometry transistors. If the grain boundary is perpendicular to the current, it impedes the current flow. In large-geometry transistors, an appreciable number of grain boundaries are present in every transistor, and the characteristics of one transistor do not differ markedly from those of another. In an array of small-geometry transistors, however, some transistors contain grain boundaries and others do not. The resulting variation in transistor characteristics (Fig. 6.25) [6.140] cannot be tolerated in high-performance circuits, and the grain boundaries must be removed from all transistors in the circuit, making control of the recrystallization process critical.

If all the grain boundaries are removed, however, the characteristics of transistor fabricated in the single-crystal films are very similar to those of transistors in bulk silicon wafers, although the floating neutral region beneath the channel leads to the *kink-effect* [6.141] and other, more subtle differences [6.142, 6.143].

Chemical-mechanical polishing: Optimum solid-state crystallization of amorphous silicon can be very time consuming. When the grain size is less critical, the silicon can be deposited in polycrystalline form. However, the surface roughness of the polysilicon degrades the device performance, and charge injection into the oxide at asperities on the rough surface degrades reliability. Chemical mechanical polishing can be used to smooth the surface and improve the transistor characteristics somewhat, while still reducing the processing time compared to solid-state crystallization [6.144, 6.145]. Polishing can reduce the rms

6.13. POLYSILICON THIN-FILM TRANSISTORS

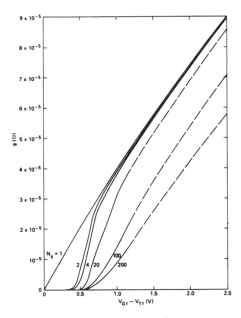

Figure 6.25: The variation in conductance between transistors containing one or a few grain boundaries ($N_g > 1$) and transistors free of grain boundaries ($N_g = 1$) makes design of VLSI circuits with polysilicon transistors difficult. From [6.140]. ©1983 IEEE. Reprinted with permission.

surface roughness by a factor of 3–6 (from 10–13 nm to 2–4 nm), with a resulting improvement in transistor performance.

Polycrystalline silicon-germanium: Because of the lower melting temperature of germanium (940°C) compared to silicon (1410°C), the crystallization temperature of an alloy of Ge and Si is lower than that of a pure Si film [6.146]. Thus, the crystallization process is more compatible with low-temperature glasses. Because of the energy position of grain-boundary states within the bandgap of Ge, the resistivity of p-type, polycrystalline silicon-germanium alloys is lower than that of similarly doped polysilicon [6.147]. Polycrystalline silicon-germanium is easily deposited, and is compatible with conventional silicon processing if the Ge content is limited to less than about 50%.

Physical vapor deposition: A limited number of devices have also been fabricated in evaporated silicon films. Diodes and transistors have been formed in initially amorphous films deposited on fused silica and then thermally crystallized to form polysilicon [6.148]. The grain size in this material appears to be a large fraction of a micrometer, and reasonable device characteristics are obtained. Other efforts have examined depositing silicon films by sputtering [1.69, 2.41, 2.42], often to optimize the deposition efficiency for large-area substrates.

6.13.3 TFTs for active-matrix, liquid-crystal displays

In an active-matrix, liquid-crystal display (AMLCD), the transparency of a pixel is modulated by applying a field across the liquid crystal within that pixel. The modulating voltage is placed on one transparent electrode [usually indium-tin oxide (ITO)] by the thin-film transistor. The counterelectrode is an unpatterned, transparent conductor on the opposite side of the liquid crystal. Color-selecting filters are placed above the liquid crystal, and polarizers are placed on both sides of the display. A light source behind the panel illuminates the display.

The thin-film transistors can be fabricated from either amorphous or polycrystalline silicon. Transistors within the pixels of a displays require low leakage current when they are off. The pixel transistors do not require high drive current when they are on; therefore, they can be fabricated with their channels in either amorphous silicon or polysilicon. Amorphous-silicon transistors have low leakage current when off, and can be fabricated at low temperatures on inexpensive glass substrates. To achieve good gray-scale performance, the pixel TFTs must have low leakage currents (<1 pA) at a typical source-to-drain bias of ~10 V. Because of the smaller bandgap of polysilicon compared to amorphous silicon, the leakage current of polysilicon TFTs is often greater than that of amorphous-silicon TFTs. For maximum brightness, the area covered by the opaque thin-film transistors should be minimum. The higher mobility of polysilicon compared to amorphous silicon allows smaller pixel transistors, increasing the fraction of the area though which light can pass (the *aperture ratio*). Obtaining a high aperture ratio is more important for high-density displays.

More importantly, the peripheral circuitry needed to address and drive the transistors in the matrix must have reasonably high mobility to provide the current for these functions. The mobility of amorphous silicon is too low (< 1 cm^2/V-s) to be used in the peripheral circuitry, and single-crystal integrated circuits must be bonded to a substrate containing the amorphous silicon matrix, increasing the cost and decreasing

6.13. POLYSILICON THIN-FILM TRANSISTORS

the reliability. The mobility of polysilicon thin-film transistors can be adequate to allow fabricating the peripheral circuitry on the same low-cost substrate as the transistors in the active matrix.

For large-area displays, the cost of the substrate is critical. For most display applications, it must be transparent. For processing temperatures above 600°C, more expensive substrates with higher softening temperatures are needed. Although expensive, fused silica ("quartz") can withstand high temperatures and can be made pure enough to be compatible with integrated-circuit processing. It can be used for physically small displays (often projection displays), where the substrate cost does not dominate. However, the thermal coefficient of expansion of fused silica is markedly lower than that of silicon. The large stresses created as the substrate and silicon film cool after deposition or other processing can cause the silicon layer or the upper regions of the substrate to fracture. Glasses can be manufactured with moderately high softening temperatures and a thermal coefficient of expansion matching that of silicon, allowing fabrication of devices on insulating substrates.

The substrate cost decreases markedly if the required softening temperature can be reduced. Therefore, significant effort is devoted to lowering the transistor processing temperatures. Processing temperatures are often limited to 600°C or less to be compatible with well developed and economically practical glass substrates. Within the temperature constraints imposed by the substrate, the goal is to achieve transistors with high mobility to improve the peripheral circuitry and low leakage for the transistors in the matrix.

Key to the successful fabrication are crystallizing the deposited amorphous silicon to form adequately large grains and depositing a gate dielectric with low interface-state density between the polycrystalline silicon and the dielectric. Forming the gate dielectric at a low temperature is especially challenging. For TFT fabrication on low-temperature substrates, the high-temperature thermal oxidation, which produces high-quality oxides and interfaces, cannot be used. Various low-temperature thermal CVD and plasma-enhanced CVD techniques have been explored. Step coverage, interface quality, and reliability must be considered. Annealing at ~600°C after oxide deposition can help to improve the oxide quality.

Obtaining an improved grain structure within the temperature constraints of the substrate is challenging. Heating with a pulsed laser is especially attractive. The laser pulse is short, so the silicon film stays at the melting temperature for less than 100 ns, limiting the heating of the underlying glass substrate (Sec. 5.5). In addition, the local heating allows crystallization of the peripheral circuitry while retaining amorphous silicon in the active matrix itself. This selective crystallization allows using polysilicon transistors with their higher mobility as the driving transistors in the peripheral circuitry and amorphous silicon transistors with their lower leakage current as the pixel transistors.

6.13.4 TFTs for static random-access memories

Cross-coupled, static random-access memory (SRAM) cells require six basic elements: a driver transistor and load element in each leg of the cross-coupled cell and two transistors to access the *data* and $\overline{data}$ lines. The load elements can be *p*-channel transistors in single-crystal silicon, but the resulting 6-transistor cell is large and less suitable for high-density SRAM circuits. They may be high-value polysilicon resistors (Sec. 6.4). However, the resistance must be high to limit the current flowing through the leg of the cell that is ON and the resulting power consumption. This limited ability to supply charge to the intermediate node degrades the stability of the cell, making it susceptible to noise and stray charge. These limitations become more serious as the supply voltage decreases.

Using polysilicon thin-film transistors as the load elements avoids most of the difficulties of the alternative solutions. The polysilicon TFTs can be placed above other circuitry, reducing the silicon area needed for the cell. The conductivity of the transistor can be modulated by the gate voltage so that it is low in the leg of circuit with the driver transistor on, and high in the opposite leg, where the driver transistor is off. This limits power consumption at the same time that it provides a source for the charge needed for cell stability. Thus, using polysilicon TFTs as the load elements of an SRAM cell simultaneously allows high density, good cell stability, and low standby power. These benefits are obtained at the expense of added process complexity.

For this application the ON/OFF current ratio must be greater than $\sim 10^5$, and good uniformity is needed from one transistor to another. To provide the area reduction, the channel length must be short, although not necessarily as short as that of the single-crystal transistors because fewer thin-film transistors are needed. To obtain lower leakage current, a drain offset can be used to reduce the electric field in and near the drain depletion region. The separation between the channel and the heavily doped region of the drain should be long enough to reduce the electric fields and lower the OFF current, but short enough so that it does not seriously increase the series resistance and reduce the ON current [6.149]. The usual connection of the polysilicon TFT to the single-crystal elements produces a parasitic diode in the polysilicon [6.150]. However, the polysilicon diode usually has a high enough leakage current that it does not severely limit the memory cell performance.

Unlike polysilicon TFTs for large-area, active-matrix LCDs on inexpensive glass substrates, TFTs for SRAMs can be processed at moderate temperatures. The temperature and time are limited by the need to avoid disturbing device elements already formed in the underlying substrate. Temperatures of approximately 800°C, and perhaps higher temperatures for short times, can be used. Therefore, the gate oxide can be formed by thermal oxidation, providing a much better polysilicon/oxide interface than can be formed at the lower temperatures compatible with the glass substrates used for AMLCDs.

6.14 Microelectromechanical Systems

Polycrystalline silicon is used extensively in the emerging field of *microelectromechanical systems* or MEMS [6.151–6.153]. The use of polysilicon in MEMS evolved from its use in piezoresistive strain sensors, but has progressed to markedly more complex structures. Polysilicon-based MEMS include accelerometers used in automotive air bags, variable capacitors, and resonant structures with potential use in filters for communication electronics. In discrete form, MEMS provide compact components that can accomplish similar functions as much larger mechanical elements. However, the great potential of MEMS lies in the ability to integrate these structures with conventional electronic elements on the same silicon chip to greatly reduce overall system cost.

MEMS using polysilicon are fabricated by *surface micromachining*, in which the mechanically moving elements are formed above the silicon substrate, rather than being etched into it, as in *bulk micromachining*. One or more layers of polysilicon, often surrounded by sacrificial oxide layers, are deposited and patterned, as shown in Fig. 6.26. The oxide layers are often heavily doped with phosphorus so that they etch rapidly in hydrofluoric-acid solutions. Removing the oxide leaves polysilicon

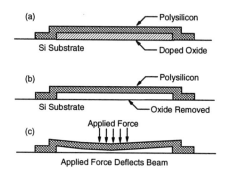

Figure 6.26: A MEMS structure can be formed by (a) depositing and patterning layers of polysilicon and sacrificial oxide. (b) The oxide is removed by wet chemical etching. (c) Applied forces deform the polysilicon. The deformation can be sensed as a change in the value of a resistor in the polysilicon, a change in the capacitance between the polysilicon and the substrate, or even a change in optical reflection.

structures anchored to the underlying substrate at only a few points — or even totally unsupported by the substrate, being constrained by a hub or pin also made of polysilicon.

6.14.1 Integrated sensors

For pressure-sensing applications, a thin layer of polysilicon is usually placed over an oxide, and resistors are formed in the polysilicon. A hole is etched through the substrate from the back, leaving a thin membrane of oxide and polysilicon. When pressure is applied to the membrane, the polysilicon distorts, and the values of the resistors change. In a typical sensor, unstrained resistors are located over the thick silicon wafer nearby, so that small changes in resistance can readily be measured by comparing strained and unstrained resistors.

Because the resistors are isolated from the substrate by an oxide layer, polysilicon sensors can operate at higher temperatures than can sensors fabricated in single-crystal silicon, which rely on p-n-junction isolation. This high-temperature capability makes them especially attractive for automotive applications. However, the sensitivity of polysilicon pressure sensors is only about half that of single-crystal sensors.

6.14. MICROELECTROMECHANICAL SYSTEMS

The piezoresistive properties of polysilicon are attributed primarily to the piezoresistive properties of the individual grains [6.154], and the lower sensitivity of polysilicon sensors is consistent with the presence of several different crystallite orientations. However, the grain boundaries may also affect the piezoresistive properties [6.155]; the piezoresistance appears to increase with increasing boron doping for polysilicon, while it *decreases* with increasing boron doping for single-crystal silicon. Although the sensitivity of polysilicon pressure sensors is somewhat lower than that of single-crystal devices, it is much greater than that of other thin-film sensors. This acceptable sensitivity and the ease of fabrication of polysilicon pressure sensors makes them useful devices.

In addition to the use of polysilicon membranes for sensing pressure, polycrystalline silicon is being applied in a wider variety of sensors. Suspended beams of polysilicon on an integrated circuit allow fabrication of accelerometers and pressure sensors with a process technology compatible with integrated-circuit fabrication. Both singly supported, *cantilever* beams and doubly supported *microbridges* can be fabricated by adding a few steps to an integrated-circuit process [6.156]. For these applications the stress in the polysilicon and any adjacent layers must be well controlled to obtain beams which are undeflected before loading. By coating a polysilicon cantilever beam or microbridge with a thin polymer film, integrated vapor sensors can be built [6.157]. The physical properties of the polymer film change as vapor is absorbed; the mass of the microbridge increases; and its resonant frequency changes. Underlying electrodes in the bulk silicon electrostatically excite and capacitively detect the vibration of the microbridge. Circuitry integrated on the same chip processes the information obtained to provide a compact, efficient sensing system.

6.14.2 Polysilicon for MEMS

Critical properties of polysilicon for MEMS include conformal deposition, stress, and etch rate [2.59]. MEMS structures are inherently nonplanar; therefore, the polysilicon must be able to fill irregular shapes during deposition. As we saw in Sec. 1.9, undoped polysilicon deposition in conventional, low-pressure reactors provides conformal coverage of very irregular, or even reentrant, features [4.28]; however for many MEMS applications, the polysilicon is doped during deposition, and the conformality degrades. Alternate doping techniques, such as diffusion from the nearby doped oxide, can be used to dope layers deposited without dopant. Even for undoped layers, the high deposition rate needed to deposit very thick layers can require operating in the mass-transport-limited regime, where deposition is expected to be much less conformal. Because the formation of MEMS structures often relies on removing sacrificial oxide layers without attacking the polysilicon elements, the etch rate of the polysilicon must be much less than that of the sacrificial layer.

Stress in the polysilicon is a serious concern for MEMS structures. When the mechanical support under a portion of the polysilicon layer is removed, compressive stress in the polysilicon may cause buckling of a beam supported at its two ends. The beam may touch the substrate, possibly shorting the plates of a capacitor. Stress gradients through the thickness of the polysilicon can cause bending of a cantilever beam supported at one end, again limiting the proper functioning of the mechanical element.

As we saw in Sec. 2.8, stress in polysilicon depends on the deposition conditions — especially temperature — and subsequent annealing treatments. Changes in the grain size can significantly affect the stress. For example, in cantilever beams, the uniformity of the stress through the thickness of the film is critical. Non-uniform stress can bend the beam with no applied force. Because of the non-uniform structure through the thickness of an as-deposited film, significant stress gradients can be present.

We saw in Sec. 2.11.3 that phosphorus-doped films are much less stable than undoped films during subsequent heat treatment. The grain size can increase significantly during annealing, reducing the stress. The marked grain growth also significantly decreases the variation of properties through the thickness of the film. When the grains extend through the entire thickness of the polysilicon, stress gradients may be decreased, along with reduced bending of cantilever beams with no applied force [6.158]. However, adding large quantities of n-type dopant can have other, less obvious, effects on the structure and mechanical properties. When phosphorus is added from a gaseous source, such as one containing $POCl_3$, the concentration, especially near the film surface, is near the solid solubility of phosphorus in crystalline silicon at the doping temperature. Subsequent heat treatment at lower temperatures causes the average dopant concentration to exceed solid solubility, and the dopant is likely to precipitate from solution. Because the grain boundaries are disordered regions, the excess phosphorus atoms are likely to migrate to the grain boundaries and precipitate there as a silicon-phosphide species [4.7]. These precipitates can change the mechanical properties of the film, and may lead to the increased compressive stress seen in heavily doped n-type films.

The effects of gain-boundary precipitation are especially important during thermal oxidation of polysilicon. If the oxidation temperature is lower than the doping temperature, excess phosphorus can precipitate at the grain boundaries, as we saw in Sec. 4.2.3. The precipitates can oxidize rapidly, leading to enhanced oxidation along the grain boundaries. The stress from the enhanced oxidation can cause severe compressive stress, especially near the top of the film. In addition to changing the stress in the film, dopant precipitation at grain boundaries can lead to unexpected chemical attack during processing. When oxide is grown on the precipitates, the resulting oxide contains a high concentration of phosphorus and is, consequently, etched rapidly, leaving an irregular surface. In the extreme case, voids can extend through the entire thickness of the polysilicon, allowing attack of underlying layers by wet chemical etching, as discussed in Sec. 2.11.3. While this is undesirable in most applications, it may be used to advantage in MEMS to rapidly remove underlying layers of sacrificial supporting oxide.

In free-standing structures both the top and the bottom of the polysilicon may be exposed during oxidation. The different properties of the polysilicon at the two surfaces can lead to different oxide thicknesses and unbalanced stresses, which can deflect a beam. These properties include (1) a larger grain size near the top of the film; (2) stronger preferred grain orientation near the top of the film; and (3) different dopant concentrations at the top and bottom of the film.

As discussed in Sec. 2.2, oxygen contamination can increase the stress markedly [2.14]. For microelectromechanical applications the oxygen content must be reduced below concentrations tolerable in polysilicon used in purely electronic devices, in which the polysilicon is supported over its entire length.

6.15 Summary

In this chapter we looked at some of the important applications of polysilicon films. Although the dominant use of polysilicon is for gate electrodes in MOS integrated circuits, the extensive development of polysilicon has led to its use in many other applications, as well. Virtually all high-performance bipolar technologies employ one, or even two, layers of polysilicon to improve transistor performance and increase packing density. The availability of polysilicon has also led to its use in floating-gate memory elements and in high-value load resistors, as well as in fusible programming links. The compatibility of polysilicon deposition with other integrated-circuit processing steps has lead to the widespread use of polysilicon in dynamic random-access memories both for trench structures below the wafer surface and for stacked structures above the wafer surface.

Devices with their active elements within a layer of polysilicon are finding select, but important applications, especially for large-area electronics. Using polysilicon thin-film transistors in active-matrix, liquid-crystal displays allows integration of the peripheral circuitry on the same, low-cost substrate as the pixel transistors.

As electronics technology continues developing, the applications of polysilicon are expanding, even as its use as the gate electrodes in MOS integrated circuits becomes restricted by its limited conductivity. Even for this application, however, the compatibility of polysilicon with integrated-circuit processing makes its continued use attractive, and it will probably be augmented by other materials, rather than being replaced in the near future. The continued use of polysilicon in existing applications and its expansion into new applications makes it a material of continuing importance in silicon and related technologies. Its use in a wide range of important applications encourages the continued and expanded study of polysilicon.

Bibliography

Chapter 1: Deposition

[1.1] S T McComas and E R G Eckert, "Combined free and forced convection in a horizontal circular tube," J. Heat Transfer, *88*, 147-153 (May 1966).

[1.2] G E Robertson, J H Seinfeld, and L G Leal, "Combined forced and free convection flow past a horizontal flat plate," AIChE Journal, *19*, 998-1008 (September 1973).

[1.3] A S Grove, *Physics and Technology of Semiconductor Devices* (Wiley, New York, 1967), p. 18.

[1.4] F C Eversteyn, P J W Severin, C H J van den Brekel, and H L Peek, "A stagnant layer model for the epitaxial growth of silicon from silane in a horizontal reactor," J. Electrochem. Soc. *117*, 925-931 (July 1970).

[1.5] American Institute of Physics Handbook, Third Edition (ed. D E Gray, McGraw-Hill, 1972), pp. 2-249—2-252.

[1.6] W Kern and G L Schnable, "Low-pressure chemical vapor deposition for very large-scale integration processing—A review," IEEE Trans. Electron Devices, *ED-26*, 647-657 (April 1979).

[1.7] W A P Claassen and J Bloem, "The growth of silicon from silane in cold wall CVD systems," Philips J. Res. *36*, 122-137 (April 27, 1981).

[1.8] R S Rosler, "Low pressure CVD production processes for poly, nitride, and oxide," Solid State Technology (April 1977), pp. 63-70.

[1.9] W A Brown and T I Kamins, "An analysis of LPCVD system parameters for polysilicon, silicon nitride and silicon dioxide deposition," Solid State Technology (July 1979), pp. 51-57, 84.

[1.10] J Witowski and E Quilantang, "Investigation of the Impact of Oxygen on Contact Resistance: The Influence of Ambient Control in a Vertical LPCVD System," Extended Abstracts, Electrochem. Soc. 1994 Spring Meeting (San Francisco, May 1994), paper 254, pp. 412-413.

[1.11] D W Foster, A J Learn, and T I Kamins, "Deposition properties of silicon films formed from silane in a vertical-flow reactor," J. Vac. Sci. Technol. B *4*, 1182-1186 (September/October 1986).

[1.12] D Foster, A Learn, and T Kamins, "Silicon films deposited in a vertical-flow reactor," Solid State Technology (May 1986), pp. 227-232.

[1.13] M Venkatesan and I Beinglass, "Single-wafer deposition of polycrystalline silicon," Solid State Technology (March 1993), pp. 49-55.

[1.14] J C Liao and T I Kamins, "Power absorption during deposition of polycrystalline silicon in a lamp-heated chemical-vapor-deposition reactor," J. Appl. Phys. *67*, 3848-3852 (April 15, 1990).

[1.15] M C Öztürk, M K Sanganeria, and F Y Sorrell, "A uniformity degradation mechanism in rapid thermal chemical vapor deposition," Appl. Phys. Lett. *61*, 2697-2699 (30 November 1992).

[1.16] W A P Claassen, J Bloem, W G J N Valkenburg, and C H J van den Brekel, "The deposition of silicon from silane in a low-pressure hot-wall system," J. Crystal Growth, *57*, 259-266 (April 1982).

[1.17] M E Coltrin, R J Kee, and J A Miller, "A mathematical model of the coupled fluid mechanics and chemical kinetics in a chemical vapor deposition reactor," J. Electrochem. Soc. *131*, 425-434 (February 1984).

[1.18] B S Meyerson and J M Jasinski, "Silane pyrolysis rates for the modeling of chemical vapor deposition," J. Appl. Phys. *61*, 785-787 (15 January 1987).

[1.19] C H J van den Brekel and L J M Bollen, "Low pressure deposition of polycrystalline silicon from silane," J. Crystal Growth, *54*, 310-322 (August 1981).

[1.20] Following M E Jones and D W Shaw, "Growth from the vapor," Chapter 5 in *Treatise on Solid State Chemistry*, Vol. 5 (ed. N B Hannay, Plenum Press, New York, 1975), pp. 283-323.

[1.21] J Bloem and W A P Claassen, "Rate determining reactions and surface species in CVD of silicon. I. The SiH_4-HCl-H_2 system," J. Crystal Growth, *49*, 435-444 (July 1980).

[1.22] A A Chernov, "Growth kinetics and capture of impurities during gas phase crystallization," J. Crystal Growth, *42*, 55-76 (December 1977).

[1.23] A A Chernov and M P Rusaikin, "Theoretical analysis of equilibrium adsorption layers in CVD systems (Si-H-Cl, Ga-As-H-Cl)," J. Crystal Growth, *45*, 73-81 (December 1978).

[1.24] M L Hitchman, J Kane, and A E Widmer, "Polysilicon growth kinetics in a low pressure chemical vapour deposition reactor," Thin Solid Films, *59*, 231-247 (1979).

[1.25] M J-P Duchemin, M M Bonnet, and M F Koelsch, "Kinetics of silicon growth under low hydrogen pressure," J. Electrochem. Soc. *125*, 637-644 (April 1978).

[1.26] K Hashimoto, K Miura, T Masuda, M Toma, H Sawai, and M Kawase, "Growth kinetics of polycrystalline silicon from silane by thermal chemical vapor deposition method," J. Electrochem. Soc. *137*, 1000-1007 (March 1990).

[1.27] B A Joyce and R R Bradley, "Epitaxial growth of silicon from the pyrolysis of monosilane on silicon substrates," J. Electrochem. Soc. *110*, 1235-1240 (December 1963).

[1.28] D W Shaw, "Chemical vapor deposition," Sec. 2.4 in *Epitaxial Growth, Part A* (ed. J W Matthews, Academic Press, New York, 1975), pp. 89-107.

[1.29] D B Meakin and W Ahmed, "LPCVD of *in-situ* phosphorus doped polysilicon from PH_3/Si_2H_6 mixtures," Fall 1986 Electrochemical Society Meeting (San Diego, October 1986), abstract 267, pp. 398-399.

[1.30] X F Hu, Z Xu, D Lim, M C Downer, P S Parkinson, B Gong, G Hess, and J G Ekerdt, "In situ optical second-harmonic-generation monitoring of disilane adsorption and hydrogen desorption during epitaxial growth on Si(001)," Appl. Phy. Lett. *71*, 1376-1378 (8 September 1997).

[1.31] K Nakazawa, "Recrystallization of amorphous silicon films deposited by low-pressure chemical vapor deposition from Si_2H_6 gas," J. Appl. Phys. *69*, 1703-1706 (1 February 1991).

[1.32] A C Adams, "Dielectric and polysilicon film deposition," in *VLSI Technology* (ed. S M Sze, McGraw-Hill, 1983), pp. 93-129.

[1.33] F C Eversteyn and B H Put, "Influence of AsH_3, PH_3, and B_2H_6 on the growth rate and resistivity of polycrystalline silicon films deposited from a SiH_4-H_2 mixture," J. Electrochem. Soc. *120*, 106-109 (January 1973).

[1.34] R F C Farrow, "The kinetics of silicon deposition on silicon by pyrolysis of silane," J. Electrochem. Soc. *121*, 899-907 (July 1974).

[1.35] P Rai-Choudhury and P L Hower, "Growth and characterization of polycrystalline silicon," J. Electrochem. Soc. *120*, 1761-1766 (December 1973).

[1.36] M L Yu and B S Meyerson, "The adsorption of PH_3 on Si(100) and its effect on the coadsorption of SiH_4," J. Vac. Sci. Technol. A *2*, 446-449 (April-June 1984).

[1.37] B S Meyerson and M L Yu, "Phosphorus-doped polycrystalline silicon via LPCVD. II. Surface interactions of the silane/phosphine/silicon system," J. Electrochem. Soc. *131*, 2366-2368 (October 1984).

[1.38] B S Meyerson and W Olbricht, "Phosphorus-doped polycrystalline silicon via LPCVD. I. Process characterization," J. Electrochem. Soc. *131*, 2361-2365 (October 1984).

[1.39] A Yeckel, S Middleman, and A K Hochberg, "The origin of nonuniform growth of LPCVD films from silane gas mixtures," J. Electrochem. Soc. *136*, 2038-2050 (July 1989).

[1.40] S Nakayama, H Yonezawa, and J Murota, "Deposition of phosphorus doped silicon films by thermal decomposition of disilane," Japan. J. Appl. Phys. *23*, L493-L495 (July 1984).

[1.41] W Ahmed and D B Meakin, "Phosphorus-doped silicon films prepared by low pressure chemical vapour deposition of disilane and phosphine," Thin Solid Films, *148*, L63-L65 (April 13, 1987).

[1.42] J Bloem and L J Giling, "Mechanisms of the chemical vapour deposition of silicon," in *Current Topics in Materials Science*, Volume I (ed. E Kaldis, North-Holland, 1978), pp. 147-342.

[1.43] L H Hall and K M Koliwad, "Low temperature chemical vapor deposition of boron doped silicon films," J. Electrochem. Soc. *120*, 1438-1440 (October 1973).

[1.44] Y Yasuda and T Moriya, "Marked effects of boron-doping on the growth and properties of polycrystalline silicon films," *Semiconductor Silicon 1973* (The Electrochemical Society, 1973), pp. 271-284.

[1.45] C-A Chang, "On the enhancement of silicon chemical vapor deposition rates at low temperatures," J. Electrochem. Soc. *123*, 1245-1247 (August 1976).

[1.46] D W. Greve, "Critical Review: Growth of epitaxial germanium-silicon heterostructures by chemical vapor deposition," Materials Science and Engineering *B18*, 22-51 (1993).

[1.47] S Nakayama, I Kawashima, and J Murota, "Boron doping effect on silicon film deposition in the Si_2H_6-B_2H_6-He gas system," J. Electrochem. Soc. *133*, 1721-1724 (August 1986).

[1.48] J Murota, "Deposition mechanism of arsenic doped polysilicon," Fall 1982 Electrochemical Society Meeting (Detroit, October 1982), abstract 226, pp. 363-364.

[1.49] R J Buss, P Ho, W G Breiland, and M E Coltrin, "Reactive sticking coefficients for silane and disilane on polycrystalline silicon," J. Appl. Phys. *63*, 2808-2819 (15 April 1988).

[1.50] D Chin, S H Dhong, and G J Long, "Structural effects on a submicron trench process," J. Electrochem. Soc. *132*, 1705-1707 (July 1985).

[1.51] V J Silvestri, "Growth kinematics of a polysilicon trench refill process," J. Electrochem. Soc. *133*, 2374-2376 (November 1986).

[1.52] W E Ham, M S Abrahams, and C J Buiocchi, "A note on the ability of CVD polysilicon to deposit nearly inaccessible areas of IC topology," J. Electrochem. Soc. *128*, 1623-1624 (July 1981).

[1.53] R P S Thakur and C Turner, "Novel method for deposition of *in situ* arsenic-doped polycrystalline silicon using conventional low pressure chemical vapor deposition systems," Appl. Phys. Lett. *65*, 2809-2811 (28 November 1994).

[1.54] C M Ransom, T N Jackson, J F DeGelormo, D Kotecki, C Graimann, and D K Sadana, "Arsenic gas-phase doping of polysilicon," J. Vac. Sci. Technol. B *12*, 1390-1393, (May/June 1994).

[1.55] T I Kamins and K L Chiang, "Properties of plasma-enhanced CVD silicon films. I: Undoped films deposited from 525 to 725°C," J. Electrochem. Soc. *129*, 2326-2331 (October 1982).

[1.56] T I Kamins and K L Chiang, "Properties of plasma-enhanced CVD silicon films. II: Films doped during deposition," J. Electrochem Soc. *129*, 2331-2335 (October 1982).

[1.57] A H Jayatissa, Y Nakanishi, and Y Hatanaka, "Preparation of polycrystalline silicon thin films by cathode-type RF glow discharge method," Japan. J. Appl. Phys. *32*, 3729-3733 (September 1993).

[1.58] A H Jayatissa, M Suzuki, Y Nakanishi, and Y Hatanaka, "Microcrystalline structure of poly-Si films prepared by cathode-type r.f. glow discharge," Thin Solid Films *256*, 234-239 (1995).

[1.59] B Moradi, V L Dalal, and R Knox, "Properties of polysilicon films deposited on amorphous substrates using reactive plasma beam deposition technique," J. Vac. Sci. Technol. A *12*, 251-254 (Jan./Feb. 1994).

[1.60] S K Kim, K C Park, and J Jang, "Effect of H_2 dilution on the growth of low temperature as-deposited poly-Si films using $SiF_4/SiH_4/H_2$ plasma," J. Appl. Phys. *77*, 5115-5118 (15 May 1995).

[1.61] H Kakinuma, M Mohri, and T Tsuruoka, "Phosphine doping effects in the plasma deposition of polycrystalline silicon films," Japan. J. Appl. Phys. *31*, L1392-L1395 (1 October 1992).

[1.62] R W Andreatta, C C Abele, J F Osmundsen, J G Eden, D Lubben, and J E Greene, "Low-temperature growth of polycrystalline Si and Ge films by ultraviolet laser photodissociation of silane and germane," Appl. Phys. Lett. *40*, 183-185 (15 January 1982).

[1.63] M Hannabusa, A Namiki, and K Yoshihara, "Laser-induced vapor deposition of silicon," Appl. Phys. Lett. *35*, 626-627 (15 October 1979).

[1.64] C P Christensen and K M Lakin, "Chemical vapor deposition of silicon using a CO_2 laser," Appl. Phys. Lett. *32*, 254-256 (15 February 1978).

[1.65] G Auvert, D Tonneau, and Y Pauleau, "Evidence of a photon effect during the visible laser-assisted deposition of polycrystalline silicon from silane," Appl. Phys. Lett. *52*, 1062-1064 (28 March 1988).

[1.66] D J Ehrlich, R M Osgood, Jr., and T F Deutsch, "Laser microreaction for deposition of doped silicon films," Appl. Phys. Lett. *39*, 957-959 (15 December 1981).

[1.67] N S J Mitchell, B M Armstrong, H S Gamble, and J Wakefield, "Low resistance polysilicon interconnects with self-aligned metal," J. Vac. Sci. Technol. B *4*, 755-757 (May/June 1986).

[1.68] M Heintze, R Zedlitz, H N Wanka, and B B Schubert, "Amorphous and microcrystalline silicon by hot wire chemical vapor deposition," J. Appl. Phys. *79*, 2699-2706 (1 March 1996).

[1.69] C Deshpandey, E Demaray, R Bar-gadda, K Law, N Turner, R Robertson, and T Takehara, "Precursor materials for polysilicon formation," Spring 1997 Materials Research Society Meeting (San Francisco, March 31-April 4, 1997), paper G3.1.

Chapter 2: Structure

[2.1] J Bloem, "Nucleation of silicon on amorphous and crystalline substrates," Proc. Seventh International Conf. on CVD (ed. T O Sedgwick and H Lydtin, The Electrochemical Society, Proc. Vol. 79-3, 1979), pp. 41-58.

[2.2] T I Kamins and T R Cass, "Structure of chemically deposited polycrystalline-silicon films," Thin Solid Films, *16*, 147-165 (May 1973).

[2.3] W A P Claassen and J Bloem, "The nucleation of CVD silicon on SiO_2 and Si_3N_4 substrates. I. The SiH_4-HCl-H_2 system at high temperatures," J. Electrochem. Soc. *127*, 194-202 (January 1980).

[2.4] T I Kamins, "Chemically vapor deposited polycrystalline-silicon films," IEEE Trans. Parts, Hybrids, and Packaging, *PHP-10*, 221-229 (December 1974).

[2.5] W A P Claassen and J Bloem, "The nucleation of CVD silicon on SiO_2 and Si_3N_4 substrates. III. The SiH_4-HCl-H_2 system at low temperatures," J. Electrochem. Soc. *128*, 1353-1359 (June 1981).

[2.6] F Hottier and R Cadoret, "*In-situ* observation of polysilicon nucleation and growth," J. Crystal Growth *56*, 304-312 (1982).

[2.7] J A Tsai and R Reif, "Polycrystalline silicon-germanium films on oxide using plasma-enhanced very-low-pressure chemical vapor deposition," Appl. Phys. Lett. *66*, 1809-1811 (3 April 1995).

[2.8] A T Voutsas and M K Hatalis, "Surface treatment effect on the grain size and surface roughness of as-deposited LPCVD polysilicon films," J. Electrochem. Soc. *140*, 282-288 (January 1993).

[2.9] B A Joyce and R R Bradley, "Epitaxial growth of silicon from the pyrolysis of monosilane on silicon substrates," J. Electrochem. Soc. *110*, 1235-1240 (December 1963).

[2.10] B A Joyce, R R Bradley, and G R Booker, "A study of nucleation in chemically grown epitaxial silicon films using molecular beam techniques. III. Nucleation rate measurements and the effect of oxygen on initial growth behaviour," Phil. Mag. *15*, 1167-1187 (June 1967).

[2.11] R Angelucci, L. Dori, and M. Severi, "Oxygen effect on the electrical characteristics of polycrystalline silicon films," Appl. Phys. Lett. *39*, 346-348 (15 August 1981).

[2.12] M Matsui, Y Shiraki, and E Maruyama, "Low-temperature formation of polycrystalline silicon films by molecular beam deposition," J. Appl. Phys. *53*, 995-998 (February 1982).

[2.13] T I Kamins and A Fischer-Colbrie, "Effect of total deposition pressure on the structure of polycrystalline-silicon films," Appl. Phys. Lett. *71*, 2322-2324 (20 October 1997).

[2.14] T I Kamins, "Deformation occurring during the deposition of polycrystalline-silicon films," J. Electrochem. Soc. *121*, 681-684 (May 1974).

[2.15] R Falckenberg, E Doering, and H Oppolzer, "Surface roughness and grain growth of thin P-doped polycrystalline Si-films," Fall 1979 Electrochem. Soc. Meeting (Los Angeles, October 1979), abstract 570, pp. 1429-1432.

[2.16] N A Haroun, "Grain size statistics," J. Materials Science, *16*, 2257-2262 (August 1981).

[2.17] F C Hull and W J Howk, "Statistical grain structure studies: Plane distribution curves of regular polyhedrons," J. Metals, *5*, 565-572 (April 1953).

[2.18] T I Kamins, "Structure and properties of LPCVD silicon films," J. Electrochem. Soc. *127*, 686-690 (March 1980).

[2.19] J V Smith, Editor, *X-Ray Powder Data File* (American Society for Testing and Materials, Philadelphia, 1960), card 5-0565.

[2.20] B D Cullity, *Elements of X-Ray Diffraction* (Addison-Wesley, Reading MA, 1956), p. 270, Eq. (9-4).

[2.21] H. Kakinuma, "Comprehensive interpretation of the preferred orientation of vapor-phase grown polycrystalline silicon films," J. Vac. Sci. Technol. A *13*, 2310-2317 (September/October 1995).

[2.22] P Joubert, B Loisel, Y Chouan, and L Haji, "The effect of low pressure on the structure of LPCVD polycrystalline silicon films,", J. Electrochem. Soc. *134*, 2541-2545 (October 1987).

[2.23] D Meakin, K Papadopoulou, S Friligkos, J Stoemenos, P Migliorato, and N A Economou, "Pressure dependence of the growth of polycrystalline silicon by low-pressure chemical-vapor deposition," J. Vac. Sci. Technol. B *5*, 1547-1550 (November/December 1987).

[2.24] D Meakin, "A new low temperature polysilicon CVD process," Semiconductor International (July 1988), pp. 74-77.

[2.25] T I Kamins, M M Mandurah, and K C Saraswat, "Structure and stability of low pressure chemically vapor-deposited silicon films," J. Electrochem. Soc. *125*, 927-932 (June 1978).

[2.26] E Kinsbron, M Sternheim, and R Knoell, "Crystallization of amorphous silicon during low pressure chemical vapor deposition," Appl. Phys. Lett. *42*, 835-837 (1 May 1983).

[2.27] A M Beers, H T J M Hintzen, H G Schaeken, and J Bloem, "CVD silicon structures formed by amorphous and crystalline growth," J. Crystal Growth, *64*, 563-571 (December 1983).

[2.28] E G Lee and J J Kim, "Investigation of microstructure and grain growth of polycrystalline silicon deposited using silane and disilane," Thin Solid Films, *226*, 123-128 (15 April 1993).

[2.29] C H Hong, C Y Park, and H-J Kim, "Structure and crystallization of low-pressure chemical vapor deposited silicon films using Si_2H_6 gas," J. Appl. Phys. *71*, 5427-5432 (1 June 1992).

[2.30] S Nakayama, H Yonezawa, and J Murota, "Deposition of phosphorus doped silicon films by thermal decomposition of disilane," Japan. J. Appl. Phys. *23*, L493-L495 (July 1984).

[2.31] S Nakayama, I Kawashima, and J Murota, "Boron doping effect on silicon film deposition in the Si_2H_6-B_2H_6-He gas system," J. Electrochem. Soc. *133*, 1721-1724 (August 1986).

[2.32] S Hasegawa, S Yamamoto, and Y Kurata, "Control of preferential orientation in *in situ* plasma supply during growth of polycrystalline silicon films," Appl. Phys. Lett. *55*, 142-144 (10 July 1989).

[2.33] S Hasegawa, S Yamamoto, and Y Kurata, "Control of preferential orientation in polycrystalline silicon films prepared by plasma-enhanced chemical vapor deposition," J. Electrochem. Soc. *137*, 3666-3674 (November 1990).

[2.34] S Hasegawa, M Arai, and Y Kurata, "Relationship between electrical properties and structure in uniaxially oriented polycrystalline silicon films," J. Appl. Phys. *71*, 1462-1468 (1 February 1992).

[2.35] T Akasaka and I Shimizu, "*In situ* real time studies of the formation of polycrystalline silicon films on glass grown by a layer-by-layer technique," Appl. Phys. Lett. *66*, 3441-3443 (19 June 1995).

[2.36] J H Comfort, L M Garverick, and R Reif, "Silicon surface cleaning by low dose argon ion bombardment for low temperature (750°C) epitaxial silicon deposition. Part I: Process considerations," J. Appl. Phys. (15 October 1987).

[2.37] L M Garverick, J H Comfort, T R Yew, R Reif, F A Baiochhi, and H S Luftman, "Silicon surface cleaning by low dose argon ion bombardment for low temperature (750°C) epitaxial deposition. Part II: Epitaxial quality," J. Appl. Phys. (15 October 1987).

[2.38] A J Mountvala and G Abowitz, "Textural characteristics and electrical properties of vacuum evaporated silicon films," Vacuum, *15*, 359-362 (July 1965).

[2.39] R M Anderson, "Microstructural analysis of evaporated and pyrolytic silicon thin films," J. Electrochem. Soc. *120*, 1539-1546 (November 1973).

[2.40] S S Chao, J Gonzalez-Hernandez, D Martin, and R Tsu, "Effects of substrate temperature on the orientation of ultrahigh vacuum evaporated Si and Ge films," Appl. Phys. Lett. *46*, 1089-1091 (1 June 1985).

[2.41] Y H Wang and J R Abelson, "Growth of polycrystalline silicon at 470°C by magnetron sputtering onto a sputtered μc-hydrogenated silicon seed layer," Appl. Phys. Lett. *67*, 3623-3625 (11 December 1995).

[2.42] Z Sun, K Y Tong, and W B Lee, "Properties of furnace crystallized polysilicon films prepared by r.f. sputtering," Thin Solid Films, *288*, 224-228 (15 November 1996).

[2.43] G Lubberts, B C Burkey, F Moser, and E A Trabka, "Optical properties of phosphorus-doped polycrystalline silicon layers," J. Appl. Phys. *52*, 6870-6878 (November 1981).

[2.44] G Harbeke, L Krausbauer, E F Steigmeier, A E Widmer, H F Kappert, and G Neugebauer, "LPCVD polycrystalline silicon: Growth and physical properties of *in-situ* phosphorus doped and undoped films," RCA Review, *44*, 287-312 (June 1983).

[2.45] M Marazzi, M E Giardini, A Borghesi, A Sassella, M Alessandri, and G. Ferroni, "Optical properties of polycrystalline silicon thin films deposited by single-wafer chemical vapor deposition," Thin Solid Films *296*, 91-93 (1997).

[2.46] G Harbeke, L Krausbauer, E F Steigmeier, A E Widmer, H F Kappert, and G Neugebauer, "Growth and physical properties of LPCVD polycrystalline silicon films," J. Electrochem. Soc. *131*, 675-682 (March 1984).

[2.47] W J Gajda, Jr., "Electrode materials for charge coupled devices," NASA Contractor Report CR-132348 (October 1973).

[2.48] K L Chiang, C J Dell'Oca, and F N Schwettmann, "Optical evaluation of polycrystalline silicon surface roughness," J. Electrochem. Soc. *126*, 2267-2269 (December 1979).

[2.49] E Ibok and S Garg, "A characterization of the effect of deposition temperature on polysilicon properties: Morphology, dopability, etchability, and polycide properties," J. Electrochem. Soc. *140*, 2927-2937 (October 1993).

[2.50] H Takahashi and Y Kojima, "Oxide-semiconductor interface roughness and electrical properties of polycrystalline silicon thin-film transistors," Appl. Phys. Lett. *64*, 2273-2275 (25 April 1994).

[2.51] G Harbeke, "Optical properties of polycrystalline silicon films," in *Polycrystalline Semiconductors. Physical Properties and Applications* (Springer-Verlag, Berlin, 1985), pp. 156-169.

[2.52] C J Dell'Oca, "Nondestructive thickness determination of polycrystalline silicon deposited on oxidized silicon," J. Electrochem. Soc. *119*, 108-111 (January 1972).

[2.53] P S Hauge, "Polycrystalline silicon film thickness measurement from analysis of visible reflectance spectra," J. Opt. Soc. Am. *69*, 1143-1152 (August 1979).

[2.54] Y-C Tai, C Mastrangelo, and R S Muller, "Thermal conductivity of heavily doped LPCVD polysilicon," *Tech. Digest* 1987 International Electron Devices Meeting (Washington DC, December 1987).

[2.55] Y-C Tai, C H Mastrangelo, and R S Muller, "Thermal conductivity of heavily doped low-pressure chemical vapor deposited polycrystalline silicon films," J. Appl. Phys. *63*, 1442-1447 (March 1988).

[2.56] C H Mastrangelo and R S Muller, "Thermal diffusivity of heavily doped, low pressure chemical vapor deposited polycrystalline silicon films," Sensors Mater. *3*, 133-142 (1988).

[2.57] K E Bean, H P Hentzschel, and D Colman, "Thermal and electrical anisotropy of polycrystalline silicon," J. Appl. Phys. *40*, 2358-2359 (April 1969).

[2.58] K E Bean, H P Hentzschel, and D Colman, "Thermal and electrical properties of polycrystalline silicon in the dielectric isolation process," in *Semiconductor Silicon 1969* (ed. R R Haberecht and E L Kern, The Electrochemical Society, 1969), pp. 747-757.

[2.59] T I Kamins, "Design properties of polycrystalline silicon," Sensors and Actuators *A21-A23*, 817-824 (1990).

[2.60] H Guckel, D W Burns, C R Rutigliano, D K Showers, and J Uglow, *Tech. Digest* 4th Int'l. Conf. Solid-State Sensors and Actuators (Tokyo, June 1987), pp. 277-282.

[2.61] L-S Fan, Y-C Tai and R S Muller, "Integrated moveable micromechanical structures for sensors and actuators," IEEE Trans. Electron Devices *ED-35*, 724-730 (June 1988).

[2.62] Y C Tai and R S Muller, "Fracture strain of LPCVD polysilicon," Proc. Solid-State Sensor and Actuator Workshop (Hilton Head Island SC, June 1988), pp. 88-91.

[2.63] L E Trimble and G K Celler, "Evaluation of polycrystalline silicon membranes on fused silica for x-ray lithography masks," J. Vac. Sci. Technol. B *7*, 1675-1679 (November/December 1989).

[2.64] B Bhushan and X Li, "Micromechanical and tribological characterization of doped single-crystal silicon and polysilicon films for microelectromechanical systems devices," J. Mater. Res. *12*, 54-63 (January 1997).

[2.65] S P Murarka and T F Retajczyk, Jr. "Effect of phosphorus doping on stress in silicon and polycrystalline silicon," J. Appl. Phys. *54*, 2069-2072 (April 1983).

[2.66] J Adamczewska and T Budzyński, "Stress in chemically vapor-deposited silicon films," Thin Solid Films, *113*, 271-285 (March 30, 1984).

[2.67] H Guckel, T Randazzo, and D W Burns, "A simple technique for the determination of mechanical strain in thin films with application to polysilicon," J. Appl. Phys. *57*, 1671-1675 (1 March 1985).

[2.68] C-L Yu, P A Flinn, S-H Lee and J C Bravman, "Stress and microstructural evolution of LPCVD polysilicon thin films during high temperature annealing," Fall 1996 Meeting Mat. Res. Soc. (Boston, 1996).

[2.69] A Benitez, J Bausells, E Cabruja, J Esteve, and J Samitier, "Stress in low pressure chemical vapour deposition polycrystalline silicon thin films deposited below 0.1 Torr." Sensors and Actuators A *37-38*, 723-726 (June-August 1993).

[2.70] F W Smith and G Ghidini, "Reaction of oxygen with Si(111) and (100): Critical conditions for the growth of SiO_2," J. Electrochem. Soc. *129*, 1300-1306 (June 1982).

[2.71] P D Agnello and T O Sedgwick, "Conditions for an oxide-free Si surface for low-temperature processing: Steady-state boundary," J. Electrochem. Soc. *139*, 2929-2934 (October 1992).

[2.72] T I Kamins and J E Turner, "Oxygen concentration in LPCVD polysilicon films," Solid State Technology, *33*, no. 4, 80-82 (April 1990).

[2.73] B A Boxall, "A change of etch rate associated with the amorphous to crystalline transition in CVD layers of silicon," Solid-State Electron. *20*, 873-874 (October 1977).

[2.74] J W Coburn, "Pattern Transfer," Solid State Technology (April 1986), p. 117.

[2.75] M A Lieberman and A J Lieberman, *Principles of plasma discharges and materials processing*, Wiley-Interscience, 1994.

[2.76] R A Gottscho, C W Jurgensen, and D J Vitkavage, "Microscopic uniformity in plasma etching," J. Vac. Sci. Tech. B *10*, 2133-2147 (September/October 1992).

[2.77] K-M Chang, T-H Yeh, I-C Deng, and H-C Lin, "Highly selective etching for polysilicon and etch-induced damage to gate oxide with halogen-bearing electron-cyclotron-resonance plasma," J. Appl. Phys. *80*, 3048-3055 (1 September 1996).

[2.78] K K Chi, H S Shin, W J Yoo, C O Jung, Y B Koh, and M Y Lee, "Effects of conductivity of polysilicon on profile distortion," Japan. J. Appl. Phys. Part I *35*, 2440-2444 (April 1996).

[2.79] S Ogino, N Fujiwara, T Maruyama, and M Yoneda, "Precise evaluation of pattern distortion with variation of the impurity concentration and conductivity of silicon films," Japan. J. Appl. Phys. Part I, *36*, 2491-2495 (April 1997).

[2.80] C Y Chang, J P McVittie, K C Saraswat, and K K Lin, "Backscattered deposition in Ar sputter etch of silicon dioxide," Appl. Phys. Lett. *63*, 2294-2296 (18 October 1993).

[2.81] M Tuda and K Ono "Mechanisms for microscopic nonuniformity in low-pressure, high-density plasma-etching of poly-si in Cl_2 and Cl_2/O_2 mixtures," Japan. J. Appl. Phys. Part I, *36*, 2482-2490 (April 1997).

[2.82] C T Gabriel and J P McVittie, "How plasma etching damages thin gate oxides," Solid State Technology, *35*, (June 1992), pp. 81-87.

[2.83] S Fang and J P McVittie, "A model and experiments for thin oxide damage from wafer charging in magnetron plasmas," IEEE Electron Device Lett. *13*, 347-349 (June 1992).

[2.84] S Fang and J P McVittie, "Charging damage to gate oxides in an O_2 magnetron plasma," J. Appl. Phys. *72*, 4865-4872 (15 November 1992).

[2.85] S Ogino, N Fujiwara, H Miyatake, and M Yoneda "Influence of poly-si potential on profile distortion caused by charge accumulation," Japan. J. Appl. Phys. Part I, *35*, 2445-2449 (April 1996).

[2.86] M Okandan, S J Fonash, O O Awadelkarim, Y D Chan, and F Preuninger, "Soft-breakdown damage in MOSFETs due to high-density plasma-etching exposure," IEEE Electron Device Lett. *17*, 388-390 (August 1996).

[2.87] O Joubert and F H Bell "Polysilicon gate etching in high-density plasmas - Comparison between oxide hard mask and resist mask," J. Electrochem. Soc. *144*, 1854-1861 (May 1997).

[2.88] M Sato and Y Arita, "Suppression of aluminum contamination in polysilicon reactive ion etching using highly purified chlorine gas," J. Electrochem. Soc. *144*, 2541-2547 (July 1997).

[2.89] Y Wada and S Nishimatsu, "Grain growth mechanism of heavily phosphorus-implanted polycrystalline silicon," J. Electrochem. Soc. *125*, 1499-1504 (September 1978).

[2.90] C Daey Ouwens and H Heijligers, "Recrystallization processes in polycrystalline silicon," Appl. Phys. Lett. *26*, 569-571 (15 May 1975).

[2.91] W J H Schins, J Bezemer, H Holtrop, and S Radelaar, "Recrystallization of polycrystalline CVD grown silicon," J. Electrochem. Soc. *127*, 1193-1199 (May 1980).

[2.92] W J H Schins and S Radelaar, "Deformation and recrystallization of polycrystalline silicon," J. Materials Science, *16*, 3153-3160 (November 1981).

[2.93] H Oppolzer, R Falckenberg, and E Doering, "Microstructure and sheet resistance of phosphorus-implanted annealed polycrystalline silicon films," Microsc. Semicond. Mater. Conf. (Oxford, April 1981), Inst. Phys. Conf. Ser. No. 60, Section 6, pp. 283-288.

[2.94] S Solmi, M Severi, R Angelucci, L Baldi, and R Bilenchi, "Electrical properties of thermally and laser annealed polycrystalline silicon films heavily doped with arsenic and phosphorus," J. Electrochem. Soc. *129*, 1811-1818 (August 1982).

[2.95] G Masetti, D Nobili, and S Solmi, "Profiles of phosphorus predeposited in silicon and carrier concentration in equilibrium with SiP precipitates," in *Semiconductor Silicon 1977* (ed. H R Huff and E Sirtl, The Electrochemical Soc., Princeton NJ, 1977), Proc. Vol. 77-2, pp. 648-657.

[2.96] L Mei, M Rivier, Y Kwark, and R W Dutton, "Grain-growth mechanisms in polysilicon," J. Electrochem. Soc. *129*, 1791-1795 (August 1982).

[2.97] T Makino and H Nakamura, "Resistivity changes of heavily-boron-doped CVD-prepared polycrystalline silicon caused by thermal annealing," Solid-State Electron. *24*, 49-55 (January 1981).

[2.98] H-J Kim and C V Thompson, "Compensation of grain growth enhancement in doped silicon films," Appl. Phys. Lett. *48*, 399-401 (10 February 1986).

[2.99] C P Ho and J D Plummer, "Si/SiO$_2$ interface oxidation kinetics: A physical model for the influence of high substrate doping levels. I. Theory," J. Electrochem. Soc. *126*, 1516-1522 (September 1979).

[2.100] N E McGruer and R A Oikari, "Polysilicon capacitor failure during rapid thermal processing," IEEE Trans. Electron Devices, *ED-33*, 929-933 (July 1986).

[2.101] C V Thompson and H I Smith, "Surface-energy-driven secondary grain growth in ultrathin (<100 nm) films of silicon," Appl. Phys. Lett. *44*, 603-605 (15 March 1984).

[2.102] M K Hatalis and D W Greve, "Large grain polycrystalline silicon by low temperature annealing of low-pressure chemical vapor deposited amorphous silicon films," J. Appl. Phys. *63*, 2260-2266 (1 April 1988).

[2.103] H-Y Kim, K-Y Lee, and J-Y Lee, "The influence of hydrogen dilution ratio on the crystallization of hydrogenated amorphous silicon films prepared by plasma-enhanced chemical vapor deposition," Thin Solid Films, *302*, 17-24 (1997).

[2.104] R B Iverson and R Reif, "Recrystallization of amorphized polycrystalline silicon films on SiO$_2$: Temperature dependence of the crystallization parameters," J. Appl. Phys. *62*, 1675-1681 (1 September 1987).

[2.105] M-K Ryu, J-W Kim, T-H Kim, K-B Kim, and C-W Hwang, "Reducing the thermal budget of solid phase crystallization of amorphous Si film using $Si_{0.47}Ge_{0.53}$ seed layer," Japan. J. Appl. Phys. *34*, L1031-L1033 (15 August 1995), and T-H Kim, M-K Ryu, J-W Kim, and K-B Kim, "Solid-phase crystallization of a-($Si_{0.69}Ge_{0.31}$/Si) and a-(Si/$Si_{0.69}Ge_{0.31}$Ge) bilayer films on SiO_2," Materials Research Soc. Spring Meeting (San Francisco, April 1996), paper H8.3.

[2.106] S W Russell, J Li, and J W Mayer, "*In situ* observation of fractal growth during a-Si crystallization in a Cu_3Si matrix," *70*, 5153-5155 (1 November 1991).

[2.107] D K Sohn, S C Park, S W Kang, and B T Ahn, "A study of Cu metal deposition of amorphous Si films from Cu solutions for low-temperature crystallization of amorphous Si films," J. Electrochem. Soc. *144*, 3592-3596 (October 1997).

[2.108] S Y Yoon, K H Kim, C O Kim, J Y Oh, and J Jang, "Low temperature metal induced crystallization of amorphous silicon using a Ni solution," J. Appl. Phys. *82*, 5865-5867 (1 December 1997).

[2.109] L Hultman, A Robertsson, H T G Hentzell, I Engström and P A Psaras, "Crystallization of amorphous silicon during thin-film gold reaction," J. Appl. Phys. *62*, 3647-3655 (1 November 1987).

[2.110] S F Gong, H T G Hentzell, A E Robertsson, L Hultman, S-E Hömström, and G Radnoczi, "Al-doped and Sb-doped polycrystalline silicon obtained by means of metal-induced crystallization," J. Appl. Phys. *62*, 3726-3732, (1 November 1987).

[2.111] G Radnoczi, A Robertsson, H T G Hentzell, S F Gong, and M-A Hasan, "Al induced crystallization of a-Si," J. Appl. Phys. *69*, 6394-6399 (1 May 1991).

[2.112] B Bian, J Yie, B Li, and Z Wu, "Fractal formation in a-Si:H/Ag/a-Si:H films after annealing," J. Appl. Phys. *73*, 7402-7406 (1 June 1993).

[2.113] T Shimizu and S Ishihara, "Effect of SiO_2 surface treatment on the solid-phase crystallization of amorphous silicon films," J. Electrochem. Soc. *142*, 298-302 (January 1995).

[2.114] Y-M Ha, S-H Lee, C-H Han, and C-K Kim, "Effects of oxygen on crystallization of amorphous silicon films and polysilicon TFT characteristics" J. Electronic Mater. *23*, 39-45 (January 1994).

[2.115] H Miura and N Okamoto, "Crystallization-induced stress in phosphorus-doped amorphous silicon thin films," J. Appl. Phys. *75*, 4747-4749 (1 May 1994).

[2.116] T S Chao, C L Lee, and T F Lei, "Multiple-angle incident ellipsometry measurement on low pressure chemical vapor deposited amorphous silicon and polysilicon," J. Electrochem. Soc. *141*, 2146-2151 (August 1994).

[2.117] J S Im and R S Sposili, "Crystalline Si films for integrated active-matrix liquid-crystal displays," Materials Research Soc. Bulletin (March 1996), pp. 39-48.

[2.118] A Marmorstein, A T Voutsas and R Solanki, "A systematic study and optimization of parameters affecting grain size and surface roughness in excimer laser annealed polysilicon thin films," J. Appl. Phys. *82*, 4303-4309 (1 November 1997).

[2.119] A Gat, L Gerzberg, J F Gibbons, T J Magee, J Peng, and J D Hong, "cw laser anneal of polycrystalline silicon: crystalline structure, electrical properties," Appl. Phys. Lett. *33*, 775-778 (15 October 1978).

[2.120] H J Kim and J S Im, "New excimer-laser-crystallization method for producing large-grained and grain boundary-location-controlled Si films for thin film transistors," Appl. Phys. Lett. *68*, 1513-1515 (11 March 1996).

[2.121] Y Komen and I W Hall, "The effect of germanium ion implantation dose on the amorphization and recrystallization of polycrystalline silicon films," J. Appl. Phys. *52*, 6655-6658 (November 1981).

[2.122] R Reif and J E Knott, "Low-temperature process to increase the grain size in polysilicon films," Electronics Lett. *17*, 586-588 (20 August 1981).

[2.123] P Kwizera and R Reif, "Solid phase epitaxial recrystallization of thin polysilicon films amorphized by silicon ion implantation," Appl. Phys. Lett. *41*, 379-381 (15 August 1982).

[2.124] S S Lau and J W Mayer, "Epitaxial growth of silicon structures—Thermal, laser- and electron-beam-induced," Chapter 3 in *Preparation and Properties of Thin Films* (ed. K N Tu and R Rosenberg), Treatise on Materials Science and Technology, vol. 24 (Academic Press, New York 1982), pp. 67-111.

[2.125] K Zellama, P Germain, S Squelard, J C Bourgoin, and P A Thomas, "Crystallization in amorphous silicon," J. Appl. Phys. *50*, 6995-7000 (November 1979).

[2.126] N A Blum and C Feldman, "The crystallization of amorphous silicon films," J. Noncryst. Solids, *11*, 242-246 (November 1972).

[2.127] R Kwizera and R Reif, "Annealing behavior of thin polycrystalline silicon films damaged by silicon ion implantation in the critical amorphization range," Thin Solid Films, *100*, 227-233 (February 18, 1983).

[2.128] K T-Y Kung, R B Iverson, and R Reif, "Seed selection through ion channeling to modify crystallographic orientations of polycrystalline Si films on SiO_2: Implant angle dependence," Appl. Phys. Lett. *46*, 683-685 (1 April 1985).

[2.129] K Egami, A Ogura, and M Kimura, "Low-temperature grain growth of initially <100> textured polycrystalline silicon films amorphized by silicon ion implantation with normal incident angle," J. Appl. Phys. *59*, 289-291 (1 January 1986).

[2.130] J W Park, D G Moon, B T Ahn, and H B Im, "Recrystallization of LPCVD amorphous Si films using F^+ implantation" Thin Solid Films *245*, 228-233 (1 June 1994).

[2.131] M Yoshimaru, J Miyano, N Inoue, A Sakamoto, S You, H Tamura, and M Ino, "Rugged surface poly-Si electrode and low temperature deposited Si_3N_4 for 64 Mbit and beyond STC DRAM cell," Tech. Digest, 1990 International Electron Devices Meeting (San Francisco, December 9-12, 1990), paper 27.4, pp. 659-662.

[2.132] H Watanabe, T Tatsumi, S Ohnishi, T Hamada, I Honma, and T Kikkawa, Tech. Digest, 1992 International Electron Devices Meeting (San Francisco, December 13-16, 1992), paper 10.1, pp. 259-262.

[2.133] B Y Tsaur and L S Hung, "Epitaxial alignment of polycrystalline Si films on (100) Si," Appl. Phys. Lett. *37*, 648-651 (1 October 1980).

[2.134] M Y Ghannam and R W Dutton, "Solid phase epitaxial regrowth of boron doped LPCVD polycrystalline silicon," Appl. Phys. Lett. (1987).

[2.135] M Ghannam and R Dutton, "Characterization and modeling of boron diffusion for polysilicon-silicon interfaces," 1986 IEEE Bipolar Circuits and Technology Meeting (Minneapolis), pp. 5-6, 110.

[2.136] Y Hirofuji and K Kugimiya, "Epitaxial regrowth of polysilicon by H_2 annealing process," JST (Japan. Semiconductor Technology) News (August 1984), vol. 3, pp. 54-55.

[2.137] M Tamura, N Natsuaki, and S Aoki, "Epitaxial transformation of ion implanted polycrystalline Si films on (100) Si substrates by rapid thermal annealing," Japan. J. Appl. Phys. *24*, L151-L154 (February 1985).

[2.138] C Y Wong, A E Michel, and R D Isaac, "The poly-single crystalline silicon interface," J. Appl. Phys. *55*, 1131-1134 (15 February 1984).

[2.139] J L Hoyt, E Crabbé, J F Gibbons, and R F W Pease, "Epitaxial alignment of arsenic implanted polycrystalline silicon films on <100> silicon obtained by rapid thermal annealing," Appl. Phys. Lett. *50*, 751-753 (23 March 1987); *50*, 1846 (22 June 1987).

[2.140] J C Bravman, G L Patton, and J D Plummer, "Structure and morphology of polycrystalline silicon–single crystal silicon interfaces," J. Appl. Phys. *57*, 2779-2782 (15 April 1985).

[2.141] N S Parekh, R V Taylor, and D O Massetti, "A simple method to control bipolar polysilicon emitter interfacial oxide," J. Electrochem. Soc. *141*, 3167-3172 (November 1994).

[2.142] R A B Devine, D Mathiot, J-B Xu, I H Wilson, M Gauneau, and W L Warren, "Grain boundary enhanced oxygen out-diffusion in annealed polycrystalline Si/SiO_2/crystalline Si structures," Thin Solid Films, *286*, 317-320 (1996).

[2.143] T P Chen, T F Lei, C Y Chang, W Y Hsieh, and L J Chen, "Investigation on the distribution of fluorine and boron in polycrystalline silicon silicon systems," J. Electrochem. Soc. *142*, 2000-2006 (June 1995).

[2.144] T Gravier, J Kirtsch, C Danterroches, and A Chantre, "Fluorine effects in n-p-n double-diffused polysilicon emitter bipolar-transistors," IEEE Electron Device Lett. *17*, 434-436 (September 1996).

[2.145] N Siabishahrivar, W Redmanwhite, P Ashburn, and H A Kemhadjian, "Reduction of 1/f noise in polysilicon emitter bipolar-transistors," Solid-State Electron. *38*, 389-400 (February 1995).

[2.146] C R Bolognesi, M B Rowlandson, "Impact of fluorine incorporation in the polysilicon emitter of npn bipolar-transistors," IEEE Electron Device Lett. *16*, 172-174 (May 1995).

[2.147] N T Quach and R Reif, "Solid-phase epitaxial growth of polycrystalline silicon films amorphized by ion implantation," Materials Letters, *2*, 362-366 (June 1984).

Chapter 3: Dopant diffusion and segregation

[3.1] R Smoluchowski, "Theory of grain boundary diffusion" Phys. Rev. *87*, 482-487 (August 1, 1952).

[3.2] P G Shewmon, *Diffusion in Solids* (McGraw-Hill, New York, 1963), Chapter 6.

[3.3] H J Queisser, K Hubner, and W Shockley, "Diffusion along small-angle grain boundaries in silicon," Phys. Rev. *123*, 1245-1254 (August 15, 1961).

[3.4] D Gupta, D R Campbell, and P S Ho, "Grain boundary diffusion," Chapter 7 in *Thin Films—Interdiffusion and Reactions* (ed. J M Poate, K N Tu, and J W Mayer, The Electrochemical Society, Princeton NJ, and Wiley-Interscience, New York, 1978), pp. 161-242.

[3.5] J C Fisher, "Calculation of diffusion penetration curves for surface and grain boundary diffusion," J. Appl. Phys. *22*, 74-77 (January 1951).

[3.6] Y M Mishin and C Herzig, "Penetration profiles for fast grain-boundary diffusion by the dissociative mechanism," J. Appl. Phys. *73*, 8206-8214 (15 June 1993).

[3.7] J W Evans, "Approximations to the Whipple solution for grain boundary diffusion and an algorithm for their avoidance," J. Appl. Phys. *82*, 628-634 (15 July 1997).

[3.8] P H Holloway, "Grain boundary diffusion of phosphorus in polycrystalline silicon," J. Vac. Sci. Technol. *21*, 19-22 (May/June 1982). See also comments by H F Mataré and reply in J. Vac. Sci. Technol. B *1*, 107-108 (January-March 1983).

[3.9] F H M Spit and H Bakker, "Diffusion of donor elements (^{125}Sb, ^{32}P, and $^{74(73)}$As) in polycrystalline silicon," Phys. Stat. Sol. (a) *97*, 135-142 (September 1986).

[3.10] D Gupta and P S Ho, "Diffusion processes in thin films," Thin Solid Films, *72*, 399-418 (October 15, 1980).

[3.11] N M Johnson, D K Biegelsen, and M D Moyer, "Grain boundaries in p-n junction diodes fabricated in laser-recrystallized silicon thin films," Appl. Phys. Lett. *38*, 900-902 (1 June 1981).

[3.12] T I Kamins, J Manoliu, and R N Tucker, "Diffusion of impurities in polycrystalline silicon," J. Appl. Phys. *43*, 83-91 (January 1972).

[3.13] L L Kazmerski, P J Ireland, and T F Ciszek, "Evidence for the segregation of impurities to grain boundaries in multigrained silicon using Auger electron spectroscopy and secondary ion mass spectroscopy," Appl. Phys. Lett. *36*, 323-325 (15 February 1980).

[3.14] A D Buonaquisti, W Carter, and P H Holloway, "Diffusion characteristics of boron and phosphorus in polycrystalline silicon," Thin Solid Films, *100*, 235-248 (February 18, 1983).

[3.15] A G O'Neill, C Hill, J King, and C Please, "A new model for the diffusion of arsenic in polycrystalline silicon," J. Appl. Phys. *64*, 167-174 (1 July 1988).

[3.16] C Hill and S K Jones, "Modelling diffusion in and from polysilicon layers," Mat. Res. Soc. Symp. Proc. vol. 182, pp. 129-140 (1990).

[3.17] K Tsukamoto, Y Akasaka, and K Horie, "Arsenic implantation into polycrystalline silicon and diffusion to silicon substrate," J. Appl. Phys. *48*, 1815-1821 (May 1977).

[3.18] H Ryssel, H Iberl, M Bleier, G Prinke, K Haberger, and H Kranz, "Arsenic-implanted polysilicon layers," Appl. Phys. *24*, 197-200 (March 1981).

[3.19] B Swaminathan, K C Saraswat, R W Dutton, and T I Kamins, "Diffusion of arsenic in polycrystalline silicon," Appl. Phys. Lett. *40*, 795-798 (1 May 1982).

[3.20] M Arienzo, Y Komem, and A E Michel, "Diffusion of arsenic in bilayer polycrystalline silicon films," J. Appl. Phys. *55*, 365-369 (15 January 1984).

[3.21] N Lewis, G Gildenblat, M Ghezzo, W Katz, and G A Smith, "Lateral diffusion of arsenic in low pressure chemical vapor deposited polycrystalline silicon," Appl. Phys. Lett. *42*, 171-172 (15 January 1983).

[3.22] Y Sato, K Murase, and H Harada, "A novel method to measure lateral diffusion length in polycrystalline silicon," J. Electrochem. Soc. *129*, 1635-1638 (July 1982).

[3.23] M M Mandurah, K C Saraswat, C R Helms, and T I Kamins, "Dopant segregation in polycrystalline silicon," J. Appl. Phys. *51*, 5755-5763 (November 1980).

[3.24] C Y Wong, C R M Grovenor, P E Batson, and D A Smith, "Effect of arsenic segregation on the electrical properties of grain boundaries in polycrystalline silicon," J. Appl. Phys. *57*, 438-442 (15 January 1985).

[3.25] W K Schubert, "Properties of furnace-annealed, high-resistivity, arsenic-implanted polycrystalline silicon films," J. Mater. Res. *1*, 311-321 (March/April 1986).

[3.26] J Miyamoto, K Shinada, S Shinozaki, and N Sekiguchi, "Application of polycrystalline silicon load for high performance bipolar memory," Tech. Digest 1980 International Electron Devices Meeting (Washington DC, December 1980), paper 3.2, pp. 50-53.

[3.27] S R Wilson, W M Paulson, R B Gregory, J D Gressett, A H Hamdi, and F D McDaniel, "Fast diffusion of As in polycrystalline silicon during rapid thermal annealing," Appl. Phys. Lett. *45*, 464-466 (15 August 1984).

[3.28] R A Powell and R Chow, "Dopant activation and redistribution in As^+-implanted polycrystalline Si by rapid thermal processing," J. Electrochem. Soc. *132*, 194-198 (January 1985).

[3.29] S Lourdudoss and S-L Zhang, "High concentration phosphorus doping of polycrystalline silicon by low temperature direct vapor phase diffusion of phosphine followed by rapid thermal annealing," Appl. Phys. Lett. *64*, 3461-3463 (20 June 1994).

[3.30] J S Makris and B J Masters, "Phosphorus isoconcentration diffusion studies in silicon," J. Electrochem. Soc. *120*, 1252-1255 (September 1973).

[3.31] D L Losee, J P Lavine, E A Trabka, S-T Lee, and C M Jarman, "Phosphorus diffusion in polycrystalline silicon," J. Appl. Phys. *55*, 1218-1220 (15 February 1984).

[3.32] J L Tandon, H B Harrison, C L Neoh, K T Short, and J S Williams, "The annealing behavior of antimony implanted polycrystalline silicon," Appl. Phys. Lett. *40*, 228-230 (1 February 1982).

[3.33] F H M Spit, H Albers, A Lubbes, Q J A Rijke, L J v Ruijven, J P A Westerveld, H Bakker, and S Radelaar, "Diffusion of Antimony (^{125}Sb) in polycrystalline silicon," Phys. Stat. Sol. (a) *89*, 105-115 (May 1985); *91*, 302 (September 1985).

[3.34] D J Coe, "The lateral diffusion of boron in polycrystalline silicon and its influence on the fabrication of sub-micron MOSTs," Solid-State Electron. *20*, 985-992 (December 1977).

[3.35] S Horiuchi and R Blanchard, "Boron diffusion in polycrystalline silicon layers," Solid-State Electron. *18*, 529-532 (June 1975).

[3.36] Y. Okazaki, S Nakayama, M Miyake, and T Kobayashi, "Characteristics of sub-1/4-μm gate surface channel PMOSFETs using a multilayer gate structure of boron-doped poly-Si on thin nitrogen-doped poly-Si," IEEE Trans. Electron Devices *41*, 2369-2375 (December 1994).

[3.37] S Nakayama and T Sakai, "Redistribution of *in situ* doped or ion-implanted nitrogen in polysilicon," J. Appl. Phys. *79*, 4024-4028 (15 April 1996).

[3.38] D Widmann and U Schwabe, "Limitations of ion implantation in MOS technology," Proc. Fourth International Conf. on Ion Implantation Equipment and Techniques (Berchtesgaden, West Germany, September 1982), (ed. H Ryssel and H Glawischnig, Springer/Verlag, Berlin, 1983), pp. 392-406.

[3.39] T E Seidel, "Channeling of implanted phosphorus through polycrystalline silicon," Appl. Phys. Lett. *36*, 447-449 (15 March 1980).

[3.40] Y Wada, S Nishimatsu, and N Hashimoto, "Arsenic ion channeling through single crystal silicon," J. Electrochem. Soc. *127*, 206-210 (January 1980).

[3.41] H Hwang, D H Lee, J-G Ahn, J S Byun, and D Yang, "Effect of channeling of halo ion implantation on threshold voltage shift of metal oxide semiconductor field-effect transistor," Appl. Phys. Lett. *68*, 938-939 (12 February 1996).

[3.42] G E Georgiou, T T Sheng, J Kovalchick, W T Lynch, and D Malm, "Shallow junctions by out-diffusion from BF_2 implanted polycrystalline silicon," J. Appl. Phys. *68*, 3707-3713 (1 October 1990).

[3.43] G E Georgiou, T T Sheng, F A Baiocchi, J Kovalchick, W T Lynch, and D Malm, "Shallow junctions by out-diffusion from arsenic implanted polycrystalline silicon," J. Appl. Phys. *68*, 3714-3722 (1 October 1990).

[3.44] I R C Post and P Ashburn, "Investigation of boron diffusion in polysilicon and its application to the design of p-n-p polysilicon emitter bipolar transistors with shallow emitter junctions," IEEE Trans. Electron Devices *38*, 2442-2451 (November 1991).

[3.45] J M Macaulay, R Hull, B Jalali, and C Magee, "Characterization of arsenic doping profile across the polycrystalline Si/Si interface in polycrystalline Si emitter bipolar transistors," Appl. Phys. Lett. *63*, 1258-1260 (30 August 1993).

[3.46] B Garben, W A Orr-Arienzo, and R F Lever, "Investigation of boron diffusion from polycrystalline silicon," J. Electrochem. Soc. *133*, 2152-2156 (October 1986).

[3.47] W J M J Josquin, P R Boudewijn, and Y Tamminga, "Effectiveness of polycrystalline silicon diffusion sources," Appl. Phys. Lett. *43*, 960-962 (15 November 1983).

[3.48] P Ashburn and B Soerowirdio, "Arsenic profiles in bipolar transistors with polysilicon emitters," Solid-State Electron. *24*, 475-476 (May 1981).

[3.49] D Tsoukalas and D Kouvatsos, "Silicon interstitial trapping in polycrystalline silicon films studied by monitoring interstitial reactions with underlying insulating films," Appl. Phys. Lett. *68*, 1549-1551 (11 March 1996).

[3.50] B Swaminathan, L Mei, A M Lin, and R W Dutton, "Enhanced diffusion in the single crystal silicon substrate during the oxidation of a deposited polysilicon doping source," 1980 Solid State Interface Specialists Conference.

[3.51] P Pichler, H Ryssel, R Ploss, C Bonafos, and A Claverie, "Phosphorus-enhanced diffusion of antimony due to generation of self-interstitials," J. Appl. Phys. *78*, 1623-1629 (1 August 1995).

[3.52] J N Burghartz, C L Stanis, and P A Ronsheim, "Dopant interactions during the diffusion of arsenic and boron in opposite directions in polycrystalline/monocrystalline silicon structures," Appl. Phys. Lett. *67*, 3156-3158 (20 November 1995).

[3.53] A Berthold, A vom Felde, M Biebl, and H von Philipsborn, "The role of point defect sources in the formation of boron polyemitters," *Tech. Digest*, 1994 International Electron Devices Meeting (San Francisco, December 1994), paper 19.7, pp. 509-512.

[3.54] J O McCaldin and H Sankur, "Diffusivity and solubility of Si in the Al metallization of integrated circuits," Appl. Phys. Lett. *19*, 524-527 (15 December 1971).

[3.55] K Nakamura and M Kamoshida, "Low-temperature diffusion of Al into polycrystalline Si," J. Appl. Phys. *48*, 5349-5351 (December 1977).

[3.56] D Pramanik and A N Saxena, "VLSI metallization using aluminum and its alloys. Part I," Solid State Technology (January 1983), pp. 127-133.

[3.57] J C M Hwang, P S Ho, J E Lewis, and D R Campbell, "Grain boundary diffusion of aluminum in polycrystalline silicon films," J. Appl. Phys. *51*, 1576-1581 (March 1980).

[3.58] N Herbots, F Van de Wiele, M Lobet, and R G Elliman, "Arsenic dopant influence upon the sintering behavior of the aluminum-polysilicon interface," J. Electrochem. Soc. *131*, 645-652 (March 1984).

[3.59] K Nakamura, S S Lau, M-A Nicolet, and J W Mayer, "Ti and V layers retard interaction between Al films and polycrystalline Si," Appl. Phys. Lett. *28*, 277-280 (1 March 1976).

[3.60] A J Learn and R S Nowicki, "Methods for minimizing silicon regrowth in thin films," Appl. Phys. Lett. *35*, 611-614 (15 October 1979).

[3.61] H Ari, S Kohda, and Y Kitano, "Lateral reaction of polycrystalline silicon film with overlapping aluminum film," J. Appl. Phys. *57*, 1143-1146 (15 February 1985).

[3.62] J R Lloyd, M J Sullivan, J L Jozwiak, and R J Murphy, "Interfacial electromigration of aluminum in thin-film polysilicon/silicide structures," Appl. Phys. Lett. *44*, 68-70 (1 January 1984).

[3.63] K Nakamura, J O Olowolafe, S S Lau, M-A Nicolet, and J W Mayer, "Interaction of metal layers with polycrystalline Si," J. Appl. Phys. *47*, 1278-1283 (April 1976).

[3.64] Ch Poisson, A Rolland, J Bernardini, and N A Stolwijk, "Diffusion of gold into polycrystalline silicon investigated by means of the radiotracer ^{195}Au," J. Appl. Phys. *80*, 6179-6187 (1 December 1996).

[3.65] A J Bevolo, F A Schmidt, H R Shanks, and G J Campisi, "Polycrystalline silicon on tungsten substrates," J. Vac. Sci. Technol. *16*, 13-19 (January/February 1979).

[3.66] T I Kamins, S S Laderman, D J Coulman, and J E Turner, "Interaction between CVD tungsten films and silicon during annealing," J. Electrochem. Soc. *133*, 1438-1442 (July 1986).

[3.67] G W Racette and R T Frost, "Deposition of polycrystalline silicon films on metal substrates under ultra-high vacuum," J. Crystal Growth, *47*, 384-388 (September 1979).

[3.68] T C Chou, C Y Wong, and K N Tu, "Enhanced grain growth of phosphorus-doped polycrystalline silicon by titanium silicide formation," Appl. Phys. Lett. *49*, 1381-1383 (17 November 1986).

[3.69] J B Lasky, J S Nakos, O J Cain, and P J Geiss, "Comparison of transformation to low-resistivity phase and agglomeration of $TiSi_2$ and $CoSi_2$," IEEE Trans. Electron Devices, *ED-38*, 262-269 (February 1991).

[3.70] L A Clevenger, R W Mann, R A Roy, K L Saenger, C Cabral Jr. and J Piccirillo, "Study of C49-$TiSi_2$ and C54-$TiSi_2$2 formation on doped polycrystalline silicon using *in situ* resistance measurements during annealing," J. Appl. Phys. *76*, 7874-7881 (15 December 1994).

[3.71] C Y Wong, L K Wang, P A McFarland, and C Y Ting, "Thermal stability of $TiSi_2$ on mono- and polycrystalline silicon," J. Appl. Phys. *60*, 243-246 (1 July 1986).

[3.72] R A Roy, L A Clevenger, C Cabral, Jr., K L Saenger, S Brauer, J Jordan-Sweet, J Bucchignano, G B Stephenson, G Morales, and K F Ludwig, Jr., "*In situ* x-ray diffraction analysis of the C49-C54 titanium silicide phase transformation in narrow lines," Appl. Phys. Lett. *66*, 1732-1734 (3 April 1995).

[3.73] T Ohguro, S-I Nakamura, M Koike, T Morimoto, A Nishiyama, Y Ushiku, T Yoshitomi, M Ono, M Saito, and H Iwai, "Analysis of resistance behavior in Ti- and Ni-salicided polysilicon films," IEEE Trans. Electron Devices *41*, 2305-2316 (December 1994).

[3.74] A Kalnitsky, P K Hurley, and A Lepert, "Dopants (P, As, and B) in the polycrystalline silicon/titanium silicide system: Redistribution and activation," J. Electrochem. Soc. *144*, 1090-1095 (March 1997).

[3.75] S P Murarka and S Vaidya, "Cosputtered cobalt silicides on silicon, polycrystalline silicon, and silicon dioxide," J. Appl. Phys. *56*, 3404-3412 (15 December 1984).

[3.76] G Ottaviani, K N Tu, P Psaras, and C Nobili, "*In situ* resistivity measurement of cobalt silicide formation," J. Appl. Phys. *62*, 2290-2294 (15 September 1987).

[3.77] S Pramanick, Y N Erokhin, B K Patnaik, and G A Rozgonyi, "Morphological instability and Si diffusion in nanoscale cobalt silicide films formed on heavily phosphorus doped polycrystalline silicon," Appl. Phys. Lett. *63*, 1933-1935 (4 October 1993).

[3.78] C R Helms, Stanford University, private communication.

[3.79] C R M Grovenor, P E Batson, D A Smith, and C Wong, "As segregation to grain boundaries in Si," Phil. Mag. A *50*, 409-423 (1984).

[3.80] H C Card and E S Yang, "Electronic processes at grain boundaries in polycrystalline semiconductors under optical illumination," IEEE Trans. Electron Devices, *ED-24*, 397-402 (April 1977).

[3.81] J. Hornstra, "Models of grain boundaries in the diamond lattice. I. Tilt about < 110 >," Physica, *25*, 409-422 (1959).

[3.82] B Swaminathan, E Demoulin, T W Sigmon, R W Dutton, and R Reif, "Segregation of arsenic to the grain boundaries in polycrystalline silicon," J. Electrochem. Soc. *127*, 2227-2229 (October 1980).

[3.83] J H Rose and R Gronsky, "Scanning transmission electron microscope microanalytical study of phosphorus segregation at grain boundaries in thin-film silicon," Appl. Phys. Lett. *41*, 993-995 (15 November 1982).

[3.84] G A Pollock, V R Deline, and B K Furman, "Secondary ion images of impurities at grain boundaries in polycrystalline silicon," in *Grain boundaries in semiconductors*, Proc. Mat. Res. Soc. Symp., vol. 5 (ed. H J Leamy, G E Pike, and C H Seager, North-Holland, 1982), pp. 71-76.

[3.85] R Angelucci, M Severi, S Solmi, and L Baldi, "Electrical properties of phosphorus-doped polycrystalline silicon films contaminated with oxygen," Thin Solid Films, *103*, 275-281 (May 20, 1983).

[3.86] J R Monkowski, J Bloem, L J Giling, and M W M Graef, "Comparison of dopant incorporation into polycrystalline and monocrystalline silicon," Appl. Phys. Lett. *35*, 410-412 (1 September 1979).

[3.87] M Y Ghannam and R W Dutton, "Resistivity of boron-doped polycrystalline silicon," Appl. Phys. Lett. *52*, 1222-1224 (11 April 1988).

[3.88] M M Mandurah, K C Saraswat, and T I Kamins, "Arsenic segregation in polycrystalline silicon," Appl. Phys. Lett. *36*, 683-685 (15 April 1980).

[3.89] A Carabelas, D Nobili, and S Solmi, "Grain boundary segregation in silicon heavily doped with phosphorus and arsenic," J. de Physique, *43*, supplement C1, C1-187–C1-192 (October 1982).

[3.90] L Baldi, G Ferla, G Queirolo, and D Beardo, "Dopant segregation in polycrystalline silicon heavily doped with arsenic," Fall 1981 Electrochemical Society Meeting (Denver, October 1981), abstract 409, pp. 992-994.

[3.91] A L Fripp, Jr., "The effect of heat treatment on the resistivity of polycrystalline silicon films," IEEE Trans. Parts, Hybrids, and Packaging, *PHP-11*, 239-240 (September 1975).

[3.92] H Uda, S Ozono, and N Owada, "Resistivity lowering phenomenon observed in heavily doped polycrystalline silicon by rapid thermal annealing," ULSI Science and Technology (Electrochem. Soc. Proc. Vol. PV 89-9), pp. 184-189 (1989).

[3.93] K Sakamoto, K Nishi, T Yamaji, T Miyoshi, and S Ushio, "Complete process modeling for VLSI multilayer structures," J. Electrochem. Soc. *132*, 2457-2462 (October 1985).

[3.94] A D Sadovnikov, "One-dimensional modeling of high concentration boron diffusion in polysilicon-silicon structures," Solid-State Electron. *34*, 969-975 (September 1991).

[3.95] A D Sadovnikov and A V Tchernyaev, "Two-dimensional model of impurity diffusion in polysilicon-silicon structures," Solid-State Electron. *35*, 193-200 (February 1992).

[3.96] M A Matsuoka and S T Dunham, "Dopant diffusion in polysilicon," in *Proc. Third International Symposium on Process Physics and Modeling in Semiconductor Technology*, (ed. G R Srinivasan, K Taniguchi, and C S Murthy,) (The Electrochem. Soc. Proc. Vol. 93-6, 1993) pp. 88-97. (Electrochem. Soc. 1993 Spring Meeting, Honolulu, May 1993.)

[3.97] H Puchner and S Selberherr, "An advanced model for dopant diffusion in polysilicon," IEEE Trans. Electron Devices, *42*, 1750-1755 (October 1995).

[3.98] X Gui, L J Friedrich, S K Dew, M J Brett, and T Smy, "Grain-boundary diffusion modeling and efficiency evaluation of thin-film diffusion barriers considering microstructure effects," J. Appl. Phys. *78*, 4438-4443 (1 October 1995).

[3.99] S Banerjee and S T Dunham, "Two stream model for dopant diffusion in polysilicon incorporating effects of grain growth," in *Proc. Fourth International Symposium on Process Physics and Modeling in Semiconductor Technology*, (ed. G R Srinivasan, C S Murthy, and S T Dunham,) (The Electrochem. Soc. 1996), pp. 92-100. (Electrochem. Soc. 1996 Spring Meeting, Los Angeles, May 1996).

Chapter 4: Oxidation

[4.1] B E Deal and A S Grove, "General relationship for the thermal oxidation of silicon," J. Appl. Phys. *36*, 3770-3778 (December 1965).

[4.2] B E Deal, "The oxidation of silicon in dry oxygen, wet oxygen, and steam," J. Electrochem. Soc. *110*, 527-533 (June 1963).

[4.3] B E Deal, "Characteristics of the surface-state charge (Q_{ss}) of thermally oxidized silicon," J. Electrochem. Soc. *114*, 266-274 (March 1967).

[4.4] T I Kamins and E L MacKenna, "Thermal oxidation of polycrystalline silicon films," Metallurgical Transactions of AIME, *2*, 2292-2294 (August 1971).

[4.5] T I Kamins, "Oxidation of phosphorus-doped low pressure and atmospheric pressure CVD polycrystalline-silicon films," J. Electrochem. Soc. *126*, 838-844 (May 1979).

[4.6] C Y Lu and N S Tsai, "Thermal oxidation of undoped LPCVD polycrystalline-silicon films," J. Electrochem. Soc. *133*, 446-447 (February 1986).

[4.7] J C Bravman and R Sinclair, "Transmission electron microscopy studies of the polycrystalline silicon-SiO_2 interface," Thin Solid Films, *104*, 153-161 (June 17, 1983).

[4.8] M Horiuchi, "Formation of thin oxide on polycrystalline silicon," J. Appl. Phys. *53*, 4943-4947 (July 1982).

[4.9] L L Kazmerski, O Jamjoum, P J Ireland, and R L Whitney, "A study of the initial oxidation of polycrystalline Si using surface analysis techniques," J. Vac. Sci. Technol. *18*, 960-964 (April 1981).

[4.10] W Shih, C Wang, S Chiao, L Chen, N Wu, T Batra, J C Kao, and O K Wu, "Structural and electrical characterization of thin polyoxides for nonvolatile memory applications," J. Vac. Sci. Technol. A *3*, 967-970 (May/June 1985).

[4.11] C P Ho and J D Plummer, "Si/SiO_2 interface oxidation kinetics: A physical model for the influence of high substrate doping levels. I. Theory," J. Electrochem. Soc. *126*, 1516-1522, (September 1979); "…. II. Comparison with experiment and discussion," J. Electrochem. Soc. *126*, 1523-1530 (September 1979).

[4.12] H Sunami, "Thermal oxidation of phosphorus-doped polycrystalline silicon in wet oxygen," J. Electrochem. Soc. *125*, 892-897 (June 1978).

[4.13] K C Saraswat and H Singh, "Thermal oxidation of heavily phosphorus-doped thin films of polycrystalline silicon," J. Electrochem. Soc. *129*, 2321-2326 (October 1982).

[4.14] J J Barnes, J M DeBlasi and B E Deal, "Low temperature differential oxidation for double polysilicon VLSI devices," J. Electrochem. Soc. *126*, 1779-1785 (October 1979).

[4.15] C C Chang, T T Sheng, and T A Shankoff, "Phosphorus depth profiles in thermally oxidized P-doped polysilicon," J. Electrochem. Soc. *130*, 1168-1171 (May 1983).

[4.16] J S Johannessen, W E Spicer, J F Gibbons, J D Plummer, and N J Taylor, "Observation of phosphorus pile-up at the SiO_2-Si interface," J. Appl. Phys. *49*, 4453-4458 (August 1978).

[4.17] T Kimura, M Hirose, and Y Osaka, "Mechanism of phosphorus pile-up in the Si-SiO_2 interface," J. Appl. Phys. *56*, 932-935 (15 August 1984).

[4.18] R W Barton, "Dopant segregation at the Si/SiO_2 interface studied with Auger sputter profiling," PhD Thesis, Stanford Electronics Laboratories, Technical Report No. J701-1, Stanford University, 1981, p. 84.

[4.19] L Baldi, G Ferla, and M Tosi, "Oxidation kinetics of As-doped polysilicon in a steam environment," Spring 1980 Electrochem. Soc. Meeting (St. Louis, May 1980), abs. 168, pp. 441-443.

[4.20] E Kinsbron, S P Murarka, T T Sheng, and W T Lynch, "Oxidation of arsenic implanted polycrystalline silicon," *130*, 1555-1560 (July 1983).

[4.21] K Suzuki, Y Yamashita, Y Kataoka, K Yamazaki and K Kawamura, "Segregation coefficient of boron and arsenic at polycrystalline silicon/SiO_2 interface," J. Electrochem. Soc. *140*, 2960-2964 (October 1993).

[4.22] M Boukezzata, D Bielle-Daspet, G. Sarrabayrouse, and F Mansour, "Characteristics of the thermal oxidation of heavily boron-doped polycrystalline-silicon thin films," Thin Solid Films, *279*, 145-154 (1996).

[4.23] E A Irene, E Tierney, and D W Dong, "Silicon oxidation studies: Morphological aspects of the oxidation of polycrystalline silicon," J. Electrochem. Soc. *127*, 705-713 (March 1980).

[4.24] H Sunami, M Koyanagi, and N Hashimoto, "Intermediate oxide formation in double-polysilicon gate MOS structure," J. Electrochem. Soc. *127*, 2499-2506 (November 1980).

[4.25] J A Appels, E Kooi, M M Paffen, J J H Schatorjé, and W H C G Verkuylen, "Local oxidation of silicon and its application in semiconductor-device technology," Philips Res. Repts. *25*, 118-132 (April 1970).

[4.26] D K Brown, S M Hu, and J M Morrissey, "Flaws in sidewall oxides grown on polysilicon gate," J. Electrochem. Soc. *129*, 1084-1089 (May 1982).

[4.27] C Y Wong, J Piccirillo, A Bhattacharyya, Y Taur, and H I Hanafi, "Sidewall oxidation of polycrystalline-silicon gates," IEEE Electron Device Lett. *10*, 420-422 (September 1989).

[4.28] T T Sheng and R B Marcus, "Advances in transmission electron microscope techniques applied to device failure analysis," J. Electrochem. Soc. *127*, 737-743 (March 1980).

[4.29] R M Levin and T T Sheng, "Oxide isolation for double-polysilicon VLSI devices," J. Electrochem. Soc. *130*, 1894-1897 (September 1983).

[4.30] H G Tompkins and B Vasquez, "A special case of using ellipsometry to measure the thickness of oxide on polysilicon: I. Theoretical considerations," J. Electrochem. Soc. *137*, 1520-1522 (May 1990).

[4.31] B Vasquez, H G Tompkins, and R B Gregory, "A special case of using ellipsometry to measure the thickness of oxide on polysilicon: II. Application," J. Electrochem. Soc. *137*, 1523-1526 (May 1990).

[4.32] T S Chao, C L Lee, T F Lei, and Y T Yen, "Poly-oxide/poly-Si/SiO$_2$/Si structure for ellipsometry measurement," Electronics Lett. *28*, 1145-1146 (4 June 1992).

[4.33] D J DiMaria and D R Kerr, "Interface effects and high conductivity in oxides grown from polycrystalline silicon," Appl. Phys. Lett. *27*, 505-507 (November 1, 1975).

[4.34] R M Anderson and D R Kerr, "Evidence for surface asperity mechanism of conductivity in oxide grown on polycrystalline silicon," J. Appl. Phys. *48*, 4834-4836 (November 1977).

[4.35] R K Ellis, "Fowler-Nordheim emission from non-planar surfaces," IEEE Electron Device Lett. *EDL-3*, 330-332 (November 1982).

[4.36] C-Y Wu and C-F Chen, "Transport properties of thermal oxide films grown on polycrystalline silicon—Modeling and experiments," IEEE Trans. Electron Devices, *ED-34*, 1590-1602 (July 1987).

[4.37] L Faraone and G Harbeke, "Surface roughness and electrical conduction of oxide/polysilicon interfaces," J. Electrochem. Soc. *133*, 1410-1413 (July 1986).

[4.38] M Sternheim, E Kinsbron, J Alspector, and P A Heimann, "Properties of thermal oxides grown on phosphorus *in-situ* doped polysilicon," J. Electrochem Soc. *130*, 1735-1740 (August 1983).

[4.39] H-H Lee and S P Marin, "Electrode shape effects on oxide conduction in films thermally grown from polycrystalline silicon," J. Appl. Phys. *51*, 3746-3750 (July 1980).

[4.40] P A Heimann, S P Murarka, and T T Sheng, "Electrical conduction and breakdown in oxides of polycrystalline silicon and their correlation with interface texture," J. Appl. Phys. *53*, 6240-6245 (September 1982).

[4.41] N Matsuo, H Fujiwara, T Miyoshi, and T Koyanagi, "Numerical analysis for conduction mechanism of thin oxide-nitride-oxide films formed on rough poly-si," IEEE Electron Device Lett. *17*, 56-58 (February 1996).

[4.42] N Matsuo and A Sasaki, "Conduction mechanism of oxide-nitride-oxide film formed on the rough polycrystalline silicon surface," Solid-State Electron. *39*, 337-342 (March 1996).

[4.43] R B Marcus, T T Sheng, and P Lin, "Polysilicon/SiO$_2$ interface microtexture and dielectric breakdown," J. Electrochem. Soc. *129*, 1282-1289 (June 1982).

[4.44] P A Heimann, P S D Lin, and T T Sheng, "Electrical conduction through polysilicon oxide: Interface texture *vs.* isolated protuberances," J. Electrochem. Soc. *130*, 2117-2119 (October 1983).

[4.45] M Hendriks and C Mavero, "Phosphorus doped polysilicon for double poly structures: I. Morphology and microstructure," J. Electrochem. Soc. *138*, 1466-1470 (May 1991); "II. Electrical Characteristics," J. Electrochem. Soc. *138*, 1470-1474 (May 1991).

[4.46] M T Duffy, J T McGinn, J M Shaw, R T Smith, R A Soltis, and G Harbeke, "LPCVD polycrystalline silicon: Growth and physical properties of diffusion-doped, ion-implanted, and undoped films," RCA Review *44*, 313-325 (June 1983).

[4.47] L Faraone, "An improved fabrication process for multi-level polysilicon structures," 1983 IEEE VLSI Tech. Symp. (Maui, September 1983), paper 8-5, pp. 110-111.

[4.48] L Faraone, R D Vibronek, and J T McGinn, "Characterization of thermally oxidized n^+ polycrystalline silicon," IEEE Trans. Electron Devices, *ED-32*, 577-583 (March 1985).

[4.49] T F Lei, J-Y Cheng, S Y Shiau, T S Chao, and C S Lai, "Improvement of polysilicon oxide by growing on polished polysilicon film," IEEE Electron Device Lett. *18*, 270-271 (June 1997).

BIBLIOGRAPHY

[4.50] L. Faraone, "Thermal SiO_2 films on n^+ polycrystalline silicon: Electrical conduction and breakdown," IEEE Trans. Electron Devices, *ED-33*, 1785-1794 (November 1986).

[4.51] M-C Jun, Y-S Kim, M-K Han, J-W Kim and K-B Kim, "Polycrystalline silicon oxidation method improving surface roughness at the oxide/polycrystalline silicon interface," Appl. Phys. Lett. *66*, 2206-2208 (24 April 1995).

[4.52] Daeje Chin, Samsung Semiconductor, private communication.

[4.53] H N Chern, C L Lee, and T F Lei, "Improvement of polysilicon oxide characteristics by fluorine incorporation," IEEE Electron Device Lett. *15*, 181-182 (May 1994).

[4.54] K Ohyu, Y Wade, S Iijima, and N Natsuaki, "Highly reliable thin silicon dioxide layers grown on heavily phosphorus doped poly-S by rapid thermal oxidation," J. Electrochem. Soc. *137*, 2261-2265 (July 1990).

[4.55] K Shinada, S Mori, and Y Mikata, "Reduction in polysilicon oxide leakage current by annealing prior to oxidation," J. Electrochem. Soc. *132*, 2185-2188 (September 1985).

[4.56] S Chiao, W Shih, C Wang, and T Batra, "Developments in thin polyoxides for non-volatile memories," Semiconductor International (April 1985), pp. 156-159.

[4.57] G Yaron and L D Hess, "Improved stacked-structure oxide by laser annealing," Appl. Phys. Lett. *36*, 284-286 (15 February 1980).

[4.58] G Yaron, L D Hess, and S Kokorowski, "Application of laser processing for improved oxides grown from polysilicon," IEEE Trans. Electron Devices, *ED-27*, 964-969 (May 1980).

[4.59] G Yaron, L D Hess, and S A Kokorowski, "Application of laser processing for improved oxides grown from polysilicon," Semiconductor International (September 1980), pp. 67-78.

[4.60] M-S Liang, N Radjy, W Cox, and S Cagnina, "Charge trapping in dielectrics grown on polycrystalline silicon," J. Electrochem. Soc. *136*, 3786-3790 (December 1989).

[4.61] H R Huff, R D Halvorson, T L Chiu, and D Guterman, "Experimental observations on conduction through polysilicon oxide," J. Electrochem. Soc. *127*, 2482-2488 (November 1980).

[4.62] Y Naito, Y Hirofuji, and H Iwaskai, "Effect of bottom oxide on the integrity of interpolysilicon ultrathin ONO (oxide/nitride/oxide) films, J. Electrochem. Soc. *137*, 635-638 (February 1990).

[4.63] C S Lai, T F Lei, and C L Lee, "The characteristics of polysilicon oxide grown in pure N_2O," IEEE Trans. Electron Devices, *43*, 326-331 (February 1966).

[4.64] H-S Lee and C-H Feng, "High-electric-field-generated electron traps in oxide grown from polycrystalline silicon," Appl. Phys. Lett. *37*, 1080-1082 (15 December 1980).

[4.65] C Hu, Y Shum, T Klein, and E Lucero, "Current-field characteristics of oxides grown from polycrystalline silicon," Appl. Phys. Lett. *35*, 189-191 (15 July 1979).

[4.66] R D Jolly, H R Grinolds, and R Groth, "A model for conduction in floating-gate EEPROMs," IEEE Trans. Electron Devices, *ED-31*, 767-772 (June 1984).

[4.67] S Mori, N Yasuhisa, T Yanase, M Sato, K Yoshikawa, and H Nozawa, "Reliability aspects of 100 Åinter-poly dielectrics for high density VLSIs," 1986 IEEE VLSI Symposium, (San Diego, May 1986), paper VI-6, pp. 71-72.

Chapter 5: Electrical properties

[5.1] K Saraswat and F Mohammadi, "Effect of scaling of interconnections on the time delay of VLSI circuits," IEEE Trans. Electron Devices, *ED-29*, 645-650 (April 1982).

[5.2] J D Joseph and T I Kamins, "Resistivity of chemically deposited polycrystalline-silicon films," Solid-State Electron. *15*, 355-358 (March 1972).

[5.3] N F Mott, "Conduction in glasses containing transition metal ions" J. Non-Cryst. Solids, *1*, 1-17 (December 1968).

[5.4] C A Dimitriadis and P A Coxon, "Hopping conduction in undoped low-pressure chemically vapor deposited polycrystalline silicon films in relation to the film deposition conditions," J. Appl. Phys. *64*, 1601-1604 (1 August 1988).

[5.5] J D Cressler, W Hwang, and T-C Chen, "On the temperature dependence of majority carrier transport in heavily arsenic-doped polycrystalline silicon thin films," J. Electrochem. Soc. *136*, 794-804 (March 1989).

[5.6] A L Fripp, "Dependence of resistivity on the doping level of polycrystalline silicon," J. Appl. Phys. *46*, 1240-1244 (March 1975).

[5.7] T I Kamins, "Hall mobility in chemically deposited polycrystalline silicon," J. Appl. Phys. *42*, 4357-4365 (October 1971).

[5.8] J Y W Seto, "The electrical properties of polycrystalline silicon films," J. Appl. Phys. *46*, 5247-5254 (December 1975).

[5.9] M Cao, T King, and K Saraswat, "Determination of the densities of gap states in hydrogenated polycrystalline Si and $Si_{0.8}Ge_{0.2}$ films," Appl. Phys. Lett. *61*, 672-674 (10 August 1992).

[5.10] G Baccarani, B Riccó, and G Spadini, "Transport properties of polycrystalline silicon films," J. Appl. Phys. *49*, 5565-5570 (November 1978).

[5.11] J P Colinge, E Demoulin, F Delannay, M Lobet and J M Temerson, "Grain size and resistivity of LPCVD polycrystalline silicon films," J. Electrochem. Soc. *128*, 2009-2014 (September 1981).

[5.12] G Baccarani, M Impronta, B Riccó, and P Ferla, "I-V characteristics of polycrystalline silicon resistors," Revue de Physique Appliqueé, *13*, 777-782 (December 1978).

[5.13] G J Korsh and R S Muller, "Conduction properties of lightly doped, polycrystalline silicon," Solid-State Electron. *21*, 1045-1051 (August 1978).

[5.14] A K Ghosh, C Fishman, and T Feng, "Theory of the electrical and photovoltaic properties of polycrystalline silicon," J. Appl. Phys. *51*, 446-454 (January 1980).

[5.15] M M Mandurah, K C Saraswat and T I Kamins, "Phosphorus doping of low pressure chemically vapor-deposited silicon films," J. Electrochem. Soc. *126*, 1019-1023 (June 1979).

[5.16] A K Ghosh, A Rose, H P Maruska, D J Eustace, and T Feng, "Hall measurements and grain-size effects in polycrystalline silicon," Appl. Phys. Lett. *37*, 544-546 (15 September 1980).

[5.17] A K Ghosh, A Rose, H P Maruska, T Feng, and D J Eustace, "Interpretation of Hall and resistivity measurements in polycrystalline silicon" J. Electronic Mat. *11*, 237-260 (March 1982).

[5.18] R H Bube, "Interpretation of Hall and photo-Hall effects in inhomogeneous materials," Appl. Phys. Lett. *13*, 136-139 (15 August 1968).

[5.19] M S Bennett, "Relationship between Hall constant and carrier densities in polycrystalline semiconductor film," J. Appl. Phys. *58*, 3470-3475 (1 November 1985).

[5.20] J W Orton, "Interpretation of Hall mobility in polycrystalline thin films," Thin Solid Films, *86*, 351-357 (December 18, 1981).

[5.21] G E Pike and C H Seager, "The dc voltage dependence of semiconductor grain-boundary resistance," J. Appl. Phys. *50*, 3414-3422 (May 1979).

[5.22] J M Andrews, "Electrical conduction in implanted polycrystalline silicon," J. Electronic Materials, *8*, 227-247 (May 1979).

[5.23] C H Seager and G E Pike, "Grain boundary states and varistor behavior in silicon bicrystals," Appl. Phys. Lett. *35*, 709-711 (1 November 1979).

[5.24] C H Seager, "Grain boundary recombination: Theory and experiment in silicon," J. Appl. Phys. *52*, 3960-3968 (June 1981).

[5.25] W B Jackson, N M Johnson, and D K Biegelsen, "Density of gap states of silicon grain boundaries determined by optical absorption," Appl. Phys. Lett. *43*, 195-197 (15 July 1983).

[5.26] W K Schubert and P M Lenahan, "Spin dependent trapping in a polycrystalline silicon integrated circuit resistor," Appl. Phys. Lett. *43*, 497-499 (1 September 1983).

[5.27] W E Spear and P G LeComber, "Electronic properties of substitutionally doped amorphous Si and Ge," Phil. Mag. *33*, 935-949 (1976).

[5.28] G Queirolo, E Servida, L Baldi, G Pignatel, A Armigliato, S Frabboni, and F Corticelli, "Dopant activation, carrier mobility, and TEM studies in polycrystalline silicon films," J. Electrochem. Soc. *137*, 967-971 (March 1990).

[5.29] V Srikant, D R Clarke, and P V Evans, "Simulation of electron transport across charged grain boundaries," Appl. Phys. Lett. *69*, 1755-1757 (16 September 1996) and Comment H F Mataré, Appl. Phys. Lett. *70*, 2055 (14 April 1997).

[5.30] C-Y Lu, N C-C Lu, and C-S Wang, "Effects of grain-boundary trapping-state energy distribution on the activation energy of resistivity of polycrystalline-silicon films," Solid-State Electron. *27*, 463-466 (May 1984).

[5.31] N C C Lu, L Gerzberg, C Y Lu, and J D Meindl, "Thermionic field emission in polycrystalline-silicon films," Fall 1980 Electrochemical Society Meeting (Hollywood FL, October 1980), abstract 483, pp. 1103-1105.

[5.32] N C-C Lu, L Gerzberg, C-Y Lu, and J D Meindl, "A conduction model for semiconductor–grain-boundary–semiconductor barriers in polycrystalline-silicon films," IEEE Trans. Electron Devices, *ED-30*, 137-149 (February 1983).

[5.33] M M Mandurah, K C Saraswat, and T I Kamins, "A model for conduction in polycrystalline silicon–Part I: Theory," IEEE Trans. Electron Devices, *ED-28*, 1163-1171 (October 1981).

[5.34] E L Murphy and R H Good, Jr., "Thermionic emission, field emission and the transition region," Phys. Rev. *102*, 1464-1473 (June 15, 1956).

BIBLIOGRAPHY

[5.35] M M Mandurah, K C Saraswat, and T I Kamins, "A model for conduction in polycrystalline silicon–Part II: Comparison of theory and experiment," IEEE Trans. Electron Devices, *ED-28*, 1171-1176 (October 1981).

[5.36] D P Joshi and R S Srivastava, "A model of electrical conduction in polycrystalline silicon," IEEE Trans. Electron Devices, *ED-31*, 920-927 (July 1984).

[5.37] J Martinez, A Criado, and J Piqueras, "Grain boundary potential determination in polycrystalline silicon by the scanning light spot technique," J. Appl. Phys. *52*, 1301-1305 (March 1981).

[5.38] E Loh, "Interpretation of dc characteristics of phosphorus-doped polycrystalline silicon films: Conduction across low-barrier grain boundaries," J. Appl. Phys. *54*, 4463-4466 (August 1983).

[5.39] N C-C Lu, and C-Y Lu, "I-V characteristics of polysilicon resistors at high electric field and the non-uniform conduction mechanism," Solid-State Electron. *27*, 797-805 (August/September 1984).

[5.40] M Taniguchi, M Hirose, Y Osaka, S Hasegawa, and T Shimizu, "Current transport in doped polycrystalline silicon," Japan. J. Appl. Phys. *19*, 665-673 (April 1980).

[5.41] M J McCarthy, M D Karim, and J A Reimer, "The properties of phosphorus in polycrystalline silicon—A nuclear magnetic resonance study," Spring 1986 Mat. Res. Soc. Meeting (Palo Alto CA, April 17, 1986), paper F5.19.

[5.42] D Ballutaud, M Aucouturier, and F Bobonneau, "Electron spin resonance study of hydrogenation effects in polycrystalline silicon," Appl. Phys. Lett. *49*, 1620-1622 (8 December 1986).

[5.43] N H Nickel, N M Johnson, and W B Jackson, "Hydrogen passivation of grain boundary defects in polycrystalline silicon," Appl. Phys. Lett. *62*, 3285-3287 (21 June 1993).

[5.44] T Makino and H Nakamura, "The influence of plasma annealing on electrical properties of polycrystalline Si," Appl. Phys. Lett. *35*, 551-552 (1 October 1979).

[5.45] D L Chen, D W Greve, and A M Guzman, "Influence of hydrogen implantation on the resistivity of polycrystalline silicon," J. Appl. Phys. *57*, 1408-1410 (15 February 1985).

[5.46] E S Cielaszyk, K H R Kirmse, R A Stewart, and A E Wendt, "Mechanisms for polycrystalline silicon defect passivation by hydrogenation in an electron cyclotron resonance plasma," Appl. Phys. Lett. *67*, 3099-3101 (20 November 1995).

[5.47] I-W Wu, A G Lewis, T-Y Huang, and A Chiang, "Effects of trap-state density reduction by plasma hydrogenation in low-temperature polysilicon TFT," IEEE Electron Device Lett. *10*, 123-125 (March 1989).

[5.48] S Ostapenko, L Jastrzebski, J. Lagowski, and R K Smeltzer, "Enhanced hydrogenation in polycrystalline silicon thin films using low-temperature ultrasound treatment," Appl. Phys. Lett. *68*, 2873-2875 (13 May 1996).

[5.49] J I Pankove, R O Wance, and J E Berkeyheiser, "Neutralization of acceptors in silicon by atomic hydrogen," Appl. Phys. Lett. *45*, 1100-1102 (15 November 1984).

[5.50] B W Liou, Y H Wu, C L Lee, and T F Lei, "Thickness effect on hydrogen plasma treatment on polycrystalline silicon thin films," Appl. Phys. Lett. *66*, 3013-3014 (29 May 1995).

[5.51] G P Pollack, W F Richardson, S D S Malhi, T Bonifield, H Shichijo, S Banerjee, M Elahy, A H Shah, R Womack, and P K Chatterjee, "Hydrogen passivation of polysilicon MOSFETs from a plasma nitride source," IEEE Electron Device Lett. *EDL-5*, 468-470 (November 1984).

[5.52] E Puppin, "Hydrogen implanted polycrystalline silicon: Resistivity and grain boundary chemistry," J. Vac. Sci. Technol. B *5*, 606-607 (March/April 1987).

[5.53] V Suntharalingam and S J Fonash, "Electrically reversible depassivation/passivation mechanism in polycrystalline silicon," Appl. Phys. Lett. *68*, 1400-1402 (4 March 1996).

[5.54] N M Johnson, D K Biegelsen, and M D Moyer "Deuterium passivation of grain-boundary dangling bonds in silicon thin films," Appl. Phys. Lett. *40*, 882-884 (15 May 1982).

[5.55] R T Young, M C Lu, R D Westbrook, and G E Jellison, Jr, "Effect of lithium on the electrical properties of grain boundaries in silicon," Appl. Phys. Lett. *38*, 628-630 (15 April 1981).

[5.56] G L Miller and W A Orr, "Lithium doping of polycrystalline silicon," Appl. Phys. Lett. *37*, 1100-1101 (15 December 1980).

[5.57] T I Kamins, "MOS transistors in beam-recrystallized polysilicon," *Tech. Digest*, 1982 International Electron Devices Meeting (San Francisco, December 1982), paper 16.1, pp. 420-423.

[5.58] S D Brotherton, D J McCulloch, J B Clegg, and J P Gowers, "Excimer-laser-annealed poly-Si thin-film transistors," IEEE Trans. Electron Devices *40*, 407-413 (February 1993).

[5.59] Y Morimoto, Y Jinno, K Hirai, H Ogata, T Yamada, and K Yoneda, "Influence of the grain boundaries and intragrain defects on the performance of poly-Si thin film transistors," J. Electrochem. Soc. *144*, 2495-2501 (July 1997).

[5.60] G K Giust and T W Sigmon, "Performance improvement obtained for thin-film transistors fabricated in prepatterned laser-recrystallized polysilicon," IEEE Electron Device Lett. *18*, 296-298 (June 1997).

[5.61] G Yaron, L D Hess, and G L Olsen, "Electrical characteristics of laser-annealed polysilicon resistors for device applications," Proc. Materials Research Society Symposium (Boston, 1979) (ed. C W White and P S Peercy, Academic Press, 1980), pp. 626-631.

[5.62] Y Wada and S Nishimatsu, "Resistivity lowering limitations of heavily doped polycrystalline silicon," Denki Kagaku, *47*, 118-123 (1979).

[5.63] S Solmi, M Severi, R Angelucci, L Baldi, and R Bilenchi, "Electrical properties of thermally and laser annealed polycrystalline silicon films heavily doped with arsenic and phosphorus," J. Electrochem. Soc. *129*, 1811-1818 (August 1982).

[5.64] N Lifschitz, "Solubility of implanted dopants in polysilicon: Phosphorus and arsenic," J. Electrochem. Soc. *130*, 2464-2467 (December 1983).

[5.65] R B Fair, "Recent advances in implantation and diffusion modeling for the design and process control of bipolar ICs," in *Semiconductor Silicon 1977* (ed. H R Huff and E Sirtl, The Electrochemical Soc., Princeton NJ, 1977), Proc. Vol. 77-2, pp. 968-987.

[5.66] R B Fair and J C C Tsai, "A quantitative model for the diffusion of phosphorus in silicon and the emitter dip effect," J. Electrochem. Soc. *124*, 1107-1118 (July 1977).

[5.67] J Murota and T Sawai, "Electrical characteristics of heavily arsenic and phosphorus doped polycrystalline silicon," J. Appl. Phys. *53*, 3702-3708 (May 1982).

[5.68] T Makino and H Nakamura, "Resistivity changes of heavily-boron-doped CVD-prepared polycrystalline silicon caused by thermal annealing," Solid-State Electron. *24*, 49-55 (January 1981).

[5.69] G Masetti, D Nobili, and S Solmi, "Profiles of phosphorus predeposited in silicon and carrier concentration in equilibrium with SiP precipitates," in *Semiconductor Silicon 1977* (ed. H R Huff and E Sirtl, The Electrochemical Soc., Princeton NJ, 1977), Proc. Vol. 77-2, pp. 648-657.

[5.70] A Lietoila, J F Gibbons, and T W Sigmon, "The solid solubility and thermal behavior of metastable concentrations of As in Si," Appl. Phys. Lett. *36*, 765-768 (1 May 1980).

[5.71] K Suzuki, N Miyata, and K Kawamura, "Resistivity of heavily doped polycrystalline silicon subjected to furnace annealing," Japan. J. Appl. Phys. *34*, 1748-1752 (April 1995).

[5.72] T I Kamins, "Resistivity of LPCVD polycrystalline-silicon films," J. Electrochem. Soc. *126*, 833-837 (May 1979).

[5.73] G Kawachi, T Aoyama, K Miyata, Y Ohno, A Mimura, N Konishi, and Y Mochizuki, "Large-area ion doping technique with bucket-type ion source for polycrystalline silicon films," J. Electrochem. Soc. *137*, 3522-3526 (November 1990).

[5.74] Y Mishima and M Takei, "Non-mass-separated ion shower doping of polycrystalline silicon," J. Appl. Phys. *75*, 4933-4938 (15 May 1994).

[5.75] R Bashir, S Venkatesan, H Yen, G W Neudeck and E P Kvam, "Doping of polycrystalline silicon films using an arsenic spin-on-glass source and surface smoothness," J. Vac. Sci. Technol. B, *11*, 1903-1905 (September/October 1993).

[5.76] B S Meyerson, F K LeGoues, T N Nguyen, and D L Harame, "Nonequilibrium boron doping effects in low-temperature epitaxial silicon films," Appl. Phys. Lett. *50*, 113-115 (12 January 1987).

[5.77] R Chow and R A Powell, "Activation and redistribution of implants in polysi by RTP," Semiconductor International (May 1985), pp. 108-113.

[5.78] W Shockley, *Electrons and Holes in Semiconductors* (John Wiley and Sons, New York 1959), pp. 318-325.

[5.79] P Panayotatos, E S Yang, and W Hwang, "Determination of the grain boundary recombination velocity in polycrystalline silicon as a function of illumination from photoconductance measurements," Solid-State Electron. *25*, 417-422 (May 1982).

[5.80] H C Card and E Yang, "Electronic processes at grain boundaries in polycrystalline semiconductors under optical illumination," IEEE Trans. Electron Devices, *ED-24*, 397-402 (April 1977).

[5.81] M A Green, "Bounds upon grain boundary effects in minority carrier semiconductor devices: A rigorous "perturbation" approach with application to silicon solar cells," J. Appl. Phys. *80*, 1515-1521 (1 August 1996).

[5.82] J E Mahan, "Threshold and memory switching in polycrystalline silicon," Appl. Phys. Lett. *41*, 479-481 (1 September 1982).

[5.83] P T Landsberg and M S Abrahams, "Effects of surface states and of excitation on barrier heights in a simple model of a grain boundary or a surface," J. Appl. Phys. *55*, 4284-4293 (15 June 1984).

[5.84] P Kenyon and H Dressel, "Negative resistance switching in near-perfect crystalline silicon film resistors," J. Vac. Sci. Technol. A *2*, 1486-1490 (October-December 1984).

[5.85] C-Y Lu, N C-C Lu, and C-C Shih, "Resistance switching characteristics in polycrystalline silicon film resistors," J. Electrochem. Soc. *132*, 1193-1196 (May 1985).

[5.86] C H Seager and G E Pike, "Anomalous low-frequency grain-boundary capacitance in silicon," Appl. Phys. Lett. *37*, 747-749 (15 October 1980).

[5.87] M Darwish and K Board, "Theory of switching in polysilicon n-p^+ structures," Solid-State Electron. *27*, 775-783 (August/September 1984).

[5.88] Y Amemiya, T Ono, and K Kato, "Electrical trimming of heavily doped polycrystalline silicon resistors," IEEE Trans. Electron Devices, *ED-26*, 1738-1742 (November 1979).

[5.89] M Tanimoto, J Murota, Y Ohmori, and N Ieda, "A Novel MOS PROM using a highly resistive poly-Si resistor," IEEE Trans. Electron Devices, *ED-27*, 517-520, (March 1980).

[5.90] O-H Kim and C-K Kim, "Effects of high-current pulses on polycrystalline silicon diode with n-type region heavily doped with both boron and phosphorus," J. Appl. Phys. *53*, 5359-5360 (July 1982).

Chapter 6: Applications

[6.1] F Faggin and T Klein, "Silicon gate technology," Solid-State Electron. *13*, 1125-1144 (August 1970).

[6.2] R S Muller and T I Kamins, *Device Electronics for Integrated Circuits*, Second Edition (Wiley, New York, 1986), pp. 418, 473.

[6.3] D Peters, "Implanted-silicided polysilicon gates for VLSI transistors," IEEE Trans. Electron Devices, *33*, 1391-1393 (September 1986).

[6.4] J F Chen and L J Chen, "Morphological stability of $TiSi_2$ on polycrystalline silicon," Thin Solid Films *293*, 34-39 (30 January 1997).

[6.5] S Pramanick, Y N Erokhin, B K Patnaik, and G A Rozgonyi, "Morphological instability and Si diffusion in nanoscale cobalt silicide films formed on heavily phosphorus doped polycrystalline silicon," Appl. Phys. Lett. *63*, 1933-1935 (4 October 1993).

[6.6] C M Osburn and E Bassous, "Improved dielectric reliability of SiO_2 films with polycrystalline silicon electrodes," J. Electrochem. Soc. *122*, 89-92 (January 1975).

[6.7] C M Osburn and N J Chou, "Accelerated dielectric breakdown of silicon dioxide films," J. Electrochem. Soc. *120*, 1377-1384 (October 1973).

[6.8] H-N Yu, A Reisman, C M Osburn, and D L Critchlow, "1 μm MOSFET VLSI technology: Part I—An Overview," IEEE Trans. Electron Devices, *ED-26*, 318-324 (April 1979).

[6.9] K Chen, M Chan, P K Ko, C Hu, and J-H Huang, "Polysilicon gate depletion effect on IC performance," Solid-State Electron. *38*, 1975-1977 (November 1995).

[6.10] N D Arora, R Rios, and C-L Huang, "Modeling the polysilicon depletion effect and its impact on submicrometer CMOS circuit performance," IEEE Trans. Electron Devices *42*, 935-943 (May 1995).

[6.11] J C Hu, et al, "Feasibility of using W/TiN as metal gate for conventional 0.13μm CMOS technology and beyond," *Tech. Digest* International Electron Device Meeting (Washington DC, December 1997), paper 33.2, pp. 825-828.

[6.12] A Chatterjee, et al, "Sub-100nm gate length metal gate NMOS transistors fabricated by a replacement gate process," *Tech. Digest* International Electron Device Meeting (Washington DC, December 1997), paper 33.1, pp. 821-824.

[6.13] H-H Tseng, M Orlowski, P J Tobin, and R L Hance, "Fluorine diffusion on a polysilicon grain boundary network in relation to boron penetration from p$^+$ gates," IEEE Electron Device Lett. *13*, 14-16 (January 1992).

[6.14] T Hosoya, K Machida, K Imai, and E Arai, "A polycide gate electrode with a conductive diffusion barrier formed with ECR nitrogen plasma for dual-gate CMOS," IEEE Trans. Electron Devices, *42*, 2111-2116 (December 1995).

[6.15] C W Chen, Y K Fang, K Y Lee, J C Hsieh, and M S Liang, "Improvement on fluorine effect under high-stress in tungsten-polycide gated metal-oxide-semiconductor field-effect transistor with oxynitride and/or reoxidized-oxynitride gate dielectric," Japan. J. Appl. Phys. *35*, 2590-2594 (May 1996).

[6.16] T S Chao and C H Chu, "Inductive-coupling-nitrogen-plasma process for suppression of boron penetration in BF2+-implanted polycrystalline silicon gate," Appl. Phys. Lett. *70*, 55-56 (6 January 1997).

[6.17] Y Okazaki, S Nakayama, M Miyake, and T Kobayashi, "Characteristics of sub-1/4-μm gate surface channel PMOSFET's using a multilayer gate structure of boron-doped poly-Si on thin nitrogen-doped poly-Si," IEEE Trans. Electron Devices *41*, 2369-2375 (December 1994).

[6.18] T Kuroi, M Kobayashi, M Shirahata, Y Okumura, S Kusunoki, M Inuishi, and N Tsubouchi, "The impact of nitrogen implantation into highly doped polysilicon gates for highly reliable and high-performance sub-quarter-micron dual-gate complementary metal-oxide-semiconductor," Japan. J. Appl. Phys. *34*, 771-775 (February 1995).

[6.19] T Sakai and S Nakayama, "The effect of nitrogen in a p^+ polysilicon gate on boron penetration through the gate oxide," J. Electrochem. Soc. *144*, 4326-4330 (December 1997).

[6.20] L Farone, "Thermal SiO_2 films on n^+ polycrystalline silicon: Electrical conduction and breakdown," IEEE Trans. Electron Devices *33*, 1785-1794 (November 1986).

[6.21] S Muramatsu, T Kubota, N Nishio, H Shirai, M Matsuo, N Kodama, M Horikawa, S-I Saito, K Arai, and T Okazawa, "The solution of over-erase problem controlling poly-Si grain size: Modified scaling principles for FLASH memory" *Tech. Digest* International Electron Devices Meeting (San Francisco, December 1994), paper 34.5, pp. 847-850.

[6.22] H Shirai, T Kubota, I Honma, H Watanabe, H Ono, and T Okazawa, "A 0.54 μm^2 self-aligned HSG floating gate cell (SAHF Cell) for 256 Mbit flash memories," *Tech. Digest* International Electron Devices Meeting (Washington, DC, December 1995), paper 27.1, pp. 653-656.

[6.23] S Tiwari, F Rana, K Chan, H Hanafi, W Chan, and D Buchanan, "Volatile and non-volatile memories in silicon with nano-crystal storage *Tech. Digest* International Electron Devices Meeting (Washington, DC, December 1995), paper 20.4, pp. 521-524.

[6.24] D W Hughes, "Polycrystalline silicon resistors for use in integrated circuits," Solid State Technology (May 1987), pp. 139-143.

[6.25] *eg*, V G McKenny, "A 5 V-only 4-K static RAM," *Digest* 1977 IEEE International Solid-State Circuits Conference (February 1977), paper WAM-1.3, pp. 16-17.

[6.26] T Ohzone, M Fukumoto, G Fuse, A Shinohara, S Odanaka, M Sasago, "Ion-implanted thin polycrystalline-silicon high-value resistors for high-density poly-load static RAM applications," IEEE Trans. Electron. Devices, *ED-32*, 1749-1756 (September 1985).

[6.27] M-K Lee, C-Y Lu, K-Z Chang, and C Shih, "On the semi-insulating polycrystalline silicon resistor," Solid-State Electron. *27*, 995-1001 (November 1984).

[6.28] T Ohzone, T Hirao, K Tsuji, S Horiuchi, and S Takayanagi, "A 2K×8-bit static RAM," *Tech. Digest* 1978 International Electron Devices Meeting (Washington DC, December 1978), paper 14.7, pp. 360-363.

[6.29] M Rodder, "Increased resistance in p-type poly-Si resistors by thermal anneal reduction of interface charge," IEEE Electron Device Lett. *12*, 160-162 (April 1991).

[6.30] M Rodder, "Effects of channel width and parasitic bipolar action of p-type poly-Si resistor characteristics," IEEE Electron Device Lett. *12*, 241-243 (May 1991).

[6.31] S Das and S K Lahiri, "Electrical trimming of ion-beam-sputtered polysilicon resistors by high current pulses," IEEE Trans. Electron Devices *41*, 1429-1434 (August 1994).

[6.32] K Kato and T Ono, "Constant voltage trimming of heavily doped polysilicon resistors," Japan. J. Appl. Phys. *34*, 48-53 (January 1995).

[6.33] D W Feldbaumer, J A Babcock, W M Mercier, and C K Y Chun, " Pulse current trimming of polysilicon resistors," IEEE Trans. Electron Devices *42*, 689-696 (April 1995).

[6.34] D W Feldbaumer and J A Babcock, "Theory and application of polysilicon resistor trimming," Solid-State Electron. *38*, 1861-1869 (November 1995).

[6.35] D L Parker and W Huang, "Polysilicon resistor trimming by laser link making," IEEE Trans. Semiconductor Manufacturing *3*, 80-83 (May 1990).

[6.36] K Kato and T Ono, "Change in temperature coefficient of resistance of heavily doped polysilicon resistors caused by electrical trimming," Japan. J. Appl. Phys. *35*, 4209-4215 (August 1996).

[6.37] S-L Jang, "A model of $1/f$ noise in polysilicon resistors," Solid-State Electron. *33*, 1155-1162 (September 1990).

[6.38] M E Lunnon and D W Greve, "The microstructure of programmed n^+pn^+ polycrystalline silicon antifuses," J. Appl. Phys. *54*, 3278-3281 (June 1983).

[6.39] J T Baek, H-H Park, S W Kang, B T Ahn, and I J Yoo, "Investigation of link formation in a novel planar-type antifuse structure," Thin Solid Films *288*, 41-44 (15 November 1996).

[6.40] J R Lloyd, M R Polcari, and G A MacKenzie, "Observation of electromigration in heavily doped polycrystalline silicon thin films," Appl. Phys. Lett. *36*, 428-430 (15 March 1980).

[6.41] M R Polcari, J R Lloyd, and S Cvikevich, "Electromigration failure in heavily doped polycrystalline silicon," Proc. 1980 IEEE International Reliability Physics Symp. (Las Vegas, April 1980), pp. 178-185.

[6.42] J R Lloyd, G S Hopper, and W B Roush, "*In-situ* IR observation of electromigration induced damage in heavily doped polycrystalline silicon resistors," Proc. 1982 IEEE International Reliability Physics Symp. (San Diego CA, March-April 1982), pp. 47-49.

[6.43] H Arai, S Kohda, and Y Kitano, "Lateral reaction of polycrystalline silicon with overlapping aluminum film," J. Appl. Phys. *57*, 1143-1146 (15 February 1985).

[6.44] D Pramanik and A N Saxena, "VLSI metallization using aluminum and its alloys. Part II," Solid State Technology (March 1983), pp. 131-138.

[6.45] J M Ford, "Al/poly Si specific contact resistivity," IEEE Electron Device Lett. *4*, 255-257 (July 1983).

[6.46] H Kamioka, K Ishii, and M Takagi, "DOPOS technology for microwave transistors," Fujitsu Scientific and Tech. Journal *8*, 147-168 (December 1972).

[6.47] *Polysilicon Emitter Bipolar Transistors* (ed. A K Kapoor and D J Roulston), IEEE Press Selected Reprint Series (IEEE, New York, 1989).

[6.48] I R C Post, P Ashburn, and G R Wolstenholme, "Polysilicon emitters for bipolar transistors: A review and re-evaluation of theory and experiment," IEEE Trans. Electron Devices *39*, 1717-1731 (July 1992).

[6.49] M Nanba, T Uchino, M Kondo, T Nakamura, T Kobayashi, Y Tamaki, and M Tanabe, "A 64-GHz f_T and 3.6-V BV_{CEO} Si bipolar transistor using *in situ* phosphorus-doped and large-grained polysilicon emitter contacts," IEEE Trans. Electron Devices *40*, 1563-1565 (August 1993); also, *Tech. Digest* International Electron Devices Meeting (Washington, DC, December 1991), paper 16.1, pp. 443-446.

[6.50] M Kondo, T Shiba, and Y Tamaki, "Analysis of emitter efficiency enhancement induced by residual stress for *in situ* phosphorus-doped polysilicon emitter transistors," IEEE Trans. Electron Devices *44*, 978-985 (June 1997).

[6.51] C A King, R W Johnson, M R Pinto, H S Luftman, and J Munanka, "*In-situ* arsenic-doped polycrystalline silicon as a low thermal budget emitter contact for $Si/Si_{1-x}Ge_x$ heterojunction bipolar transistors," Appl. Phys. Lett. *68*, 226-228 (8 January 1996).

[6.52] B Cunningham, P A Ronsheim, and B H Yun, "Microstructural effects of emitter size on polysilicon-emitter bipolar transistors," J. Appl. Phys. *70*, 5318-5322 (15 November 1991).

[6.53] J N Burghartz, J Y-C Sun, C L Stanis, S R Mader, and J D Warnock, "Identification of perimeter depletion and emitter plug effects in deep-submicrometer, shallow-junction polysilicon emitter bipolar transistors," IEEE Trans. Electron Devices *39*, 1477-1489 (June 1992).

[6.54] K Kikuchi, S Kameyama, M Kajiyama, M Nishio, and T Komeda, "A high-speed bipolar LSI process using self-aligned double diffusion polysilicon technology," *Tech. Digest* International Electron Devices Meeting (Los Angeles, December 1986), paper 16.4, pp. 420-423.

[6.55] C I Drowley, *et al,* "STRIPE - A high-speed VLSI bipolar technology featuring self-aligned single-poly base and submicron emitter contacts," Symp. VLSI Technology, Digest of Technical Papers, pp. 53-54 (Honolulu, June 1990).

[6.56] S-Y Chiang, D Pettengill, and P Vande Voorde, "Bipolar device design for circuit performance optimization," Proc. IEEE Bipolar Circuits and Technology Meeting (Minneapolis MN, September 1990), paper 8.1, pages, 172-179.

[6.57] H C de Graaff and J G de Groot, "The SIS tunnel emitter: A theory for emitters with thin interface layers," IEEE Trans. Electron Devices, *26*, 1771-1776 (November 1979).

[6.58] T Yamaguchi, Y-C S Yu, V F Drobny, and A M Witkowski, "Emitter resistance and performance tradeoffs of submicrometer self-aligned double-polysilicon bipolar devices," IEEE Trans. Electron Devices *35*, 2397-2405 (December 1988).

[6.59] S L Wu, C L Lee, T F Lei, C F Chen, L J Chen, K Z Ho, and Y C Ling, "Enhancement of oxide break-up by implantation of fluorine in poly-Si emitter contacted p^+-n shallow junction formation," IEEE Electron Device Lett. *15*, 120-122 (April 1994).

[6.60] N E Moiseiwitsch and P Ashburn, "The benefits of fluorine in pnp polysilicon emitter bipolar transistors," IEEE Trans. Electron Devices *41*, 1249-1256 (July 1994).

[6.61] N E Moiseiwitsch, J F W Schiz, C D Marsh, P Ashburn, and G R Booker, "Increased current gain and reduced emitter resistance in F-implanted low thermal budget polysilicon emitters for SiGe HBTs," Proc. IEEE Bipolar/BiCMOS Circuits and Technology Meeting (Minneapolis MN, September 1996), paper 11.2, pp. 177-180.

[6.62] N Siabi-Shahrivar, W Redman-White, P Ashburn and I Post, "Modeling and characterization of noise of polysilicon emitter bipolar transistors," Proc. IEEE Bipolar Circuits and Technology Meeting (Minneapolis MN, September 1990), paper 10.2, pp. 236-238.

[6.63] G R Wolstenholme, N Jorgensen, P Ashburn, and G R Booker, "An investigation of the thermal stability of the interfacial oxide in polycrystalline silicon emitter bipolar transistors by comparing device results with high-resolution electron microscopy observations," J. Appl. Phys. *61*, 225-233 (1 January 1987).

[6.64] J L Egley and J L Gray, "Demonstration of the importance of the oxide breakup in polysilicon-contacted-emitter modeling," IEEE Trans. Electron Devices *38*, 2112-2117 (September 1991).

[6.65] E F Chor, P Ashburn, and A Brunnschweiler, "Emitter resistance of arsenic- and phosphorus-doped polysilicon emitter transistors," IEEE Electron Device Lett. *6*, 516-518 (October 1985).

[6.66] K K Ng and H C Card, "Asymmetry in the SiO_2 tunneling barriers to electrons and holes," J. Appl. Phys. *51*, 2153-2157 (April 1980).

[6.67] P Ashburn and B Soerowirdjo, "Comparison of experimental and theoretical results on polysilicon emitter bipolar transistors," IEEE Trans. Electron Devices, *31*, 853-860 (July 1984).

[6.68] D J Doyle, J D Barrett, W A Lane, M O'Neill, D Bain, R Baker, and P J Mole, "Comparison of bipolar NPN polysilicon emitter interface formation at three different manufacturing sites," IEEE Trans. Semiconductor Manufacturing *5*, 241-247 (August 1992).

[6.69] T H Ning and R D Isaac, "Effect of emitter contact on current gain of silicon bipolar devices," IEEE Trans. Electron Devices, *27*, 2051-2055 (November 1980).

[6.70] P H Yeung and W C Ko, "Current gain in polysilicon emitter transistors," IEEE Trans. Electron Devices, *30*, 593-597 (June 1983).

[6.71] G L Patton, J C Bravman, and J D Plummer, "Physics, technology, and modeling of polysilicon emitter contacts for VLSI bipolar transistors," IEEE Trans. Electron Devices, *33*, 1754-1768 (November 1986).

[6.72] A Neugroschel, M Arienzo, Y Komem, and R D Isaac, "Experimental study of the minority-carrier transport at the polysilicon-monosilicon interface," IEEE Trans. Electron Devices, *32*, 807-816 (April 1985).

[6.73] P A Potyraj, D-L Chen, M K Hatalis, and D W Greve, "Interfacial oxide, grain size, and hydrogen passivation effects on polysilicon emitter transistors," IEEE Trans. Electron Devices *35*, 1334-1343 (August 1988).

[6.74] B Jalali and E S Yang, "A general model for minority carrier transport in polysilicon emitters," Solid-State Electron. *32*, 323-327 (April 1989).

[6.75] T Matsushita, N Oh-uchi, H Hayashi, and H Yamoto, "A silicon heterojunction transistor," Appl. Phys. Lett. *35*, 549-550 (1 October 1979).

[6.76] K Nakazato, T Nakamura, and M Kato, "A 3GHz lateral PNP transistor," *Tech. Digest* International Electron Devices Meeting (Los Angeles, December 1986), paper 16.3, pp. 416-419.

[6.77] U S Davidsohn and F Lee, "Dielectric isolated integrated circuit substrate processes," Proc. IEEE, *57*, 1532-1537 (September 1969).

[6.78] H J A van Dijk and J de Jonge, "Preparation of thin silicon crystals by electrochemical thinning of epitaxially grown structures," J. Electrochem. Soc. *117*, 553-554 (April 1970).

[6.79] R L Meek, "Electrochemically thinned N/N^+ epitaxial silicon—Method and application," J. Electrochem. Soc. *118*, 1240-1246 (July 1971).

[6.80] T I Kamins, "A new dielectric isolation technique for bipolar integrated circuits using thin single-crystal silicon films," Proc. IEEE, *60*, 915-916 (July 1972).

[6.81] I Kobayashi, "A new technology for high-power IC," IEEE Trans. Electron Devices, *ED-18*, 45-49 (January 1971).

[6.82] I Kobayashi, "Technology for monolithic high-power integrated circuits using polycrystalline Si for collector and isolation walls," IEEE Trans. Electron Devices, *20*, 399-404 (April 1973).

[6.83] J A Schoeff, "Polycrystalline silicon technology for bipolar integrated circuits," *Tech. Digest* International Electron Device Meeting (Washington DC, October 1970), paper 11.6, pp. 74-76.

[6.84] J I Raffel and S E Bernacki, "Dielectric isolation using shallow oxide and polycrystalline silicon," *Tech. Digest* International Electron Devices Meeting, (Washington DC, December 1976), paper 24.3, pp. 601-604.

[6.85] R L Guldi, B McKee, G M Damminga, C Y Young, and M A Beals, "Characterization of poly-buffered LOCOS in manufacturing environment," J. Electrochem. Soc. *136*, 3815-3820 (December 1989).

[6.86] P U Kenkare and J R Pfiester, "Extending LOCOS-based approaches for Sub-0.5 μm CMOS," Semiconductor International (May 1994), pp. 64-68.

[6.87] J Nagel, M Reiche, S Hopfe, and D Katzer, "Stress-induced void formation in interlevel polysilicon films during polybuffered local oxidation of silicon," J. Electrochem. Soc. *140*, 2356-2359 (August 1993).

[6.88] V K Mathews, P C Fazan, and R L Maddox, "Residues, polycrystalline voids, and active area damage with the polycrystalline silicon buffered local oxidation of silicon isolation process," Appl. Phys. Lett. *64*, 94-96 (3 January 1994).

[6.89] Y S Obeng, D C Brady, S C Vitkavage, J A Taylor, K J Hanson, and B J Sapjeta, "Pitting of the silicon layer of poly buffered LOCOS stack," J. Electrochem. Soc. *142*, 1680-1688 (May 1995).

[6.90] T Kobayashi, S Nakayama, M Miyake, Y Okazaki, and H Inokawa, "Nitrogen in-situ doped poly buffer LOCOS: Simple and scalable isolation technology for deep-submicron silicon devices," IEEE Trans. Electron Devices *43*, 311-317 (February 1996).

[6.91] R D Rung, H Momose, and Y Nagakubo, "Deep trench isolated CMOS device," *Tech. Digest* International Electron Devices Meeting (San Francisco, December 1982), paper 9.6, pp. 237-240.

[6.92] H Sunami, "Cell structure for future DRAMs," *Tech. Digest* International Electron Devices Meeting (Washington DC, December 1985), paper 29.1, pp. 694-697.

[6.93] V J Silvestri, "Si selective epitaxial trench refill," Fall Electrochemical Society Meeting (San Diego CA, October 1986), abstract 269, pp. 402-403.

[6.94] V J Silvestri, "Growth kinematics of a polysilicon trench refill process," J. Electrochem. Soc. *133*, 2374-2376 (November 1986).

[6.95] H Sunami, T Kure, N Hashimoto, K Itoh, T Toyabe, and S Asai, "A corrugated capacitor cell (CCC)" IEEE Trans. Electron Devices, *31*, 746-753 (June 1984).

[6.96] C Koburger, F R White, L Nesbit, and S D Emmanuel, "Process-dependent properties of three-dimensional capacitors," IEEE Trans. Electron Devices, *33*, 766-771 (June 1986).

[6.97] N Lu, P Cottrell, W Craig, S Dash, D Critchlow, R Mohler, B Machesney, T Ning, W Noble, R Parent, R Scheuerlein, E Sprogis, and L Terman, "The SPT cell—A new substrate-plate trench cell for dRAMs," *Tech. Digest* International Electron Devices Meeting (Washington DC, December 1985), paper 29.8, pp. 771-772.

[6.98] S Nakajima, K Miura, K Minegishi, and T Morie, "An isolation-merged vertical capacitor cell for large capacity DRAM," *Tech. Digest* 1984 International Electron Devices Meeting (San Francisco, December 1984), paper 9.4, pp. 240-243.

[6.99] M Taguchi, S Ando, N Higaki, G Goto, T Ema, K Hashimoto, T Yabu, and T Nakano, "Dielectrically encapsulated trench capacitor cell," *Tech. Digest* 1986 International Electron Devices Meeting (Los Angeles, December 1986), paper 6.3, pp. 136-139.

[6.100] W F Richardson, D M Bordelon, G P Pollack, A H Shah, S D S Malhi, H Shichijo, S K Banerjee, M Elahy, R H Womack, C P Wang, J Gallia, H E Davis, and P K Chatterjee, "A trench transistor cross-point dRAM cell," *Tech. Digest* 1985 International Electron Devices Meeting (Washington DC, December 1985), paper 29.6, pp. 714-717.

[6.101] P C Fazan and A Ditali, "Electrical characterization of textured interpoly capacitors for advanced stacked DRAMs." *Tech. Digest* International Electron Devices Meeting (San Francisco, December 1990), paper 27.5, pp. 663-666.

[6.102] H Watanabe, T Tatsumi, S Ohnishi, T Hamada, I Honma, and T Kikkawa, "A new cylindrical capacitor using hemispherical grained Si (HSG-Si) for 256 Mb DRAMs," *Tech. Digest* International Electron Devices Meeting (San Francisco, December 1992), paper 10.1, pp. 259-262.

[6.103] H Watanabe, T Tatsumi, S Ohnishi, H Kitajima, I Honma, T Ikarashi, and H Ono, "Hemispherical grained Si formation on *in-situ* phosphorus doped amorphous-Si electrode for 256Mb DRAM's capacitor," IEEE Trans. Electron Devices *42*, 1247-1253 (July 1995).

[6.104] P C Fazan, V K Mathews, N Sandler, G Q Lo, and D L Kwong, "A high-C capacitor (20.4 fF/μm^2) with ultrathin CVD–Ta$_2$O$_5$ films deposited on rugged poly-Si for high density DRAMs," *Tech. Digest* International Electron Devices Meeting (San Francisco, December 1992), paper 10.2, pp. 263-266.

[6.105] Y-C Jeon, J-M Seon, J-H Joo, K-Y Oh, J-S Roh, J-J Kim, and D-S Kim, "Thermal stability of Ir/polycrystalline-Si structure for bottom electrode of integrated ferroelectric capacitors," Appl. Phys. Lett. *71*, 467-469 (28 July 1997).

[6.106] J Manoliu and T I Kamins, "*p-n* junctions in polycrystalline-silicon films," Solid-State Electron. *15*, 1103-1106 (October 1972).

[6.107] M Dutoit and F Sollberger, "Lateral polysilicon *p-n* diodes," J. Electrochem. Soc. *125*, 1648-1651 (October 1978).

[6.108] K Okada, K Aomura, M Suzuki, and H Shiba, "PSA—A new approach for bipolar LSI," IEEE J. Solid-State Circuits, *13*, 693-698 (October 1978).

BIBLIOGRAPHY

[6.109] M Dutoit, "Influence of impurities on lateral polysilicon pn diodes," J. Electrochem. Soc. *130*, 967-968 (April 1983).

[6.110] A Kalnitsky, J Li, and C E D Chen, "Experimental characterization of the diode-type polysilicon loads for CMOS SRAM," IEEE Trans. Electron Devices *40*, 358-363 (February 1993).

[6.111] C Cohen, "Vertical MOS FET has diode protection," Electronics (October 6, 1982), pp. 70-72.

[6.112] E Schibli and A G Milnes, "Effects of deep impurities on n^+p junction reverse-biased small-signal capacitance," Solid-State Electron. *11*, 323-334 (March 1968).

[6.113] M Dutoit and F Sollberger, "Lateral polysilicon pn diodes in CMOS technology," Fall 1977 Electrochem. Soc. Meeting (Atlanta, October 1977), abs. 300, pp. 802-804.

[6.114] H C de Graaff, M Huybers, and J G de Groot, "Grain boundary states and the characteristics of lateral polysilicon diodes," Solid-State Electron. *25*, 67-71 (January 1982).

[6.115] D W Greve, P A Potyraj, and A M Guzman, "Field-enhanced emission and capture in polysilicon pn junctions," Solid-State Electron. *28*, 1255-1261 (December 1985).

[6.116] A Aziz, O Bonnaud, H Lhermite, and F Raoult, "Lateral polysilicon PN diodes: Current-voltage characteristics between 200 K and 400 K using a numerical approach," IEEE Trans. Electron Devices *41*, 204-211 (February 1994).

[6.117] A W De Groot and H C Card, "Charge emission from interface states at silicon grain boundaries by thermal emission and thermionic-field emission—Part II: Experiment," IEEE Trans. Electron Devices, *31*, 1370-1376 (October 1984).

[6.118] D-L Chen, A M Guzman, and D W Greve, "Effect of hydrogen implantation on polysilicon p-n junctions," IEEE Trans. Electron Devices, *33*, 270-274 (February 1986).

[6.119] K Sagara and Y Tamaki, "Effect of grain structure on the electrical characteristics of platinum silicide polysilicon Schottky barrier diodes," J. Electrochem. Soc. *138*, 616-619 (February 1991).

[6.120] T L Chu, S S Chu, G A van der Leeden, C J Lin, and J R Boyd, "Polycrystalline silicon p-n junctions," Solid-State Electron. *21*, 781-786 (May 1978).

[6.121] A Criadi, E Calleja, J Martinez, J Piqueras, and E Muñoz, Solid-State Electron. *22*, 693-700 (August 1979).

[6.122] S D S Malhi, H Shichijo, S K Banerjee, R Sundaresan, M Elahy, G P Pollack, W F Richardson, A H Shah, L R Hite, R H Womack, P K Chatterjee, and H W Lam, "Characteristics and three-dimensional integration of MOSFETs in small-grain LPCVD polycrystalline silicon," IEEE Trans. Electron Devices, *32*, 258-281 (February 1985).

[6.123] S W Depp, A Juliana, and B G Huth, "Polysilicon FET devices for large area input/output applications," *Tech. Digest* 1980 International Electron Devices Meeting (Washington DC, December 1980), paper 27.2, pp. 703-706.

[6.124] C H Fa and T T Jew, "The poly-silicon insulated-gate field-effect transistor," IEEE Trans. Electron Devices, *13*, 290-291 (February 1966).

[6.125] T I Kamins, "Field-effects in polycrystalline-silicon films," Solid-State Electron. *15*, 789-799 (July 1972).

[6.126] I-W Wu, T-Y Huang, W B Jackson, A G Lewis, and A Chiang, "Passivation kinetics of two types of defects in polysilicon TFT by plasma hydrogenation," IEEE Electron Device Lett. *12*, 181-183 (April 1991).

[6.127] M K Hatalis and D W Greve, "High-performance thin-film transistors in low-temperature crystallized LPCVD amorphous silicon films," IEEE Electron Device Lett. *8*, 361-364 (August 1987).

[6.128] T Noguchi, H Hayashi, and T Ohshima, "Advanced superthin polysilicon film obtained by Si^+ implantation and subsequent annealing," J. Electrochem. Soc. *134*, 1771-1777 (July 1987).

[6.129] M Cao, S Talwar, K J Kramer, T W Sigmon, and K C Saraswat, "A high-performance polysilicon thin-film transistor using XeCl excimer laser crystallization of pre-patterned amorphous Si films," IEEE Trans. Electron Devices *43*, 561-567 (April 1996).

[6.130] I-W Wu, "High-definition displays and technology trends in TFT-LCDs" J. Soc. for Information Display *2*, 1-14 (April 1994).

[6.131] C-T Liu, C-H D Yu, A Kornblit, and K-H Lee, "Inverted thin-film transistors with a simple self-aligned lightly doped drain structure," IEEE Trans. Electron Devices *29*, 2803-2809 (December 1992).

[6.132] T I Kamins and P J Marcoux, "Hydrogenation of transistors fabricated in polycrystalline-silicon films," IEEE Electron Device Lett. *1*, 159-161 (August 1980).

[6.133] S D S Malhi, R R Shah, H Shichijo, R F Pinizzotto, C E Chen, P K Chatterjee, and H W Lam, "Effects of grain boundary passivation on the characteristics of p-channel MOSFETs in LPCVD polysilicon," Electronics Lett. *19*, 993-994 (10 November 1983).

BIBLIOGRAPHY

[6.134] H J Singh, K C Saraswat, J D Shott, J P McVittie, and J D Meindl, "Hydrogenation by ion implantation for scaled SOI/PMOS transistors," IEEE Electron Device Lett. *6*, 139-141 (March 1985).

[6.135] N D Young, "The formation and annealing of hot-carrier-induced degradation in poly-Si TFT's, MOSFET's, and SOI devices, and similarities to state-creation in αSi:H," IEEE Trans. Electron Devices *43*, 450-456 (March 1966).

[6.136] A F Tasch, Jr., T C Holloway, K F Lee, and J F Gibbons, "Silicon-on-insulator MOSFETs fabricated on laser-annealed polysilicon on SiO_2," Electronics Lett. *15*, 435-436 (5 July 1979).

[6.137] K F Lee, J F Gibbons, K C Saraswat, and T I Kamins, "Thin film MOSFET's fabricated in laser-annealed polycrystalline silicon," Appl. Phys. Lett. *35*, 173-175 (15 July 1979).

[6.138] T I Kamins, K F Lee, J F Gibbons, and K C Saraswat, "A monolithic integrated circuit fabricated in laser-annealed polysilicon," IEEE Trans. Electron Devices, *27*, 290-293 (January 1980).

[6.139] S Seki, O Kogure, and B Tsujiyama, "Effects of crystallization on trap state densities at grain boundaries in polycrystalline silicon," IEEE Electron Device Lett. *8*, 368-370 (August 1987).

[6.140] J G Fossum and A Ortiz-Conde, "Effects of grain boundaries on the channel conductance of SOI MOSFETs," IEEE Trans. Electron Devices, *30*, 933-940 (August 1983).

[6.141] J Tihanyi and H Schlotterer, "Properties of ESFI MOS transistors due to the floating substrate and the finite volume," IEEE Trans. Electron Devices, *22*, 1017-1023 (November 1975).

[6.142] N Sasaki, "Charge pumping in SOS-MOS transistors," IEEE Trans. Electron Devices, *28*, 48-52 (January 1981).

[6.143] H K Lim and J G Fossum, "Transient drain current and propagation delay in SOI CMOS," IEEE Trans. Electron Devices *31*, 1251-1258 (September 1984).

[6.144] M Cao, S Kuehne, K C Saraswat, and S S Wong, "A-low-thermal budget polysilicon thin-film transistor using chemical mechanical polishing," Conf. Record of 14th International Display Research Conf. (Monterey, CA, October 1994,) pp. 294-298.

[6.145] C-Y Chang, H-Y Lin, T F Lei, J-Y Cheng, L-P Chen, and B-T Dai, "Fabrication of thin film transistors by chemical mechanical polished polycrystalline silicon films," IEEE Electron Device Lett. *17*, 100-102 (March 1996).

[6.146] T-J King and K C Saraswat, "A low-temperature ($\leq 550°C$) silicon-germanium MOS thin-film transistor technology for large-area electronics," Tech. Digest International Electron Devices Meeting (Washington, DC, December 1991), paper 20.4, pp. 567-570.

[6.147] D S Bang, M Cao, A Wang, K C Saraswat, and T-J King, "Resistivity of boron and phosphorus doped polycrystalline $Si_{1-x}Ge_x$ films," Appl. Phys. Lett. *66*, 195-197 (9 January 1995).

[6.148] C Feldman and R Plachy, "Vacuum deposited silicon devices on fused silica substrates," J. Electrochem. Soc. *121*, 685-688 (May 1974).

[6.149] J D Hayden, R C Taft, P Kenkare, C Mazuré, C Gunderson, B-Y Nguyen, M Woo, C Lage, B J Roman, S Radhakrishna, R Subrahmanyan, A R Sitaram, P Pelley, J-H Lin, K Kemp, and H Kirsch, "A quadruple well, quadruple polysilicon BiCMOS process for fast 16 Mb SRAM's," IEEE Trans. Electron Devices *41*, 2318-2325 (December 1994).

[6.150] Parasitic polysilicon diode in SRAM S Ikeda, *et al*, A polysilicon transistor technology for large capacity SRAMs," *Tech. Digest*, 1990 International Electron Devices Meeting (San Francisco, CA, December 1990), paper 18.1, pp. 469-472.

[6.151] J M Jaffe, "Monolithic polycrystalline-silicon pressure transducer," Electronics Lett. *10*, 420-421 (3 October 1974).

[6.152] J C Erskine, "Polycrystalline silicon-on-metal strain gauge transducers," IEEE Trans. Electron Devices, *30*, 796-801 (July 1983).

[6.153] J Suski, V Mosser, and G Le Roux, "The piezoresistive properties of polycrystalline silicon films," Fall 1986 Electrochemical Society Meeting (San Diego, October 1986), abstract no. 586, p. 870.

[6.154] J Y W Seto, "Piezoresistive properties of polycrystalline silicon," J. Appl. Phys. *47*, 4780-4783 (November 1976).

[6.155] H Mikoshiba, "Stress-sensitive properties of silicon-gate MOS devices," Solid-State Electron. *24*, 221-232 (March 1981).

[6.156] R T Howe and R S Muller, "Polycrystalline silicon micromechanical beams," J. Electrochem. Soc. *130*, 1420-1423 (June 1983).

[6.157] R T Howe and R S Muller, "Resonant-microbridge vapor sensor," IEEE Trans. Electron Devices, *33*, 499-506 (April 1986).

[6.158] T A Lober, J H Huang, M A Schmidt, and S D Senturia, "Characterization of the mechanisms producing bending moments in polysilicon micro-cantilever beams by interferometric deflection measurements," Proc. Solid-State Sensor and Actuator Workshop (Hilton Head Island SC, June 1988), pp. 92-95.

Index

absorption coefficient, light **88**
absorption length, x-ray 78
accelerometers 311
access transistor 286
accumulation-mode 301
activated conduction 207, 212
active matrix 44, 56, 114, 297, **308**
adsorbed moisture 197
adsorption 4, 37, 42
adsorption sites 28
AFM 93, 180
aluminum **145** 268
aluminum oxide 256
AMLCD **308**
amorphous
    films 64, 72, **110**
    silicon 54, 87, 209, 215
    silicon, conduction **198**
    surfaces **58**
amorphous-silicon transistor 114
anisotropic 96
anisotropic diffusion 137, 160
anisotropic etching 101
annealing 98
antimony
    diffusion 138
    segregation 155
aperture ratio 308
arsenic 107, 118, 145, 172, 230, 237, 258, 273
    segregation 152
    diffusion 135
arsine 45
asperity 105, 182, 184
atmospheric pressure 33

atomic-force microscope 93, 180
automated loading 18

backside gettering 268
bandgap 199
bar 5
barium-strontium titanate 291
barrier height 202, 203
barriers 145, 201
    grain-boundary **219**
    potential **201**, 219
base current 277
base pressure 23
batch reactor **33**
bicrystal 126
bipolar transistor **270**
bird's beak 178
bond angle 73
boron 49, 107, 119, 173, 221, 230
    diffusion 138
    penetration 259
    segregation 155
boundary layer 3, **6**, 26
    diffusion 4
Bragg's law 76
Bravais theory 86
breakdown 102
    field 189, 196
Brillouin zone 90
buffer layer 283
bulk micromachining 311
byproduct 5

cantilever 313
capacitor

cylindrical 290
finned 290
stacked **289**
trench **287**
carrier
    gas 13, 17, 67
    injection 191
    mobility 231
    transport **205**
carrier trapping **200**, 219
    model 203, 221
    oxide **191**
carriers, minority **238**
channeling, implant 140
charge emission 241
charge neutral grains 204
charge trapping 241
charge-coupled devices 260
charging 101
chemical effect 149
chemical oxide 119
chemical mechanical polishing 93, 187, 306
chlorinated silanes 13
clean room 18
cleaning 63
cluster processing 24
cluster tool 24, 31
clustering, arsenic 159
CMOS **248**
CMP 93, 187, 306
cobalt silicide 148, 255
coefficient of friction 97
cold-wall reactor **24**, **33**
columnar grains 78, 130
columnar structure 82, 259
complementary MOS **248**
computer modeling 279
    diffusion **159**
    etching 101
conduction 195
    activated 207
    hopping 199

    moderately doped polysilicon 199
    oxide 262
    undoped films **196**
conductive filament 242
conformal deposition **51**, 185, 313
constant, Hall 210
contacts **268**, **270**
contamination 68, 79, **99**, 295
    oxygen 315
control gate 261
convection **5**
cooling rate 158, 236
correction factor 77
counterdoping 251, 253
critical cluster 58
critical dopant concentration 203, 223
crystal structure 27
crystal texture 78
crystallization 110, 165, 215, 305
    laser 114
    modeling 161
    selective 310
crystallized films 90
current flow, nonuniform 223
cut-off frequency 277
CVD dielectrics **192**
cw laser 228
cylindrical capacitor 290

dangling bonds 201, 214, 226, 299
deactivation, dopant 236
decomposition 4, **35**, 42
deep states 201, 299
deformation 97
depletion regions 201
deposited dielectrics 187
deposition **1**
    chamber **25**
    efficiency 22
    rate **39**
desorption 4, 42, 58
deuterium 228

INDEX

diborane 45, 49
dielectric isolation **281**
dielectrics, CVD **192**
diffusion **123**
    diffusion coefficient 10
    coefficient 126
    current 293
    kinetics (A,B,C) 128
    length 38, 67, 126
    limited 12
    from polysilicon **269**
    in polysilicon 129
    anisotropic 137, 160
    antimony 138
    boron 138
    computer modeling **159**
    grain growth 161
    grain-boundary 78
    lateral 265
    phosphorus 137
diluent gas 13, 41
diode 239
    gate-protection 293
    polysilicon **292**, 311
disilane **44**, **48**, 50, **84**, 112
distortion, wafer 282
dopant
    deactivation 236
    diffusion 28, **123**, 169
    diffusivity 126
    precipitation 167, 170
    segregation **149**, 170, 219, 232
doped films **45**
doped oxides 234
doping
    in situ 235
    methods **232**
    nitrogen 140
    nonuniform 235
drain offset 310
DRAM **285**
driving force 9
dynamic random-access memory 116, **285**

EEPROM 163, 261
effective
    diffusivity 132, 139
    mass 87, 182, 220
    mobility **209**
effective-medium 63
electric field, local 288
electrical properties **195**
electromigration 267
electron
    affinity 250
    diffraction 72
    injection 182, 274
    spin resonance 215, 226
electronics, large-area 297
electrostatic **51**
elemental silicon 35
ellipsometry 62, 180
embedded DRAM 287
emissivity 29
emitter
    injection efficiency 274
    plug 273
    resistance 275
energy 257
    distribution 215
enhanced deposition **53**
    oxidation 165
entropy 150
epitaxial alignment **117**, 142
epitaxy 64
equi-axed grains 131
equilibrium constant 3
etching 281, 290
etching **100**
electron volt 10
evaluation **90**
    techniques **70**
evaporated films **85**, 307
extrinsic base 270

Fermi level 51, 212, 214, 217, 239, 250
field oxidation 284

field-assisted trap emptying 242
field-effect mobility 298, 300
field-enhanced emission 295
field-programmable gate array 267
filament, conductive 242
film thickness **94**
    finite 169
filters, color 308
finned capacitor 290
flash memory 261
flatband voltage 250
floating gate 261
fluorine 55, 120, 189, 259, 275
forced convection 4, 5, 26, 28
forming gas 227
Fowler-Nordheim injection 182
FPGA 267
fracture 97
free carriers 87, 200
free convection 5
friction 97
fused silica 16
fusible links **267**

gain, transistor 275
gas depletion **19**, 28
gas-phase
    diffusion 7
    doping 199, 232
    nucleation 45
    processes **5**
    reactions 36
gate dielectric 309
gate electrode, conductivity type **253**
gate oxide, void 177
gate protection 293
gate stack 24
germanium 50, 63, 86, 113, 307
gettering **268**
glass 44, 53
    flow 260
    substrate 113, 297, 309
glow discharge 215

gold 146
grain boundary 73, 123
    barriers **219**
    carrier trapping 200
    diffusion 78, 124
    diffusivity 132
    high-angle 125
    migration 108
    modification **225**
    movement 133
    oxidation 190, 315
    passivation **226**
    precipitation 170, 234
    segregation **123**, **149**, 170, 266
    sliding 178
    traps 203
    recombination velocity 239
grain
    grooving 109
    orientation **76**
    size **73**
    structure **72**
graphite 26
gray room 18
gray scale 308
growth texture 82
growth-controlled 66
Gummel number 276

H_3PO_4
Hall mobility **210**
hard mask 102
HCl 274, 296
    oxidation 189
HDP 102
heavily doped films 140, **230**
heavy metals 268
hemispherical-grain polysilicon **116**, 183, 262, 291
heterogeneous decomposition 35
high permittivity 291
high-density plasma 102
highest occupied molecular
    orbital 198

INDEX

holding voltage 241
hole injection 274
HOMO 198
homogeneous decomposition **35**
hopping conduction 197, 199
horizontal reactor 18
horns 179
hot-point-probe 196
hot-wall reactor **15**
HSG **116**, 183, 262, 291
hydrofluoric acid 63
hydrogen
    evolution 228
    implantation 227
    passivation 305
    stability 228
hydrogenation 226
    amorphous silicon 112

ideality factor 293, 296
impurities 80
impurity drag 103
inclusion 184, 185
    in oxide 179
incorporation 4, 42
incubation 59
index of refraction **87**
indium-tin oxide 308
input protection 293
in situ doping 235
interconnections 195, 230, **254**, 270
interdiffusion 145
interface recombination
        velocity 278
interface segregation 142, 277
interfacial oxide 142, 274, 275
interference 30, 180
interstitials 118, 123, 143
intrinsic base 270
intrinsic gettering 268
inversion layer 288
inversion mode 301
ion 54
ion bombardment 84

ion implantation 107, 114, 173,
       199, 227, 233, 269
    channeling 140
    damage 141
iridium 292
isolation
    device **281**
    trench **284**
ITO 308

junction spiking **268**

kilocalorie 10
kinetic 3
kink 38
kink effect 303

lamp-heated 25
large-area 44
    electronics 297
laser 228, 309
    crystallization 114
    melting 191, 237, 306
    recrystallization 228
    pulsed 229
latch-up, CMOS 285, 297
lateral diffusion 136
lateral junction 292
lateral pnp transistor **279**
LDD 252, 304
leakage current 303
    diode 292
    transistor 249
ledge 38
lifetime 294
    minority carrier **238**, 276
light scattering 86, 88
lightly doped drain 252, 304
limitations of model 221
limiting resistivity 254
linear-parabolic oxidation 171
links, fusible **267**
liquid-crystal display 44, 56,
       114, 297, **308**

lithium 228
load lock 31, 69
load, SRAM 310
local electric field 288
local oxidation of silicon 178, 251
localized states 198
LOCOS 178, 251
    poly-buffered **283**
low-pressure reactor **15**
lowest unoccupied molecular orbital 198
LPCVD 15
LUMO 198

masks 252
mass-transport limited 12
mean free path 6, 52
mechanical properties **97**
mechanical support 282
membrane 312
memory switching 242
memory, dynamic **285**
MEMS **311**
metal gate 246, 258
metal silicide **147**, 238
metal-induced crystallization 113
metals **144**
microbridge 313
microelectromechanical systems **311**
micromachining 311
microprocessor 261
microtwins 73
mid-gap states 214
Miller feedback capacitance 247
minority carriers **238**
    lifetime **238**
mobility 237, 302
    carrier 231
    effective **209**
    field effect 298, 300
    Hall **210**
    minimum 210
model limitations 221

modification, grain-boundary **225**
molecular-beam deposition 69
MOS transistor 129, **246**
    metal-gate 258
multi-point injection 21

N_2O 296
    oxidation 191
native oxide 93, 109, 119, 142, 186
nickel silicide 147
nitrogen 259
    doping 140
noise 267
nonuniform current flow 223
nonuniform doping 235
nonvolatile memories **261**
nucleating layer 83
nucleation 4, 103, 110
    controlled 66
    grains 305
    spontaneous 115
    texture 82
nuclei 59

offset gate 304
Ohm's law 209
ohmic conduction 197
open-flow 2, 5
operating sequence **23**
optical
    interference 30
    microscopy 70
    properties **86**
    pyrometer 29
overhead 24
    time bf 31
oxidation **163**
    arsenic-doped films **172**
    boron-doped films **173**
    device geometry **177**
    doped films 166
    grain-boundary 190, 315
    phosphorus-doped films **171**
    undoped films 164

INDEX

oxide 93, 119, 197
    breakup 120
    breakup, modeling 161
    charging 101
    conduction 185, **181**, 262
    sacrificial 311
    stress-relief 283
oxygen 86
    contamination **99**, 114, 315

parasitic diodes 292
particles 35, 45
pascal 5
passivation 54
    grain-boundary **226**
    hydrogen 305
PECVD 54, 65, 215, 309
permittivity 87, 193
PH_3 232
phosphine 45, 47, 232
phosphorus 46, 84, 107, 171, 189, 221, 230, 237, 253, 268, 273, 314
    diffusion 137
    segregation 154
photo-enhanced deposition 55
photoconductivity decay 239
photolysis 55
physical vapor deposition 307
piezoresistance 311, 313
pixel 297
plasma 54, 63
    hydrogenation 227
    nitride 227
    passivation 227
plasma-enhanced CVD 54, 65, **84**, 309
plasmons 94
platinum 292
$POCl_3$ 168, 232, 253
point defects 109, 118, 123, 168, 269
Poisson's equation 202
polarity, voltage 188

polishing 186, 282, 306
poly-buffered LOCOS **283**
polysilicon depletion 257
polysilicon-emitter transistor 117, 121, 141, 271, **274**
Poole-Frenkel effect 295
potential barriers **201**, 219
power adsorbed 30
precipitates 234, 315
precipitation 167
pressure 42, 68, 312
primary recrystallization 104
process compatibility **259**
profilometer 94
protrusions 109
protuberances 184
pulsed laser 229, 309
purity 32, 34
pyrolysis 55
pyrometer 29

quality factor, diode 293
quartz 16
quasi-Fermi level 239

radical 54
random-access memory **285**
rapid thermal
    annealing 121
    processing 137, 143, 279
    processor **29**, 189r
reactant flux 8
reactant gas 67
reaction **10**, **34**
reaction-rate limited 12
reactive-sticking coefficient 39, 51
reactive-ion etching 100
rearrangement, subsurface **70**
recombination current 293
recombination velocity 239
recrystallization **103**, **228**, 238, 306
recrystallized grain 134
redundancy 267
reflectance 71, 180

376 INDEX

surface **90**
reflectivity 30, 229
   patterned layers 229
refractive index **87**
reliability, gate-oxide **255**
residence time 35
resistivity 229
   heavily doped **230**
   stability **235**
   temperature dependence 211
   thickness variation 209
   undoped **196**
resistor 129, **262**
reverse injection 274
Reynold's number 6
roughened surface 291
roughness **90**
ruthenium 292

sacrificial oxide 311
salicide 148, 255
saturation 233
   density 59
scaling 287
scanning-electron microscopy 71
Schottky diodes 296
secondary recrystallization 105, 110
seeding, recrystallization 230
segregation 109, **123**, **149**, 242
   coefficient 173
selective
   crystallization 310
   deposition 285
   etching 101
self-alignment 271
SEM 71
semi-insulating polysilicon 279
sensors, integrated **312**
sensors, strain 311
series gates 303
shallow junction 141
Sherwood number 13
shower-head injector 26

sidewall-contact transistor 280
SiF_4 55
SiH_2
SiH_3PH_2 48
silane 13
silicide 146, **147**, 238, 254
silicon carbide 26
silicon monoxide 63
silicon phosphide 176
silicon-gate transistor 146, **246**
   limitations **257**
silicon-germanium 307
silver 146
silylene 49
single-crystal surfaces **64**
single-wafer reactor **24**
SIPOS 279
size effect 149
solar cells 296
solid solubility 145, 171, 176, 190, 231, 233, 237, 266, 314
solid-state crystallization 83
spiking 268
spontaneous nucleation 112, 115
sputtered films **85**, 308
sputtering 56, 101
stability, resistivity **235**
stability, structural **102**
stabilizing oxide 109, 186
stacked capacitor **289**
stagnant layer 8
static RAM 263, 292, **310**
statistical variation 229, 306
sticking coefficient 39, 51
storage capacitor 286
storage region 288
strain sensors 311
strained bonds 299
stress 98, 278, 284, 309, 314
stress gradient 100, 314
stress-relief oxide 283
structure **57**
substitutional dopant 200
subthreshold 303

INDEX

swing 303
supply function 220
surface
    adsorption 37
    diffusion 4, 38, 42, 52, 99
    diffusion **65**
    diffusion length 67
    energy 110
    irregularity 182
    micromachining 311
    oxide 93
    processes **5**
    reflectance **90**
    roughness **90**, 185, 262, 306
susceptor 26
switching **240**
    memory 242
    time 257
    threshold 240
    voltage 241
system integrity 34, 69
system pressure 68

Ta_2O_5
tail states 299
tantalum silicide 146
TCE 17
TCR 266
TEM 71, 72
temperature coefficient of resistance 266
temperature dependence, resistivity 211
temperature gradient 20, 196
texture 78
texture, transition 80
thermodynamics **2**
thermal
    coefficient of expansion 17, 309
    conductivity **96**
    mass 29, 32
thermionic emission 205, 218, 295
thermionic-field emission **218**, 295, 303

thickness **94**
thin-film transistor 93, 110, 114, 187, 265, 292, **296**, **298**
threshold switching 240
threshold voltage 141, 247, **249**, 253, 299
time-to-breakdown 255
titanium 146, 255
titanium silicide 147, 148, 255
Torr 5
transfer chamber 31
transient-enhanced diffusion 269, 273
transistor
    bipolar **270**
    silicon-gate **246**
    thin-film **296**
transition texture 80
transmission-electron microscopy 71, 72
transport, carrier **205**
trap
    concentration **213**
    density 208, 225, 239
    distribution **213**
    emptying 242
trench
    287
    filling **52**
    isolation 52, **284**
trimming, resistor 266
trivalent silicon 214
tungsten silicide 147
tunneling 181, 205, 218, 237, 275
turbulent flow 6
twin boundary 154
twins 73t
two-step deposition 259
two-stream model 159

ultraviolet 180
undercutting 179
undoped films, conduction **196**
uniformity 16, 19, 44, 48

vacuum-evaporated films 85
vanadium 146
vapor sensor 313
vertical reactor 18
vertical-flow reactor 22
viscosity 6
viscous flow 6
viscous forces 7
void 284, 315
voltage, Hall 210

wafer cage 49
wafer distortion 282
warpage 282
wear-out-time 256
WKB approximation 220
work function 86, 250

x-ray analysis 152
x-ray diffraction 71, 76

Young's modulus 97